Immunocytochemistry in Diagnostic Cytology

Immunocytochemistry in Diagnostic Cytology

Pranab Dey MBBS MD MIAC FRCPath
Professor
Department of Cytology and Gynecologic Pathology
Post Graduate Institute of Medical Education and Research
Chandigarh, India

JAYPEE

JAYPEE BROTHERS MEDICAL PUBLISHERS
The Health Sciences Publisher
New Delhi | London

Jaypee Brothers Medical Publishers (P) Ltd

Headquarter

Jaypee Brothers Medical Publishers (P) Ltd
4838/24, Ansari Road, Daryaganj
New Delhi 110 002, India
Phone: +91-11-43574357
Fax: +91-11-43574314
Email: jaypee@jaypeebrothers.com

Overseas Office

J.P. Medical Ltd
83 Victoria Street, London
SW1H 0HW (UK)
Phone: +44 20 3170 8910
Fax: +44 (0)20 3008 6180
Email: info@jpmedpub.com

Website: www.jaypeebrothers.com
Website: www.jaypeedigital.com

Inquiries for bulk sales may be solicited at: jaypee@jaypeebrothers.com

Immunocytochemistry in Diagnostic Cytology / Pranab Dey

First Edition: **2021**

ISBN: 978-93-90020-57-7

Printed at Replika Press Pvt. Ltd.

Dedicated To

Shree Shree Satyananda Giri
Shree Shree Paramahansha Yogananda
Rini and Madhumanti

PREFACE

Pranab Dey MBBS MD MIAC FRCPath
Professor
Department of Cytology and Gynecologic Pathology
Post Graduate Institute of Medical Education and Research
Chandigarh, India

This book Immunocytochemistry in Diagnostic Cytology deals with the practical aspect of immunocytochemistry which is essential for day-to-day work. The book is in two sections: Practical (Section 1) and Diagnostic Application (Section 2). Section 1 deals with the basic theory of immunochemistry, cell block preparation, staining procedure, automation in immunostaining, and quality control. Section 2 highlights the diagnostic applications of immunocytochemistry. The text in this part is provided with numerous tables, boxes, and illustrations. I hope that this book will help both, the students and the pathology practitioners.

ACKNOWLEDGMENTS

I am thankful to Shri Jitendar P Vij (Group Chairman), Mr Ankit Vij (Managing Director), Mr MS Mani (Group President), Dr Richa Saxena (Associate Director-Professional Publishing), Ms Nedup Denka Bhutia (Development Editor) and Other staffs of M/s Jaypee Brothers Medical Publishers (P) Ltd, New Delhi, India. The primary source of inspiration for this book is Mr Jitendar P Vij, who first time requested me to write such a book.

I am also thankful to Professor Arvind Rajwanshi, Head of the Department of Cytology and Gynecology.

My sincere thanks to all my colleagues in the department for always supporting me.

I wish to give my special thanks to my wife Rini and daughter Madhumanti whose constant encouragement helps me to accomplish new projects.

I am grateful to God for his immense blessings.

Pranab Dey

ABBREVIATIONS

Full form	Short form
3,3′-Diaminobenzidine	DAB
3-Amino-9-ethylcarbazole	AEC
4-Chloro-1-naphthol	CN
Activated B-cell like	ABC
Acute lymphoblastic lymphoma	ALL
Alkaline phosphatase	AP
Alkaline phosphatase–antialkaline phosphatase	APAAP
Alpha-fetoprotein	AFP
Anaplastic large cell lymphoma	ALCL
Angioimmunoblastic T-cell lymphoma	AITL
Antibody	Ab
Antigen	Ag
Antigen retrieval	AR
Avidin and biotin complex	ABC
Biotin–Streptavidin	B–S
Carbonic anhydrase IX	CA IX
Carcinoembryonic antigen	CEA
Caudal-related homeobox 2	CDX-2
Chromogranin	CHG
Classical Hodgkin lymphoma	CHL
Cytokeratin	CK
Desmoplastic small round cell tumors	DSRT
Epithelial membrane antigen	EMA
Estrogen receptors	ER
Ewing's sarcoma	ES
Extranodal marginal zone lymphoma	EZML
Fluorescein isothiocyanate	FITC
Follicular adenoma	FA
Follicular carcinoma	FC
Follicular lymphoma	FL
Forkhead box transcription factor L2	FOXL2
Friend leukemia virus integration site 1	Fli-1

Full form	Short form
Gastrointestinal stromal tumor	GIST
Gastrointestinal tract	GIT
Germinal center B-cell like	GCB
Glucose transporter isoform	GLUT1
Gross cystic disease fluid protein	GCDFP
Hairy cell leukemia	HCL
Heat-induced antigen retrieval	HIAR
Hector Battifora Mesothelial cell 1	HBME-1
Hepatocellular carcinoma	HCC
Hepatocyte nuclear factor β	HNF1β
Hepatocyte paraffin 1	HepPar1
Horseradish peroxidase	HRP
Human chorionic gonadotropin	HCG
Human epidermal growth factor receptor 2	HER-2
Human Melanoma Black 45	HMB45
Hypoxanthine-guanine phosphoribosyltransferase	HGPRT
Immunohistochemistry	IHC
Immunocytochemistry	ICC
Kidney injury molecule-1	KIM-1
Langerhans cell histiocytosis	LCH
Leukocyte common antigen	LCA
Liquid-based cytology	LBC
Lymphocyte and histiocytic	L&H
Lymphoepithelial sialadenitis	LESA
Malignant peripheral nerve sheath tumor	MPNST
Malignant round cell tumors	MRCT
Mammary analogue secretory carcinoma	MASC
Mantle cell lymphoma	MCL
Monoclonal antibody	MoAb
Mucoepidermoid carcinoma	MEC
Nasopharyngeal carcinoma	NPC
Neuroblastoma	NB

Full form	Short form
Neuroendocrine tumors	NET
Neuron-specific enolase	NSE
Neutral buffered formalin	NBF
Nitroblue tetrazolium chloride	NBT
Nodal marginal zone lymphoma	NMZL
Nodular lymphocyte-predominant Hodgkin lymphoma	NLPHL
Non-Hodgkin Lymphoma	NHL
Octamer binding transcription factor	OCT
Pancreatic and duodenal homeobox-1	PDX-1
Papillary thyroid carcinoma	PTC
Parathyroid hormone	PTH
Paired box gene 8	PAX-8
Peripheral T-cell lymphoma	PTCL
Peroxidase-anti-peroxidase	PAP
P-iodonitrotetrazolium chloride	INT
Placental alkaline phosphatase	PLAP
Plasmacytoid lymphoma	PL
Polymorphous low-grade adenocarcinoma	PLGA
Primary effusion lymphoma	PEL
Primitive neuroectodermal tumors	PNET
Progesterone receptor	PR
Prostate-specific antigen	PSA

Full form	Short form
Reed Sternberg	RS
Renal cell carcinoma	RCC
Rhabdomyosarcoma	RMS
Salivary duct carcinoma	SDC
Small cell carcinoma	SCC
Small lymphocytic lymphoma	SLL
Smooth muscle actin	SMA
Solid and cystic papillary neoplasm	SCPN
Special AT-rich sequence binding protein 2	SATB2
Splenic marginal zone lymphoma	SMZL
Squamous cell carcinoma	SQCC
Synaptophysin	SNP
Thyroid transcription factor	TTF-1
Transducer-like enhancer of split 1	TLE-1
Tris-buffered saline	TBS
Tyramine signal amplification	TSA
Undifferentiated carcinoma	UDC
Urothelial cell carcinoma	UCC
V-ETS avian erythroblastosis virus oncogene homolog	ERG
von Willebrand factor	vWF
Wilms' tumor gene 1	WT-1

CONTENTS

1

SECTION

Practical Aspects of Immunocytochemistry

CHAPTER 1

Basic Principles of Immunocytochemistry

INTRODUCTION

There is a long chain of development in the technique of immunohistochemistry (IHC). In the year of 1942, Coons et al. first time demonstrated antigen–antibody (Ag-Ab) reaction with the help of fluorescent-tagged primary antibody on frozen section.[1] Nakane and Pierce,[2] in the year of 1967, first time overcame the problems of fluorescent microscopy and applied enzyme-linked primary antibody. Later on, Taylor et al. in early 1970, introduced IHC in the formalin-fixed paraffin section of the routine sample. Subsequently, the chromogen-tagged secondary antibody, monoclonal antibody (MoAb), avidin-biotin, and biotin–streptavidin methods were introduced.[3-6] Presently, tyramide[7] and extremely sensitive polymer-labeled visualization system[8] are now widely used in many laboratories.

In the last few decades, a large number of antibodies are available for the formalin-fixed tissue section. Previously, it was assumed that the epitopes of the antigens are not preserved in formalin. However, it is now proved that the epitopes of the antigen are never lost in formalin fixation, they are only masked. Heat retrieval technique helps to unmask the antigenic epitope and the demonstration of various antigens in formalin-fixed tissue is now feasible.

Presently, IHC/immunocytochemistry (ICC) is now an integral technical part in every pathology laboratory. ICC has various useful applications that include diagnosis, classification, and prognostic prediction. This technique can assay the antigen in the tissue section that may have an essential part in the management of the disease. For the successful application of ICC, proper standardization and reproducibility are mandatory. Strict quality control is necessary in this technique. Recently, many laboratories have introduced automated immunostaining that needs special attention.

BASIC PRINCIPLES AND THEORY OF IMMUNOCYTOCHEMISTRY

Each antigen contains a three-dimensional structure known as the epitope or antigen-binding site. The epitope binds with the antibody because the epitope and the antibody both have the complimentary geometric shape. The detection system then demonstrates this Ag-Ab reaction.

Antigen

Antigen is a chemical substance that can generate an immunogenic response. It has an epitope or antigenic determinant site. The epitope is the specialized three-dimensional charge-shaped arrangement of amino acid or monosaccharide that provokes the formation of an antibody. Each antigen may have one or more such unique epitopes that may generate the antibody. As mentioned before, this epitope reacts explicitly with the corresponding paratope of the antibody.

Antibody

Antibody is also known as "immunoglobulin." It is generated due to the immunogenic response against the corresponding antigen. They have the unique property to bind with the corresponding antigen. The antibodies are produced by the plasma cells. The basic unit of immunoglobulin is the monomeric type. However, it can be dimeric or trimeric or tetrameric type. Each immunoglobulin consists of a pair of light and heavy chains. There are two types of light chains: Kappa and lambda. Each immunoglobulin has either kappa or lambda light chain. Heavy chains are of five kinds: α, γ, δ, ε, and μ and the corresponding immunoglobulins are IgA, IgG, IgD, IgE, and IgM. The immunoglobulin IgG type is widely used in ICC. The shape of immunoglobulin is Y like (**Fig. 1**; **Box 1**). The ends of the two arms of Y are the antigen-binding sites. These sites have a variable sequence of amino acids and are labeled as "variable domain." The junction of the two arms of Y is known as the hinge region. The immunoglobulin is a protein molecule, and the specific part of it may have an antigenic determinant site and thereby can produce an antibody against it. So the immunoglobulin may also behave as an antigen and a second antibody can bind with the first antibody. This fact is widely used in ICC.

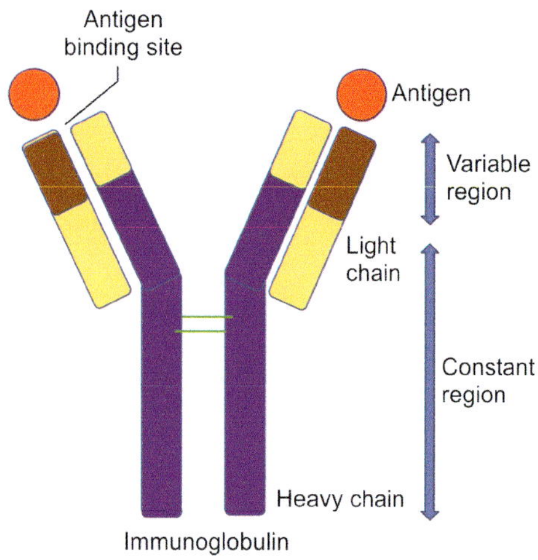

FIG. 1: Schematic diagram of the immunoglobulin.

BOX 1 Antibody (immunoglobulin).

- Antibody (Ab) is produced as an immunogenic response against an antigen
- Ab is against the epitope of antigen
- Ab is generated by the plasma cells
- Ab is composed of a pair of light and heavy chains
- The light chains: Kappa or lambda
- Heavy chains: α (alpha), γ (gamma), δ (delta), ε (epsilon), and μ (mu)
- The respective immunoglobulins are IgA, IgG, IgD, IgE, and IgM
- Immunoglobulin is Y-shaped
- The ends of the two arms of Y are the antigen-binding sites

Affinity

Affinity means the binding power of the paratope of the antibody with the respective antigenic epitope. It depends on the fitting of the three-dimensional epitope of the Ag with the corresponding paratope of the Ab.

Avidity

Avidity means how strongly an Ab binds with the corresponding Ag molecule. It reflects the overall stability of the Ag-Ab complex. The multivalent antibody has more avidity than a monovalent antibody.

The avidity of an Ab depends on the following factors (**Fig. 2**):

- Avidity is directly proportional to the number of valency of Ab.
- The affinity of the epitope of the Ag and paratope of the Ab.

The differences between affinity and avidity are highlighted in **Table 1**.

Specificity of an Antibody

Antibody specificity means the selective binding of the Ab with the unique epitope on the Ag. It helps in the detection of the target antigen. The different antigen may have the same epitope and therefore, may react with the same antibody.

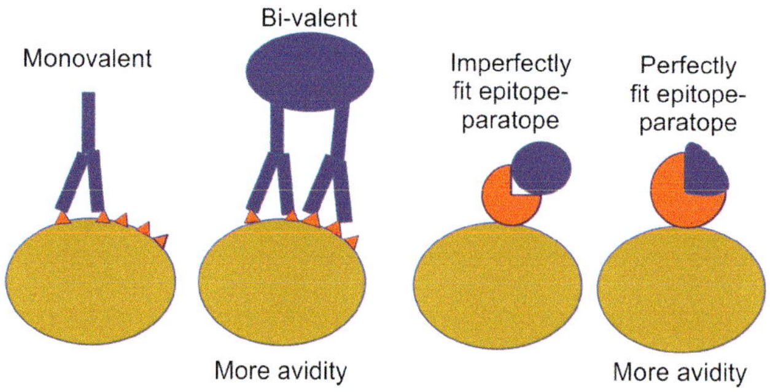

FIG. 2: Schematic diagram showing how avidity is affected.

TABLE 1: Difference between affinity and avidity.

Affinity	Avidity
Only the binding force between the epitope and paratope	The total binding strength between the antigen and antibody
Depends only between geometrical and chemical interaction between the epitope and paratope	Depends on the valency of the antibody and also the affinity of the antibody *Example*: Multivalent antibody binds with the multiple sites of a single antigen or multiple antigen and has higher avidity than monovalent antibody

Sensitivity

Sensitivity of the ICC is related to the lowest amount of antigen that can be detected by the ICC technique. The highly sensitive technique will show more intense signal than the inferior sensitive technique.

Antibody Production

Polyclonal Antibody

The production of polyclonal antibody is simple and cost-effective and massive amount of the Ab can be produced. The polyclonal antibody is produced by introducing the antigen or immunogen in the animal. The antibody is produced against the different epitopes of the antigen. A mixture of different antibodies are produced by the various clones of the plasma cells. These antibodies are known as polyclonal as these antibodies are produced against different epitopes of the antigen so they may cross-react with other closely similar epitopes of the other antigens and therefore, the polyclonal antibody has low specificity. Usually, mammalian and avian species are used to produce the polyclonal antibody. The white rabbits are the most suitable for the generation of the polyclonal antibody. However, a rat or a guinea pig can also be used for the generation of polyclonal antibody. For large scale production of these antibodies, donkeys or sheep are used. The steps of polyclonal antibodies are as follows (**Fig. 3**):

- Introduction of immunogens (antigen) to the animal in the skin or peritoneum.
- Engulfing of the antigen by the antigen-presenting cells.
- Transfer of Ag to the B cell.

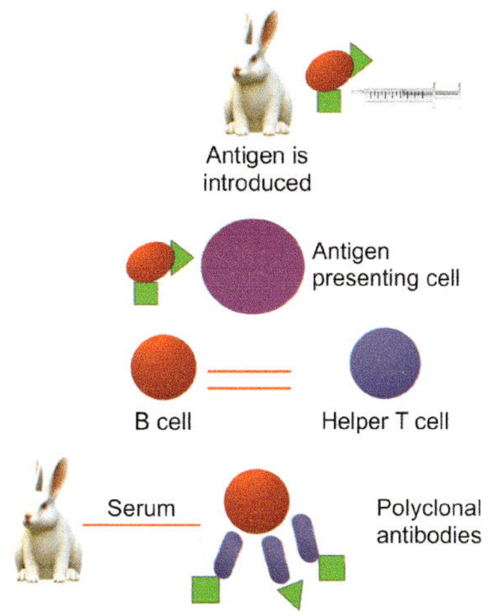

FIG. 3: The basic technique of polyclonal antibody production.

- Interaction of B cell and T helper cells and cytokine secretion from the T cell to stimulate B cells.
- Production of antibodies by the B cell. The antibodies are produced against different epitopes of the Ag.

Monoclonal Antibody

Monoclonal antibody is produced from a single clone of B cell that is destined to generate Ab against the single epitope of the Ag. The production of the MoAb is a revolutionary change in the ICC as the pure and specific antigen can be developed in large quantity with immense diagnostic importance.

Major advantages of MoAb include the following:
- It is specific to the single epitope of the Ag.
- Excellent affinity
- Consistent result
- No cross-reactivity
- Unlimited amount of MoAb production is possible by the hybridoma technique.

Disadvantages:
- Needs skilled technology
- A minor change of epitope may cause gross reduction of binding performance with the epitope.

Production of MoAb

Monoclonal antibody is produced by the hybridoma technique (**Fig. 4**), and the steps include:[9]
- *Immunization*: The mouse is immunized by introducing the antigen of interest.
- After 72 hours, the spleen of the mouse is taken out, and a free cell suspension is made.
- The plasma cells of the mouse are fused with myeloma cells (the immortal cancer cell). These myeloma cells have two essential characteristics: (1) The cells should not generate their own antibodies and (2) the cells contain the specific wild type HGPRT (hypoxanthine guanine phosphoribosyl transferase) gene.
- The fused cells are cultured in HAT medium (containing hypoxanthine, aminopterin and thymidine) and aminopterin. These hybridoma cells can only proliferate in the HAT medium. The myeloma cells cannot survive with aminopterin and B cells die as they cannot proliferate incessantly.

Table 2 shows the differences between monoclonal and polyclonal antibodies. The differences are also highlighted in **Figure 5**.

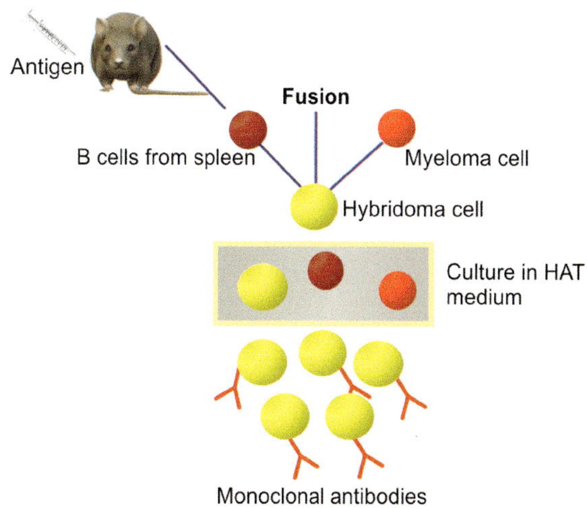

(HAT: hypoxanthine, aminopterin and thymidine)

FIG. 4: The basic technique of monoclonal antibody production.

TABLE 2: **Monoclonal versus polyclonal antibody.**

Features	Monoclonal antibody	Polyclonal antibody
Specificity	Very high	Low
Epitope binding	Produced against a single epitope	Produced against multiple epitopes
Technical skill needed	High	Low
Cost	High	Low
Variation	No batch-to-batch variation	Batch-to-batch variation is common

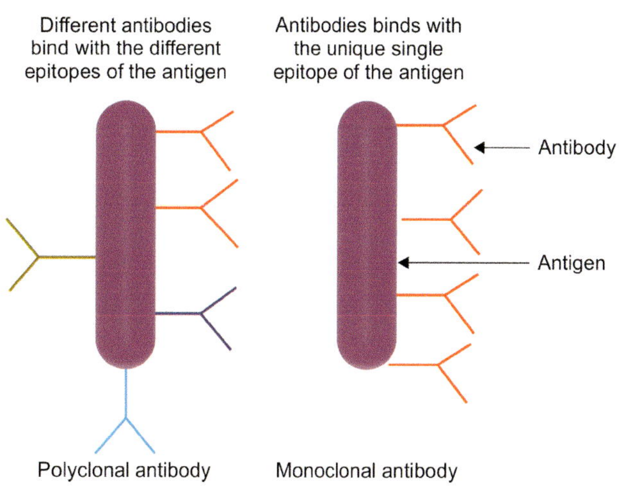

FIG. 5: Schematic diagram showing the difference between monoclonal and polyclonal antibody.

Labeling of Antibody

The antibodies are not visible in the light microscope. Therefore, we need to tag either the primary or the secondary antibody to the label in such a way that the antigen and the antibody reaction can be visible by light microscopy (**Box 2**). The same principle is applicable to visualize the Ag-Ab reaction by electron microscopy if the product is electron-dense. The two most frequent labels of the antibody are enzymatic labels and fluorescent labels.

Enzymatic Labeling of Antibody

The antibody is conjugated with enzymes that reacts with the substrate and produces the chromogenic product (**Fig. 6**).

Enzyme + Substrate = Chromogenic product

These chromogenic products are insoluble and precipitate in the Ag-Ab reaction sites. Usually, the secondary antibody is conjugated with the enzymes rather than the primary antibody because it increases the amplification of the signal as several secondary antibodies labeled with multiple enzymes react to a primary antibody. The most commonly used enzymes are horseradish peroxidase (HRP) and alkaline phosphatase (AP). These enzymes are less expensive and also stable products.

BOX 2 Labeling of antibody.

- Either primary or secondary antibody are labeled to visualize the antigen–antibody reaction
- *Two types of labeling:*
  - Enzymatic labeling
  - Fluorescent labeling
- *Enzymatic labeling:*
  - In case of enzymatic labeling of an antibody, a substrate is used to produce a colored product and the antigen–antibody reaction is visualized by light microscope
  - Frequently used enzymes are horseradish peroxidase (HRP) and alkaline phosphatase (AP)
  - Advantages of enzyme labeling: Stable reaction, easy to do, and ordinary light microscope can be used
- *Fluorescent labeling:*
  - In fluorescent labeling, antibodies are labeled with fluorescent dye and fluorescent microscope is needed to visualize the reaction
  - *Advantages of fluorescent labeling*: Bright signal and different colored dye can be used

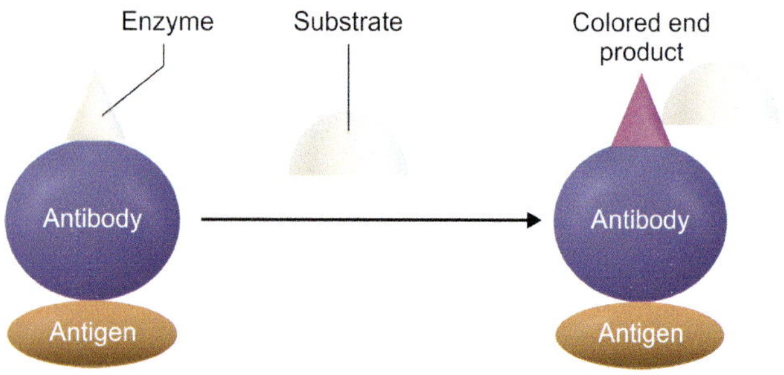

FIG. 6: Enzyme-substrate reaction to visualize the antigen–antibody reaction.

Horseradish Peroxidase

Horseradish peroxidase enzyme is extracted from the root of the horseradish plant. This enzyme binds with the Ab by the covalent or noncovalent bond. The HRP enzyme reacts with hydrogen peroxide (H_2O_2) and liberates water and atomic oxygen. This atomic oxygen oxidizes several substrates to generate a chromogenic product.

Substrates for Horseradish Peroxidase

3,3′-diaminobenzidine (DAB): DAB is oxidized by HRP in the presence of H_2O_2 to generate a brown-colored product. This end product is insoluble in alcohol. The smear can be dehydrated by alcohol and dibutylphthalate polystyrene xylene (DPX) mounting is possible. DAB is carcinogenic and needs careful handling.

3-amino-9-ethylcarbazole (AEC): AEC generates red-colored product that is soluble in alcohol. So, it requires aqueous counterstaining and mounting medium. It fades slowly in the sunlight as it is further oxidized. Therefore, the smear should be stored in the dark.

4-chloro-1-naphthol (CN): CN gives blue-colored end product that is highly soluble in alcohol. The chromogenic product is soluble in alcohol and therefore aqueous counterstain and aqueous medium should be used.

Alkaline Phosphatase

Alkaline phosphatase enzyme removes the phosphate group from its substrate naphthol phosphate and produces phenolic compounds and phosphate. The soluble colorless phenolic compound couples with the diazonium salts such as NBT (nitro blue tetrazolium chloride) or INT (P-iodonitrotetrazolium chloride) to form colored insoluble precipitate in the Ag-Ab reaction site.

Substrates for Alkaline Phosphatase

Naphthol AS-MX phosphate: This substrate is used with the chromogen Fast Red BB or Fast Red TR with the production of blue or red color, respectively. The chromogenic product is soluble in alcohol, and so aqueous counterstain and mounting medium are needed.

Naphthol AS-BI phosphate: This substrate is used with the chromogen new fuchsin to produce a red-colored chromogenic product that is insoluble in alcohol. So, dehydration by alcohol and DPX mounting is possible.

Table 3 shows the details of some frequently used enzymes and their substrates.

Fluorescence Labeling

Antibodies are labeled with a fluorescent dye such as fluorescein isothiocyanate (FITC). Usually, secondary antibodies are labeled with the fluorescein dye as it is of low cost and technique is flexible. Once the photon of the light hits the fluorochrome molecule, the electron from the outer shell absorbs energy and jumps to the higher orbit. The dye molecule is converted from the stable state (S_0) to excited singlet state (S_2) (**Fig. 7**). This excited state of the dye molecule is unstable and lasts only for a few nanoseconds. This unstable fluoro-

TABLE 3: Commonly used enzymes and the corresponding substrates.

Procedure	Color of the end product	Solubility in alcohol	Counterstain and mounting media
Peroxidase enzyme			
DAB: 3,3′-diamniobenzidine	Brown	Insoluble	Dehydration before putting the coverslip and DPX mounting possible
4-CN: 4-chloro-1-napthol	Blue	Soluble	Aqueous counterstain and aqueous mounting
AEC: 3-amino-9-ethylcarbazole	Red	Soluble	Aqueous counterstain and aqueous mounting
Alkaline phosphatase			
Fast blue BB	Blue	Soluble	Aqueous counterstain and aqueous mounting
Fast red TR	Red	Soluble	Aqueous counterstain and aqueous mounting
New fuchsin	Red	Insoluble	Dehydration before the putting the coverslip and DPX mounting possible

(DPX: dibutylphthalate polystyrene xylene)

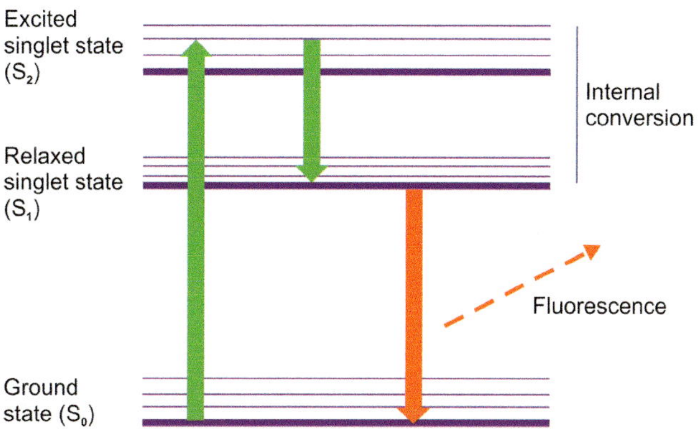

FIG. 7: Schematic diagram showing the fluorochrome excitation.

chrome molecule undergoes an internal conformational change which is known as "internal conversion." Here, energy is lost in the form of heat energy. This state of the molecule is known as a relaxed singlet (S_1). The electron of the molecule comes back to the lower level. The molecule now rapidly comes back to the stable ground state (S_0) by releasing energy in the form wave of light. This emission of energy is low, and therefore, the wavelength of the light is longer than the excited beam of light. The difference of the wavelength between the maximum absorption and emission spectra is called as "stokes shift."

Advantages

- Bright signal
- The different colored dye can be used.

Disadvantages

- The fluorescence may be unstable and may fade in the course of time.
- The fluorescent microscope is needed.
- The excitation filter beam should be appropriately selected for the specific dye.

Antibody Storage

The antibodies should be stored properly to prevent the aggregation because of hydrophobic interaction and also to maintain its biological activity. The optimum temperature to store the antibody is 2–8°C. The manufacturers always mention the "shelf life" and optimum storage temperature.

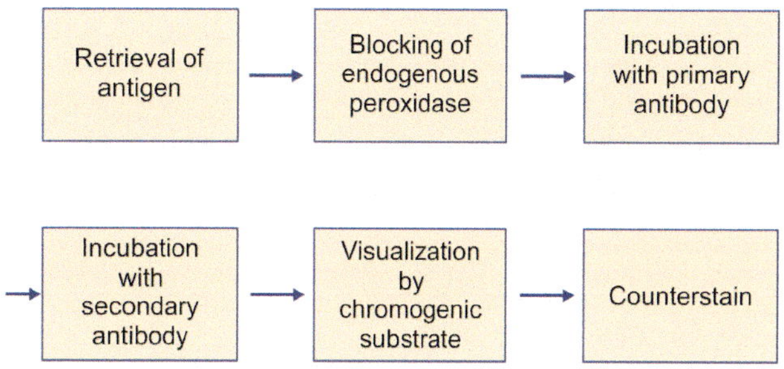

FIG. 8: Steps of immunocytochemistry.

Necessary Steps of Immunocytochemistry

The principle steps of immunocytochemistry are as follows (**Fig. 8**):

- *Retrieval of antigen*: This is essential to unmask the epitope of the antigen that is masked due to fixation. This can be done by heating or enzymatic method.
- *Endogenous peroxidase (or other enzyme) blocking*: This step helps to block the endogenous peroxide enzyme (if immunoperoxidase method is used).
- *Incubation with primary antibody*: Selected Ab to detect the corresponding antigen.
- *Incubation with secondary antibody*: Ab against the primary Ab tagged with enzyme to visualize the reaction.
- *Visualization by chromogen substrate*: The chromogenic substrate to react with the enzyme and produce visible end product to localize Ag-Ab reaction.
- *Counterstaining*: Counterstaining the cells to visualize the tissue and the spatial distribution

REFERENCES

1. Coons AH, Creech HJ, Jones RN. Immunological properties of an antibody containing a fluorescent group. Exp Biol Med. 1941;47(2):200-2.
2. Nakane PK, Pierce GB. Enzyme-labeled antibodies for the light and electron microscopic localization of tissue antigens. J Cell Biol. 1967;33(2):307-18.

3. Taylor CR, Burns J. The demonstration of plasma cells and other immunoglobulin-containing cells in formalin-fixed, paraffin-embedded tissues using peroxidase-labelled antibody. J Clin Pathol. 1974; 27(1):14-20.
4. McMichael AJ, Pilch JR, Galfre G, Mason DY, Fabre JW, Milstein C. A human thymocyte antigen defined by a hybrid myeloma monoclonal antibody. Eur J Immunol. 1979;9(3):205-10.
5. Erber WN, McLachlan J. Use of APAAP technique on paraffin wax embedded bone marrow trephines. J Clin Pathol. 1989;42(11):1201-5.
6. Cordell JL, Falini B, Erber WN, Ghosh AK, Abdulaziz Z, MacDonald S, et al. Immunoenzymatic label of monoclonal antibodies using immune complexes of alkaline phosphatase and monoclonal anti-alkaline phosphatase (APAAP) complexes. J Histochem Cytochem. 1984;32(2):219-29.
7. Sabattini E, Bisgaard K, Ascani S, Poggi S, Piccioli M, Ceccarelli C, et al. The EnVision++ system: a new immunohistochemical method for diagnostics and research. Critical comparison with the APAAP, ChemMate, CSA, LABC, and SABC techniques. J Clin Pathol. 1998;51(7):506-11.
8. Toda Y, Kono K, Abiru H, Kokuryo K, Endo M, Yaegashi H, et al. Application of tyramide signal amplification system to immunohistochemistry: A potent method to localize antigens that are not detectable by ordinary method. Pathol Int. 1999;49(5):479-83.
9. Tomita M, Tsumoto K. Hybridoma technologies for antibody production. Immunotherapy. 2011;3(3):371-80.

CHAPTER 2

Visualization of an Antigen-antibody Reaction

INTRODUCTION

Antigen (Ag) and antibody (Ab) reaction cannot be visualized by light microscope. The detection system is needed by visualizing this reaction. The enzymes or fluorescent probes are used to demonstrate the Ag-Ab reaction. In the case of enzyme labelling a suitable substrate is used to develop an insoluble chromogenic product.

The various types of visualization system are described here (**Box 1**).

BOX 1 Different technique of visualization of antigen-antibody reaction.

Direct technique:

- The primary Ab is tagged with the enzymes
- Rarely used

Indirect method:

- The enzyme-conjugated secondary Ab is used against the unlabeled primary antibody

Peroxidase-antiperoxidase (PAP) method:

- The PAP reagent is a stable complex molecule that contains horseradish peroxidase (HRP) Ag and Ab against it. The bridging secondary Ab is used that binds with both the primary and PAP complex

Avidin and biotin method:

- The secondary Ab is chemically attached with biotin that reacts with the primary Ab
- The avidin conjugated to HRP binds with the biotinylated secondary Ab

Avidin and biotin complex (ABC) technique:

- Primary Ab, biotinylated secondary Ab, and ABC are used
- The biotinylated secondary Ab attaches with avidin of ABC

Biotin–streptavidin method:

- Primary Ab, biotinylated secondary Ab, and enzyme-conjugated streptavidin enzyme conjugated streptavidin
- Streptavidin binds with biotin

Polymer-based labelling method:

- A polymer is used as backbone with embedded multiple enzymes and secondary Abs

Tyramine signal amplification:

- The biotinylated tyramine is activated by HRP in the presence of hydrogen peroxide
- The activated biotinylated tyramide forms covalent bond with tyramine in the tissue and deposited in the vicinity of HRP
- The biotinylated tyramide is further demonstrated by HRP-conjugated streptavidin-biotin technique with 3,3'-diaminobenzidine (DAB) as substrate

DIRECT METHOD

In the direct staining method, the primary Ab is tagged with the enzyme (**Fig. 1**). This Ab combines directly with the epitope of the Ag. Subsequently, with the help of chromogenic substrate, the Ag-Ab reaction is visualized.

Advantages

- Rapid technique
- Relatively simple

Disadvantages

- The Ab should be pure and highly specific otherwise erroneous staining my happen.
- The sensitivity of the direct staining technique is much lower than the indirect method.
- The various Ab should be labeled separately.
- A higher concentration of primary Ab is needed, and therefore, it is costly.

INDIRECT METHOD

Here the conjugated (enzyme/fluorochrome) secondary Ab is used against the unlabeled primary Ab. The Ag-Ab reaction

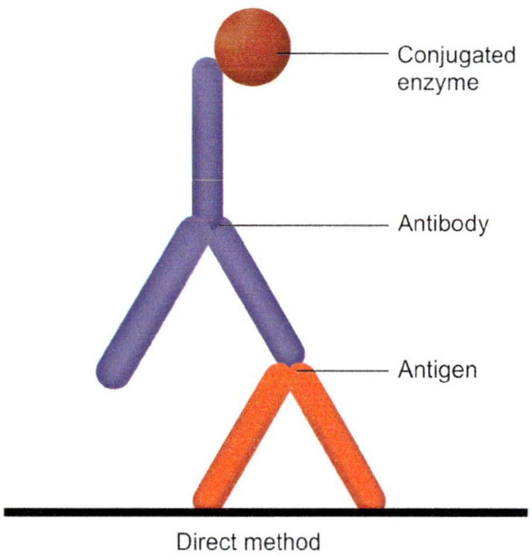

FIG. 1: Schematic diagram showing the direct technique of immunostaining.

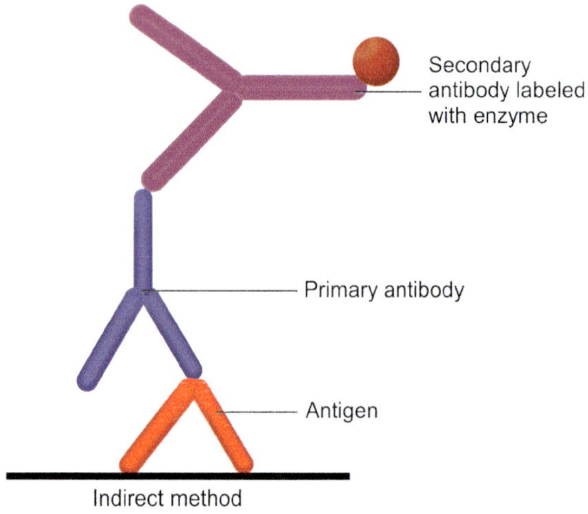

FIG. 2: Enzyme labeled secondary antibody against the primary antibody.

is demonstrated by a suitable substrate forming a colored product (**Fig. 2**).

Advantages

- The primary Ab is used in higher dilution, and so the cost is less.
- Easy to conjugate a single type of secondary Ab.
- The secondary Ab can be produced easily with high specificity against the primary Ab.
- Easy to have negative control by omitting the primary Ab.

PEROXIDASE-ANTIPEROXIDASE METHOD

The peroxidase-antiperoxidase (PAP) technique is introduced by Sternberger.[1] The PAP reagent is a stable complex molecule that contains horseradish peroxidase (HRP) Ag and Ab against it. The bridging secondary Ab is used that binds with both the primary and PAP complex (**Fig. 3**). The primary Ab and PAP complex must be of the same species, or they may have an almost similar antigenic determinant site. The secondary Ab is developed from another species and has specificity against both the primary Ab and Ab present within the PAP complex.

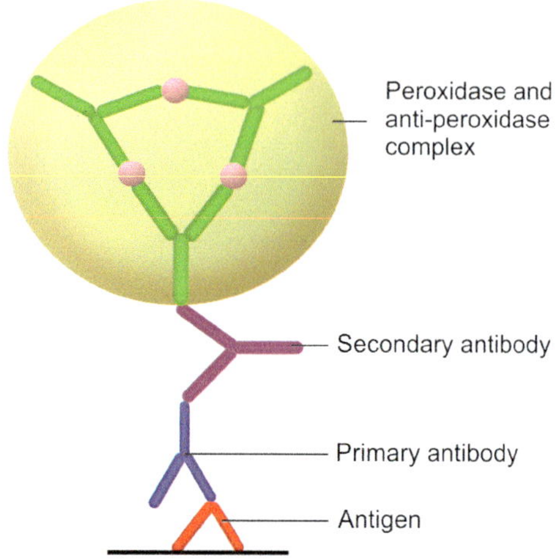

FIG. 3: Schematic diagram of peroxidase-antiperoxidase.

Advantages

Peroxidase-antiperoxidase method is a very sensitive technique.

AVIDIN AND BIOTIN METHOD

Avidin and biotin have high affinity with each other, and this is exploited in this technique. The secondary Ab is chemically attached with biotin that reacts with the primary Ab. Now the avidin conjugated to HRP is used that binds with the biotinylated secondary Ab (**Fig. 4**). The reaction is visualized by adding the substrate.

Advantages

- Relatively rapid technique
- Good sensitivity

Disadvantages

- The normal tissue may contain endogenous biotin and may produce false-positive reaction by binding with HRP-conjugated avidin. This reaction can be prevented by using a suitable blocking agent.
- The binding affinity between avidin-biotin may differ from batch to batch and thereby, may affect the sensitivity and reproducibility of the technique.

AVIDIN AND BIOTIN COMPLEX TECHNIQUE

The avidin and biotin complex (ABC) method is the modification of the previously described avidin-biotin technique.[2]

The essential ingredients of this technique are: (1) Primary Ab, (2) Biotinylated secondary Ab, and (3) ABC.

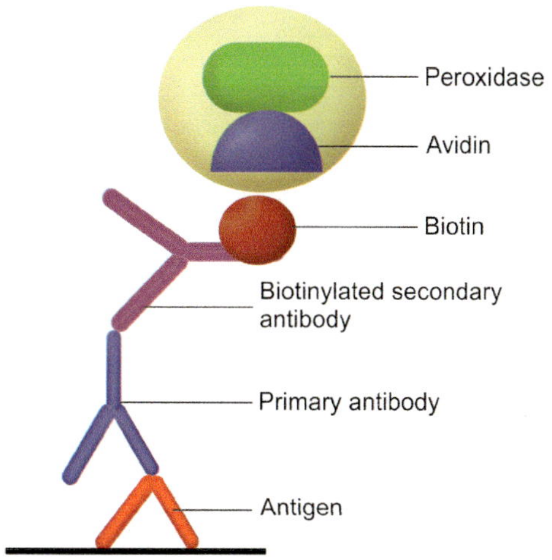

FIG. 4: Schematic diagram showing avidin-biotin technique.

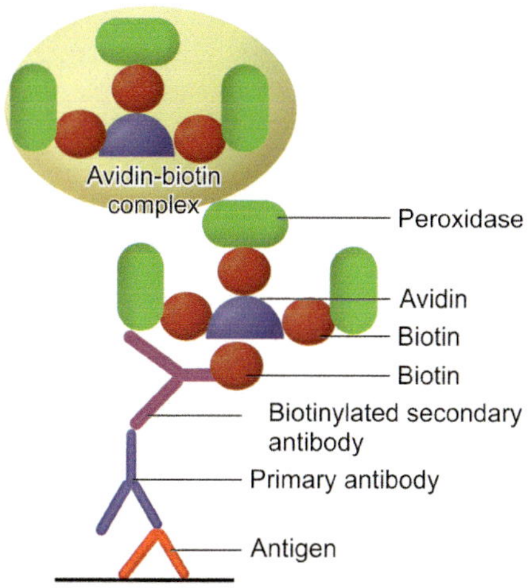

FIG. 5: Avidin-biotin complex technique.

The preformed ABC contains avidin and biotin-labeled peroxidase. This large lattice-like molecule of ABC has a strong affinity with biotin molecule. The secondary Ab is also tagged with biotin that attaches with avidin of ABC as the ABC molecule contains multiple peroxidase enzymes, so the signal intensity is high (**Fig. 5**).

Advantages

- Sensitivity is many-fold higher than simple avidin-biotin technique.
- High dilution of primary Ab can be used.

Disadvantages

Endogenous biotin in tissue may also react with ABC.

BIOTIN–STREPTAVIDIN METHOD

In the biotin-streptavidin (B-S) technique instead of avidin, we use enzyme-conjugated streptavidin that has a very high affinity with biotin (**Fig. 6**). The B-S technique is better than the avidin-biotin method due to its higher sensitivity.

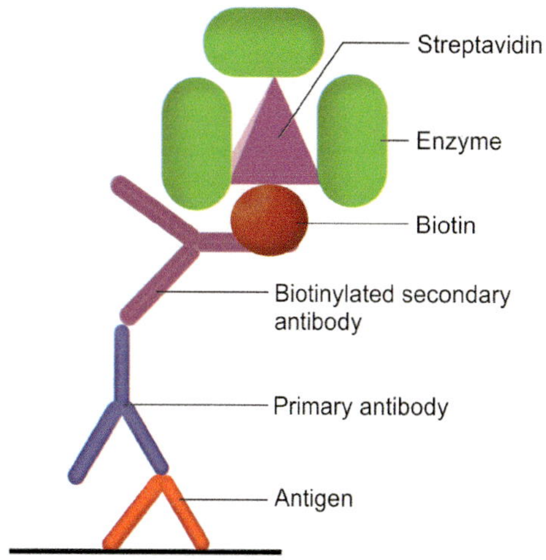

FIG. 6: The schematic diagram showing the streptavidin and biotin technique.

Advantages

- Streptavidin does not contain any carbohydrate, so there is no cross-reaction with a lectin-like molecule that is present in liver, kidney, etc.
- The streptavidin-enzyme complex is more stable, and therefore, the diluted form can be stored a long time and immediately usable.
- No nonspecific staining develops in this technique.

ALKALINE PHOSPHATASE–ANTIALKALINE PHOSPHATASE METHOD[3]

The alkaline phosphatase–antialkaline phosphatase (APAAP) technique is not used nowadays. Here, similar to PAP technique, we apply a complex of APAAP along with a secondary Ab that binds both the complex and primary Ab. The primary Ab and the Ab in the APAAP complex should be from the same species, and secondary Ab should be from some other species.

Advantages

- The APAAP technique is useful in the tissue sections that are rich in endogenous peroxidase.
- Dual labelling with PAP technique.
- The APAAP complex is relatively stable.

POLYMER-BASED LABELING METHOD

Polymer-based labeling method is a simple and very sensitive technique. Here a polymer is used as a backbone with multiple embedded enzymes and secondary Abs.[4] The polymer can be made of dextran, dendrimers or polypeptides. Recently micropolymer of very small size is used to prevent steric hindrance. This system is extremely sensitive because each polymer contains a large number of secondary Abs and

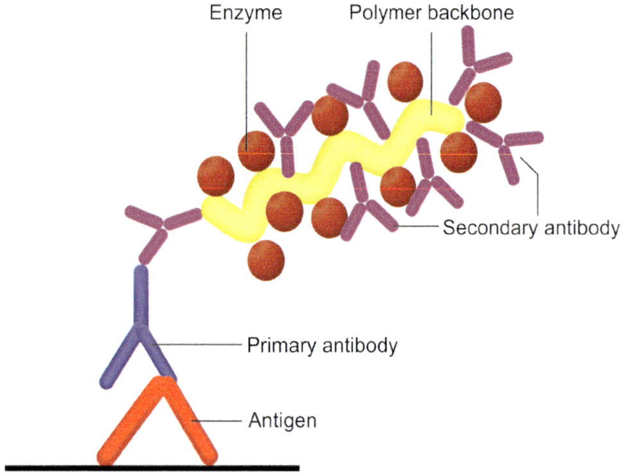

FIG. 7: Polymer-based labelling method.

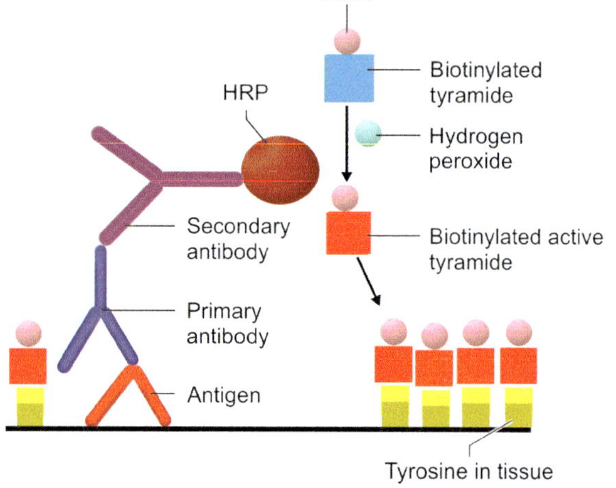

(HRP: horseradish peroxidase)

FIG. 8: Tyramine signal amplification method.

enzymes (**Fig. 7**). As no biotinylated Ab is used so there is no need to use any blocking agent to prevent cross-reaction with tissue biotin.

Advantages

- Extremely sensitive technique as a large number of enzymes are embedded in polymer.
- Rapid, simple two steps method.
- No cross staining as no biotin or avidin is used.

TYRAMINE SIGNAL AMPLIFICATION

Tyramine signal amplification (TSA) method is also called a catalyzed reporter deposition technique. It is based on the deposition of the activated biotinylated tyramine.[5] The biotinylated tyramine is activated by the oxidizing agent HRP in the presence of hydrogen peroxide. Subsequently, this activated-biotinylated tyramide forms a covalent bond with electron rich tyramine in the tissue and deposited in the vicinity of HRP (**Fig. 8**). The biotinylated tyramide is further demonstrated by HRP-conjugated streptavidin-biotin technique with 3,3'-diaminobenzidine (DAB) as a substrate.

Steps of TSA:

- At first, complete the step up to the secondary Ab with HRP.
- Now add biotinylated tyramine.
- After proper wash add HRP attached streptavidin.
- Develop color by adding DAB substrate.

MULTIPLE IMMUNOCYTOCHEMISTRY IN A SINGLE SECTION

Multiple immunocytochemistry in a single tissue section helps to detect the various Ag in the tissue localized in the different cell or the same cell in the different subcellular compartment. It provides crucial information about the distribution of the Ags and the character of the neighboring cells.

The prerequisite before the designing of the multiple staining:

- *The target section*: Formalin-fixed paraffin-embedded or cytology smear
- *Reagents available*: Primary Ab of the same species or different species and also the secondary Ab against the primary Ab from the same or different species.
- *Sequential or simultaneous staining*: The staining of all the Ags at the same time happens in the simultaneous staining, and it requires less time whether the staining of the Ag in a stepwise manner is sequential and needs more time.
- Nature of the target antigen.

Selection of primary Ab: The selection of the primary Ab from the different host species is the simplest technique.

Selection of secondary Ab: The secondary Abs are usually raised against the primary Ab of the host species. However, these different secondary Abs are developed in the same species. The normal serum of the host species of the secondary Ab is used as blocking agent to prevent the nonspecific immunostaining.

Selection of the enzymes and substrates: The enzyme-substrate combination should be designed in such a way to have red-blue combination than red-brown combination because red-blue combination of the final color is easy to distinguish. To develop blue color chromogenic product alkaline phosphatase and fast blue BB is the most popular combination, whereas, for the red color development, HRP and AEC (3-amino-9-ethylcarbazole) is the desirable combination.

Steps:

- After Ag retrieval, please wash the sections
- Apply primary Abs
- Wash
- Endogenous biotin blocking
- Wash
- Apply secondary Ab
- Peroxidase blocking by hydrogen peroxide
- Wash
- Incubate with New fuchsin chromogen
- Wash
- DAB
- Wash
- Mount and see

REFERENCES

1. Sternberger LA. Immunocytochemistry. Englewood Cliffs, NJ: Prentice Hall; 1974.
2. Hsu SM, Raine L, Fanger H. A comparative study of the peroxidase-antiperoxidase method and an avidin-biotin complex method for studying polypeptide hormones with radioimmunoassay antibodies. Am J Clin Pathol. 1981;75(5):734-8
3. Erber WN, McLachlan J. Use of APAAP technique on paraffin wax embedded bone marrow trephines. J Clin Pathol.1989;42(11):1201-5.
4. Kämmerer U, Kapp M, Gassel AM, Richter T, Tank C, Dietl J, et al. A new rapid immunohistochemical staining technique using the EnVision antibody complex. J Histochem Cytochem. 2001;49(5):623-30.
5. Toda Y, Kono K, Abiru H, Kokuryo K, Endo M, Yaegashi H, et al. Application of tyramide signal amplification system to immunohistochemistry: A potent method to localize antigens that are not detectable by ordinary method. Pathol Int. 1999;49(5):479-83.

CHAPTER 3

Basic Preparation for Immunostaining

INTRODUCTION

The common cytology specimens for immunocytochemistry (ICC) include cell block specimen, liquid-based cytology (LBC) smear, cytospin smear, and the direct cytology smear from the imprint or fine-needle aspiration cytology (FNAC). Out of all these specimens, the cell block is the preferable and the most commonly used specimen for ICC.

CELL BLOCK

The cell block is prepared to have paraffin-embedded tissue. The cell block can be made from the FNAC material and fluid specimens such as effusion fluid or bronchoalveolar lavage. The section from the cell block can be used for ICC, hematoxylin, and eosin staining and also various special stains.
The advantages of cell block are as follows:

- A panel of antibody can be applied from the multiple sections.
- Minimal background stain.
- Relatively consistent cellularity.
- Tissue can be preserved for the archival use.

Methods of Cell Block Preparation

- Collect sample in 10% neutral buffered formalin (NBF): FNAC material can be put directly in the formalin and the fluid sample can be centrifuged and then be kept in 10% NBF.
- Keep the sample in NBF for 4 hours for cellular fixation.
- Centrifuge the fluid for 10–15 minutes at 1,500 rounds per minute.
- Take the sediment and wash it in phosphate buffer solution.
- Add plasma (100 μL) and thrombin (30 μL).
- Remove the clot and put it on the filter paper.
- Process the clot in the tissue processor.

Compact Cell Block Technique

Compact cell block technique can also be done as described by Young et al.[1] *Advantages of compact cell block technique are as follows:*

- The cells are in a relatively smaller area, and so more cells are obtained in a single section.
- The tissue section is entirely devoid of red blood cells (RBCs).

Steps

- Centrifuge approximately 40 mL of fluid at 3,000 rounds per minute for 8 minutes.
- Add CytoRich Red solution (1:1 ratio) and keep it for 2 minutes. It will lyse the RBCs and dissolve the proteins.
- Add plasma (four drops) and three drops of thrombin (5,000 UI/10 mL).
- Gently agitate the solution to get a clot and wait for 2 minutes to complete the process.
- Transfer the clot with embedded cells in a lens paper.
- Wrap the clot and fix it in the formalin.
- Formalin-fixed tissue is now processed in the tissue processor.

LIQUID-BASED CYTOLOGY

Commercially available LBC technique can produce multiple monolayer smears which can be used for ICC. The smears should be fixed in a suitable fixative for ICC.

Advantages

- Background of the smear is usually free of debris and RBCs.
- Multiple smears can be produced in LBC.
- The smear is in a small circular area.

Direct Smear

The smear(s) can be prepared from the FNAC or sediment from the centrifuged fluid sample for ICC. This type of direct smear is substandard for the ICC.

Advantages

- This is the easiest sampling technique, and no extra effort is needed.

Disadvantages

- The smear often gives background artifact.
- Only a limited number of slides can be prepared, and so the choice of antibody is limited.

Cytospin Preparations

Multiple smears are easily made with the help of cytospin technique of fluid, particularly effusion sample and FNAC material collected in the fluid.

Advantages

- Relatively multiple smears can be prepared.
- Multiple antibodies can be applied.

Disadvantages

- Mucoid and blood-rich specimen cannot be processed.
- Specimen with scanty cellularity cannot be processed.

Fixation of Tissue

The fixation of tissue or cell sample is the most crucial factor to preserve the antigen and its demonstration (**Box 1**). The following factors regarding fixation are essential to get a good result in ICC:

- The proper concentration and pH of the fixative.
- Immediate fixation
- Duration of fixation
- Type of fixative

There is no effect on ICC on the fixative and tissue ratio or the duration after fixation.

There were various recommendations to improve the result of immunohistochemistry for estrogen receptor assessment in breast carcinoma.[2]

These recommendations are equally useful for ICC on cytology materials.

The summary of some crucial suggestions is given below.

BOX 1 Fixation of tissue.

Basic purpose: To preserve the cell morphology and the epitope of the antigen

Universal fixative: 10% neutral buffered formalin (NBF)

Duration: At least 8 hours

Major recommendations:

- Fix the sample without delay
- Use only 10% NBF
- Note the time of fixation
- Apply adequate duration

Mechanism of fixation of NBF: By crosslinking of aldehyde group with lysine and forming methylene bridge

Major advantages of NBF:

- Most of the epitope of antigen is preserved
- Cheap
- Bactericidal effect

Disadvantages:

- Antigen retrieval needed
- NBF is carcinogenic

Recommendation on Fixation

- Fix the specimen as quick as possible.
- All the samples should be fixed in an identical manner.
- Only 10% NBF should be used as a fixative.
- The minimum fixation time is 8 hours.
- Alcoholic fixation should be avoided in cytology sample and formalin should be the choice of fixation.

Formalin Fixation

Neutral Buffered Formalin

Formaldehyde is a gas and when it is dissolved in water, methylene hydrate is formed. The compound forms polymers with each other, and the solution is called formalin. This formalin solution consists of 37–40% formaldehyde and the remaining 60–63% water. This formalin is considered as 100% formaldehyde. For the routine laboratory use, 10% of the formalin is prepared.

Mechanism of Fixation

The formaldehyde has the aldehyde group that binds with the lysine of the protein to form a methylene bridge. Subsequently, the crosslinking of the molecules produces an insoluble product.

Preparation

- Take 100 mL of formaldehyde (40%).
- Add 900 mL distilled water.
- Add sodium dihydrogen phosphate: 4 g.
- Add disodium dihydrogen phosphate: 6.5 g.

Advantages

- It is the most cost-effective fixative.
- Individual cellular morphology is intact.
- Formalin has the sterilization property and prevents bacterial degradation.
- Formalin prevents any diffusion of protein from the cell.
- NBF is an excellent fixative for hormones.

Disadvantages

- It is unsuitable for any ribonucleic acid (RNA)-based ICC.
- Formalin often masks the epitope, and therefore demasking or antigen retrieval (AR) is needed.[3]
- Formalin is carcinogenic.

Alcohol Fixation

Ethyl alcohol (95%) and methyl alcohol are often used as fixative.

Mechanism: Coagulation of protein.

Advantages

- Quick fixation than NBF.
- No AR is needed.

Other fixatives for ICC: Paraformaldehyde, glutaraldehyde, osmium tetroxide, acetone, etc.

Antigen Retrieval

The aldehyde group of formaldehyde causes methylene bridge with the protein that hinders the access of the antibody to the epitope. Therefore, there is a great need to unmask the epitope of the antigen for the antigen–antibody reaction. The process of unmasking the antigenic epitope is called as AR. It has been noted that formalin does not cause a change of the epitope configuration and by simply heating or treating the tissue in alkali can reverse the reaction of formalin and protein.[4] The two most common methods of AR are: (1) Heat-induced AR, and (2) Enzymatic method (**Box 2**).

Major Factors of Antigen Retrieval

- *Heat*: Higher temperature heating provides better AR. However, to preserve the cell morphology, it is better to apply a lower temperature (90°C) with prolonged time. The degree of temperature and heating time are inversely proportional.
- *pH of the AR solution*: The pH of the AR solution may grossly affect AR. The best pH of AR solution is 8–9.
- *Composition of AR solution*: The chemical constituents of the AR solution may affect AR. Nowadays, the vendor of the automated immunostainer recommends their own AR solution with secretive chemical composition.

Heat-induced Antigen Retrieval

Mechanism

Exact mechanism is not understood correctly. Possibly, heat breaks down the methylene bridge of the protein and formalin and helps to penetrate the antibody to the epitope.

Microwave, pressure cooker, or water bath can be used for heat-induced AR.

Microwave retrieval: Jessup et al.[5] has described microwave retrieval technique in 1994. It is one of the most efficient methods. However, to get a consistent, reproducible result, one should take care of the various factors such as:

- The wattage of the microwave (750–800 W).
- *AR buffer solution*: Citrate buffer (0.01M, pH 6).
- *The volume of AR buffer solution*: Approximately 600 mL for 25 slides.
- *Number of slides*: Usually 25 slides; however, it depends on the container.
- *Time*: Usually 10–15 minutes. Nuclear Ag may take more time.

BOX 2 Antigen retrieval.

- Antigen retrieval (AR) is essential to unmask the antigenic determinant site (epitope)
- The two most common methods of AR are: (1) Heat-induced AR, and (2) Enzymatic method

Heat-induced AR:

- Microwave retrieval: The most efficient technique
- Pressure cooker method
- Water bath heating

Enzyme method of AR: Less popular

Method:

- At first, dewax the paraffin section and rehydrate it.
- Put the slides in a microwave oven-resistant plastic slide rack and then place it in the incubation container filled with an adequate amount of AR buffer solution to cover the slides fully.
- Trisodium citrate buffer is used as an AR solution:
    - Add 2.94 g trisodium citrate in 1 L of ultrapure water.
    - Adjust pH at 6 with concentrated hydrochloric acid.
- Put a lid over the container to prevent evaporation.
- *Incubate the slides in the microwave oven*: 15 minutes in 800 W.
- Remove the container with slides and cool it at room temperature by running cold tap water for 20 minutes.
- Proceed for ICC protocol.

Precautions:

- Always prevent drying of the slides.
- Always avoid metallic container.

Pressure cooker method: It is also a useful technique as it gives uniform heating to the slides.

Methods:

- At first, dewax the paraffin section and rehydrate it.
- Pour 1.5 L AR buffer solution in the pressure cooker without lid and heat the pressure cooker until the AR solution is boiling.
- Now carefully immerse plastic holder with slides within the AR solution.
- Seal the lid and heat the pressure cooker to generate full pressure and incubate the slides for 2 minutes.
- Cool the cooker by running cold water for 20 minutes.
- Open the lid carefully and allow the slides to cool.
- Proceed for ICC protocol.

Precautions:

- Always prevent drying of the slides.
- The AR buffer solution should be enough to immerse the slides.

Water bath heating: This technique is easy as most of the laboratories have the facility of the water bath.

Method:

- At first, dewax the paraffin section and rehydrate it.
- Fill the Coplin jar with AR solution.
- Put the jar in the water bath and heat it till the temperature of the bath is 98°C.
- Immerse the slide in this preheated AR solution and cover the lid.
- Incubate the slides in the water bath for 30 minutes.
- Remove the container and allow it to cool.
- Wash the slides and proceed for ICC protocol.

Enzyme Method of Antigen Retrieval

The enzymatic method of AR is relatively less popular as it often gives the inconsistent result, and strict supervision is needed. The incubation time of this technique depends on various factors such as slide load, pH of the enzyme solution, and temperature. The commonly used enzymes are trypsin and proteinase K.

The composition of enzyme solution is as follows:
- 0.1 g chymotrypsin
- 0.1 g calcium chloride
- 1,000 mL distilled water
- Maintain pH 7.8 by using sodium hydroxide solution.

Method
- Dewax the paraffin section and rehydrate it.
- Keep the slides in distilled water at 37°C.
- Incubate the slides in the freshly prepared enzyme solution for 10 minutes at 37°C.
- Take out the slides and wash them in cold running water for 3–5 minutes.
- Proceed for ICC protocol.

BLOCKING NONSPECIFIC BACKGROUND STAINING

Nonspecific background staining is caused by the following (**Box 3**):
- Nonspecific antigen binding.
- Endogenous enzymatic activity.

Nonspecific Antibody Binding

Causes

- Primary antibody, particularly polyclonal antibody, binds with the unwanted antigen.
- The antibodies are highly charged molecules and may be attracted by the corresponding reciprocal charged molecules in the tissue such as collagen and reticulin. This binding of the primary antibody with the connective tissue is nonimmunological. Now the addition of the secondary tagged antibody will bind both the primary antibody binds with connective tissue and also the specific antigen. It will produce false-positive localization of the antigen.

BOX 3 Non-specific background staining.

It is caused by:
- Nonspecific antigen binding
- Endogenous enzymatic activity

Nonspecific antigen binding:
- Primary antibody binds with the unwanted antibody of the tissue
- The highly charged antibodies molecule binds with the corresponding reciprocal charged molecules in the tissue (collagen and reticulin) producing false positive localization of the antigen

Blocking:
- Prior treatment of the section with normal serum to neutralize and saturate the charged sites of the connective tissue
- The hydrophobic bonding of the primary antibody is minimized by adding the detergent or highly concentrated salt solution to the buffer

Endogenous enzymatic activity:
- Endogenous enzymes of the tissue react with the substrate used in the visualization system

Blocking:
- Prior treatment of the tissue with the peroxidase (for endogenous peroxidase) or levamisole (alkaline phosphatase)

Blocking

This nonspecific binding can be blocked by:
- Prior treatment of the section with normal serum causes binding of the serum protein with the charged sites. It will neutralize and saturate the charged sites of the connective tissue and therefore may decrease the nonspecific binding of the antibody. Usually, the normal serum from the same animal species is used in which the secondary antibody is raised.
- The hydrophobic bonding of the primary antibody and tissue can be minimized by adding the detergent (e.g., Triton X) or highly concentrated salt solution to the buffer.

Endogenous Enzymatic Activity

Endogenous enzymatic activity is another cause of the background false positive staining in ICC. The normal tissue contains endogenous peroxidase or alkaline phosphatase enzyme. This enzyme may not be inactivated by routine processing. Alkaline phosphatase enzyme is completely inactivated by routine formalin fixation and paraffin embedding. However, the endogenous peroxidase enzymes retain their activity and may react with the substrate used in the visualization system (**Figs. 1A** and **B**), if we use the secondary antibody tagged with a similar enzyme. Therefore, it is mandatory to block the endogenous tissue enzyme before the addition of secondary antibody.

Blocking

The blocking of endogenous peroxidase enzymes is done by:
- Incubation of the tissue section by methanol-hydrogen peroxide for 10 minutes (absolute methanol with 0.5% hydrogen peroxide).

The blocking of endogenous alkaline phosphatase enzymes is done by:
- 1 mM levamisole in 0.5 M HCl solution.

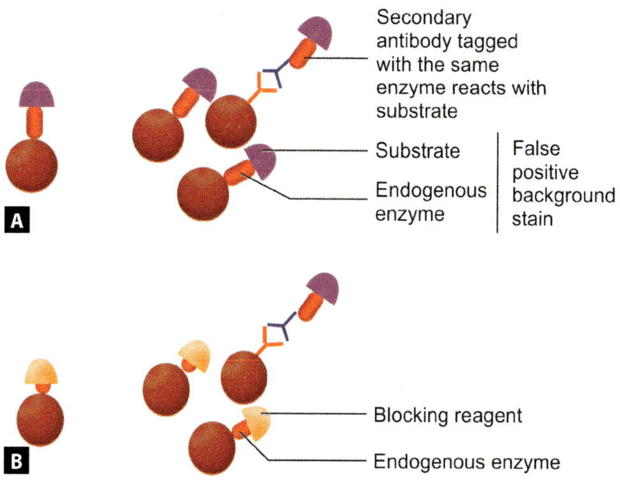

FIGS. 1A AND B: (A) Background false positive staining due to the reaction of endogenous enzyme and substrate. (B) Blocking of the endogenous enzyme before incubation with secondary antibody prevents background staining.

SELECTION OF PRIMARY ANTIBODY

In case of selection of primary antibodies, the following factors are important (**Box 4**):

- *Clinical application*: What is the exact use of the antibody?
- *Sensitivity and specificity*: The lower expression of the target antigens may need the antibody with higher sensitivity. For the diagnostic purpose, the antibody should not cross-react with the other antigen, and highly specific monoclonal antibody is needed.
- *The type of specimen to use this antibody*: Smear or formalin-fixed paraffin-embedded tissue or flow cytometry.
- *In which species the antibody is raised*: This is important for using the secondary antibody.
- *AR*: It is essential to know the necessity of AR and the preferable procedure.
- *Antibody dilution*: In which concentration the antibody should be used to have the desired result.
- *The choice of positive control*: Using a positive control is mandatory, and so one should have an idea about the preferable positive control for the antibody.

It is recommended to purchase the antibody from the well-reputed commercial company for the routine diagnostic use.

The details of the antibody are always available in the manufacturer's website and the following information is usually given:

- Conjugated or not
- Clone
- Biological source
- Application of formalin-fixed paraffin-embedded tissue or immunofluorescence or flow cytometry/species reactivity
- Molecular weight
- Storage temperature
- Antibody product type (primary or secondary antibody)
- Contains (preservative)

Antibodies are usually supplied in the concentrated form, either solid or liquid. During application, one has to dilute the antibody ranging from 1 to 20 µg/1 mL. It is important to remember that the more the concentration of the primary antibody at the time of incubation, the more is the risk of getting the background staining.

BOX 4 Primary antibody.

The important points for selection of primary antibody:

- *Clinical application*: The exact use of the antibody
- *Sensitivity and specificity*: The antibody should not cross react with the other antigen
- The type of specimen to use this antibody
- In which species the antibody is raised
- *Antigen retrieval*: Needed or not
- *Antibody dilution*: The optimum concentration the antibody
- The choice of positive control for using the antibody

Optimizing the dilution of primary antibody:

- To get maximum intensity and minimum background staining
- The optimum incubation time of the antibody
- The appropriate incubation temperature of the antibody

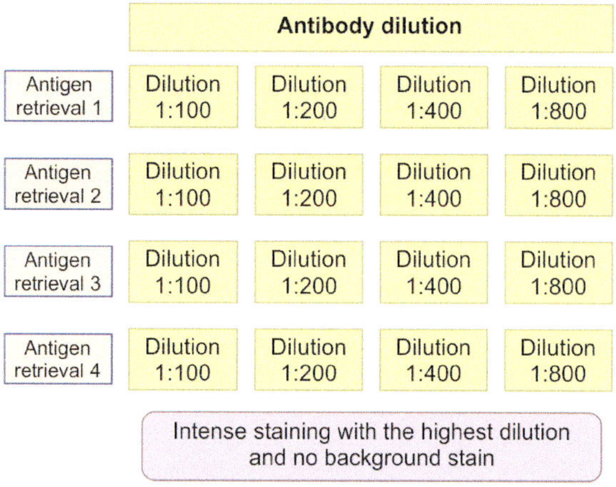

FIG. 2: Schematic diagram showing the optimization of primary antibody.

Optimizing the Dilution of Primary Antibody

The optimization of the primary antibody indicates:

- The most suitable AR protocol.
- The optimum dilution of the antibody to get the maximum intensity and minimum background staining.
- The optimum incubation time.
- The appropriate incubation temperature.

Antigen retrieval technique is the crucial factor to avoid false-negative immunostaining. For the initial standardization of the antibody, at least four different AR protocols should be used (**Fig. 2**). Each protocol should follow four different dilutions from the lowest to highest dilution. Usually, the range of 1–5 µg/mL dilution is recommended for the initial incubation. It is essential to follow the absolute concentration than simply the dilution. For the initial test, the incubation should be overnight, and if it works then the incubation time can be reduced to 30 minutes to 1 hour. To compare the result of the different dilution of antibody, we should follow consistent time and temperature. The best working highest dilution of antibody should be used.

REFERENCES

1. Yang GC, Wan LS, Papellas J, Waisman J. Compact cell blocks. Use for body fluids, fine needle aspirations and endometrial brush biopsies. Acta Cytol. 1998;42(3):703-6.
2. Yaziji H, Taylor CR, Goldstein NS, Dabbs DJ, Hammond EH, Hewlett B, et al. Consensus recommendations on estrogen receptor testing in breast cancer by immunohistochemistry. Appl Immunohistochem Mol Morphol. 2008;16(6):513-20.
3. Shi SR, Cote RJ, Taylor CR. Antigen retrieval techniques: current perspectives. J Histochem Cytochem. 2001;49(8):931-7.
4. Shi S-R, Key ME, Kalra KL. Antigen retrieval in formalin-fixed, paraffin-embedded tissues: An enhancement method for immunohistochemical staining based on microwave oven heating of tissue sections. J Histochem Cytochem. 1991;39(6):741-8.
5. Jessup E. Antigen retrieval technique for the demonstration of immunoglobulin light chain in formalin fixed paraffin embedded section. UK NEQAS Newsletter. 1994:4;12-16.

CHAPTER 4

Practical Technique, Quality Control and Troubleshooting in Immunocytochemistry

INTRODUCTION

The necessary steps of immunochemistry (ICC) have been outlined in Chapter 1. To have a good successful immunostaining, one should be familiar with the nature of localization of the antigen, the type of primary antibody (Ab) and its working nature (dilution, incubation time, temperature, etc.), the chromogen and substrate.

CONTROL

Selection of control: This is one of the essential parts of ICC as it gives the validation of the test. The control provides information about the specificity of the primary Ab and also the sensitivity of the technique.

Positive tissue control: Positive tissue control means that the tissue contains the antigen in the stable and demonstrable levels. The positive control indicates that the stain is working correctly. It is preferable to take the same type of section or smear for the control which means that for the cytology smear, the tissue section should not be taken as control.

Negative tissue control: Negative tissue control is needed to assess the specificity of the Ab. A tissue with the absence of the corresponding antigen can be used. However, usually, the primary Ab is omitted in the negative control.

Absorption control: This is mainly needed to test the specificity of a new Ab. The primary Ab is absorbed by the immunogens against which the Ab is formed. It may be assumed that the immunogens completely neutralize the primary Ab, and it should not react with the tissue antigen in the section. Now, the absorbed Ab is used for the ICC. If the positive stain occurs, then it indicates that the staining is nonspecific.

PRACTICAL STEPS OF IMMUNOCHEMISTRY

1. *Antigen retrieval (AR):* The details of AR have been already described in Chapter 3.
2. *Blocking nonspecific background staining:* Blocking of the endogenous enzymes and also nonspecific Ab binding have been described in Chapter 3.
3. *Incubation with primary antibody:* The primary Ab should be optimally diluted to provide the best result. The vendor usually mentions the optimal dilution; however, as described before, the optimal dilution of the Ab can be checked as mentioned in Chapter 3. Similarly, temperature and incubation time are also critical. The following practical factors should be noted regarding the incubation of the primary Ab:
   a. The slides should be kept in the humid chamber, and the sections should not be dried under any circumstances.
   b. The slides should be in a level so that the Ab cannot be drained out.
   c. The tissue section should be preferably in the center of the slide.
   d. Two slides should be placed with a reasonable distance to avoid cross-contamination of the Ab.
4. *Incubation with labeled secondary antibody:* Secondary Ab is raised against the primary Ab, and it is conjugated with the enzyme or enzyme complex. Usually, 30 minutes' incubation of the secondary Ab is necessary.
5. *Chromogenic substrate:* Now, appropriate chromogen is used. Chapter 1 describes details of the combination of the enzyme and chromogenic substrate. Proper dilution of the chromogenic substrate is mandatory as mentioned by the company. Proper care should be taken to cover the section or smear by the substrate solution.
6. *Counterstaining and mounting the slides:* This is the final step of ICC. Light hematoxylin stain is preferable for the nuclear stain to visualize the nuclear antigen. Usually, it takes 10–20 seconds to stain the nuclei. However, the fresh hematoxylin stain solution may take less time. The alcohol-fast chromogen product such as HRP-DAB (horseradish peroxidase-3,3′-diaminobenzidine) is mounted by DPX (dibutylphthalate polystyrene xylene), whereas the alcohol-soluble chromogenic product such as AP-Fast Blue BBN should be mounted in aqueous mounting media.

PREPARATION FOR STAINING

Reagent Preparation

Tris-buffered Saline (TBS)

- Take 6 g Trisaminomethane and 8.76 g Sodium chloride in 500 mL distilled water and adjust pH to 7.6 by 1M hydrochloric acid (HCl).
- Make the solution up to 1,000 mL.

You can keep it in 4°C for 3 months.

3,3′-Diaminobenzidine Solution

- Take 6 mg DAB in 10 mL of 0.05 M TBS.
- Now add 0.1 mL of 3% hydrogen peroxide.

3% Hydrogen Peroxide

- Add 100 µL of 30% hydrogen peroxide in 10 mL distilled water.

Antigen Retrieval Solution

Trisodium citrate buffer AR solution:

- Add 2.94 g Trisodium citrate in 1 L of ultrapure water.
- Adjust pH at 6 with concentrated HCI.

Primary Antibody Dilution Buffer

Reagents to mix:

- 15 mL sodium chloride
- 10 mL Tris-HCl
- 2.5 mL Triton X
- 10 g bovine serum albumin
- 2.5 mL 20% NaN_3

Now add distilled water up to 500 mL, filter it and keep it at 4°C.

Secondary Antibody Dilution Buffer

- Take 0.01M phosphate buffer solution and keep pH 7.2.
- Add 0.05% sodium azide as preservative.

IMMUNOCYTOCHEMISTRY PROTOCOLS

Peroxidase Technique

Manual ICC technique:

- Dewax the tissue.
- Bring the tissue in water by serial changes in graded alcohol.
- Rinse the section in TBS.
- *Antigen retrieval:* Do AR if needed.
- *Endogenous enzyme block:* Apply 3% hydrogen peroxide in methanol for 10 minutes in case of peroxidase blocking.
- Wash three times by TBS.
- *Nonspecific blocking:* Block nonspecific immunostaining by applying 3% bovine serum albumin for 30 minutes.
- Wipe excess solution.
- *Primary antibody:* Incubate the section with a primary Ab with proper dilution for 30–60 minutes.
- Rinse with TBS thoroughly.
- *Secondary antibody:* Apply enzyme-conjugated secondary Ab for 30 minutes.
- *Substrate:* Apply DAB solution for 10 minutes.
- Wash by running water.
- *Counterstain:* Hematoxylin for 10 seconds
- Dehydrate, clear, and mount by DPX.

Alkaline Phosphatase—Antialkaline Phosphatase (APAAP) Technique[1]

- Dewax the tissue.
- Bring the tissue in water by serial changes in graded alcohol.
- Rinse the section in TBS.
- *Antigen retrieval:* Do AR if needed.
- Rinse in TBS.
- Wipe excess solution and incubate in 10% casein for 10 minutes.
- *Primary antibody:* Incubate the section with optimally diluted primary Ab for 30 minutes.
- Rinse in TBS.
- *Secondary antibody:* Apply optimally diluted bridging secondary Ab for 30 minutes.
- Rinse in TBS.
- *Enzyme complex:* Incubate the section in a properly diluted APAAP complex solution for 30 minutes.
- *Substrate:* Apply the substrate solution (fast blue BB or fast red, as per choice) for 10 minutes.
- Aqueous counterstain and aqueous mounting.

Avidin–biotin Complex (ABC) Technique

- Dewax the tissue.
- Bring the tissue in water by serial changes in graded alcohol.
- Rinse the section in TBS.
- *Antigen retrieval:* Do AR if needed.
- Rinse in TBS.
- *Primary antibody:* Incubate the section with optimally diluted primary Ab for 60 minutes.
- Rinse in TBS.
- *Secondary antibody:* Apply optimally diluted biotinylated secondary Ab for 30 minutes.
- Rinse in TBS.
- *Enzyme complex:* Incubate the section in properly diluted ABC for 30 minutes as per vendor's recommendation.
- Rinse in TBS.
- *Substrate:* Apply the substrate solution (DAB for peroxidase) for 10 minutes.
- Wash by running water.
- *Counterstain:* Hematoxylin for 10 seconds
- Dehydrate, clear, and mount by DPX.

Polymer-based Technique[2]

- Dewax the tissue.
- Bring the tissue in water by serial changes in graded alcohol.
- Rinse the section in TBS.

- *Antigen retrieval:* Do AR if needed.
- Rinse in TBS.
- *Primary antibody:* Incubate the section with optimally diluted primary Ab for 60 minutes.
- Rinse in TBS.
- Now incubate the section in EnVision polymer reagent for 30 minutes.
- Rinse in TBS.
- Incubate in DAB solution for 10 minutes.
- Wash by running water.
- *Counterstain:* Hematoxylin for 10 seconds
- Dehydrate, clear, and mount by DPX.

QUALITY CONTROL

Quality control (QC) and quality assurance (QA) are two essential components of ICC. QC comes under the broad heading of QA.

The term QC means the collection of processes to identify and correct the shortcomings of the analytical process.[3] It is necessary to reduce the variability of the test and to reproduce the best quality of the test. The QC in ICC can be divided into three phases:[4]

1. Preanalytical
2. Analytical
3. Postanalytical

Preanalytical

The preanalytical step mainly includes:

- *Fixation of the tissue:* Optimum fixation is one of the crucial parts of the QC. An ideal fixative is neutral buffered formalin, and at least 8 hours' fixation is must for ICC.
- *Slide drying time and temperature:* The slide should be dried for a minimum of 1 hour in 50°C.

Analytical

The following factors are related to the analytical phase:

- *Selection of the primary antibody:* This is the crucial part of ICC. The choice of the Ab depends on various factors such as (1) Clinical application of the Ab, (2) Sensitivity and specificity of the Ab, (3) Clonality, (4) Volume of the Ab, and (5) Use in an exact tissue section.
- *Antigen retrieval, dilution, and incubation time:* A battery of tests are needed to have the optimum AR, dilution, and incubation time which have been described in Chapter 3.
- *Antibody validation:* Each laboratory must validate the Ab before its use. At least 10 positive and 10 negative control slides should be assessed for the analytic test and 20 positive and negative controls each should be studied for the predictive markers.

Postanalytical

The postanalytical phase includes positive and negative controls, interpretation of the slides, and performance of the pathologists.

EXTERNAL QUALITY ASSURANCE

Daily routine screening of the slides and maintaining the laboratory standard of ICC are essential. Similarly, it is also very much necessary to have an interlaboratory comparison to see the reproducibility and standard of the ICC. The unstained slides could be sent to different laboratories for the series of immunostaining. Those stained slides should be reviewed with the in-house standard ICC and should be scored accordingly.

Troubleshooting

The best way to have good immunostaining is to follow these basic rules:

- Check the expiry date of each reagent and also the storage condition.
- Check the constituents of the buffer.
- Check the compatibility of all the reagents.
- Rigorously follow each step of the ICC.
- Check the available data of the primary Ab regarding the type of tissue and optimal specimen fixation of the sample.

Table 1 shows the possible causes of trouble and its solution.

TABLE 1: Troubleshooting in an immunocytochemistry technique.

No signal in both test and control sections	
Possible causes	**Remedies**
Primary or secondary antibody is out of order	• Check the expiry date of the antibodies • Check the storage condition of the antibodies and use fresh antibodies
Forgot to add the antibody	• Do the test again
Too much diluted primary antibody	• Change the dilution and use the concentrated primary antibody
The primary antibody is incubated for too less time	• Optimal incubation time needed
The primary antibody is incubated in too low or too high temperature	• Follow the optimum temperature
The reagent is incorrectly made	• Follow the vendor's instruction
Enzyme and substrate are incompatible	• Use compatible enzyme and substrate

Continued

Continued

No signal in both test and control sections	
Possible causes	**Remedies**
Antigenic determinant site is destroyed at the time of processing	• Maintain optimum melting point of the wax
Improper fixation	• Use proper fixative
Improper dewaxing	• Keep the sections in xylene for more time with preferably fresh xylene
The antigen is too less in the tissue	• Use a more sensitive detection technique
The chromogenic product is alcohol soluble	• Use aqueous mounting medium
Nuclear overstain by the counterstain is too dark to visualize the Ag-Ab reaction	• Spend less time for counterstain
Improper use of the buffer solution	• Omit sodium azide when doing peroxidase staining • Check the pH of the buffer
Positive tissue control is stained properly but weak or absent signal in test section	
Inadequately fixed test tissue	• Fix in proper fixative with optimum time
Too low antigen in the test section	• Use a more sensitive detection technique
The epitope of the antigen is not unmasked in the test tissue	• Proper antigen retrieval needed
Necrotic or damaged test tissue	• Better sample is needed
Background staining in both control and test sections	
Too much incubation time for antibody or chromogenic substrate	• Optimize the incubation time
Too much concentration of the primary antibody	• Optimize the concentration of the primary antibody
Nonspecific cross-reaction of the secondary antibody with the endogenous background immunoglobulin	• Use normal serum to block the cross-reacting antigen • Use a preadsorbed secondary antibody
The improper blocking serum is applied	• Apply the serum of the species in which the secondary Ab is developed
Improper blocking of endogenous peroxidase	• Do appropriate blocking of the enzymes
Section is dried during ICC	• Avoid drying of the tissue by using a humidified environment
Chromogenic substrate is not completely dissolved	• Dissolve the chromogenic substrate
Improper rinsing	• Rinse the sections properly
Positive tissue control and test samples have background staining. However, no background stain in negative control	
Too much concentrated primary antibody	• Optimally dilute the primary antibody
Negative tissue control also showing signal	
Forgot to omit the primary antibody	• Repeat the ICC properly
Secondary antibody is contaminated	• Use fresh pipette and reagents

REFERENCES

1. Cordell JL, Falini B, Erber WN, Ghosh AK, Abdulaziz Z, MacDonald S, et al. Immunoenzymatic labeling of monoclonal antibodies using immune complexes of alkaline phosphatase and monoclonal anti-alkaline phosphatase (APAAP complexes). J Histochem Cytochem. 1984;32(2):219-29.
2. Sabattini E, Bisgaard K, Ascani S, Poggi S, Piccioli M, Ceccarelli C, et al. The EnVision++ system: a new immunohistochemical method for diagnostics and research. Critical comparison with the APAAP, ChemMate, CSA, LABC, and SABC techniques. J Clin Pathol. 1998;51(7):506-11.
3. College of American Pathologists. Standard for Laboratory Accreditation. Skokie, IL: College of American Pathologists; 1987.
4. Lin F, Chen Z. Standardization of diagnostic immunohistochemistry: literature review and geisinger experience. Arch Pathol Lab Med. 2014;138(12):1564-77.

CHAPTER 5

Automated Immunostaining

INTRODUCTION

In the last few decades, there are massive innovative developments in the field of immunocytochemistry (ICC) and automated immunostaining is one of those. It was first time introduced in 1992, and since then there was continuous development in this area.[1] The automated ICC performs most of the tedious task such as antigen retrieval, incubating the secondary antibody, addition of the chromogenic substrate, counterstaining, etc. The various advantages of the automated ICC have been documented in different reviews.[2,3]

Advantages of Automated Immunostaining

- Less workforce is needed. Automation is most effective in those laboratories where the workload is high.
- Extensive knowledge of the troubleshooting and in-depth theory of immunostaining are not needed for the technologists.
- The barcoded slides and reagents assure the possibility of the correct staining.
- The instrument also gives a signal in case of incorrect temperature, incompatible reagents, the inadequate volume of the reagents, etc.
- The automated immunostainer also keeps a track about the expiry date and stock of the reagents.

Disadvantages of Automated Immunostaining

- There is no flexibility of the staining compared to manual staining.
- Laboratory staffs are entirely dependent on the mercy of the instrument and respective company. The sophisticated machine has more chance of malfunctioning and therefore repeated maintenance is needed.
- The less experienced and less knowledgeable person can work on automated immunostainer. Therefore, if there is a suboptimal stain, the person may not be able to solve the issue.

"OPEN" AND "CLOSED" SYSTEMS

There are two distinct types of automated immunostaining machine—open and closed systems. The open systems are more flexible regarding the antigen retrieval program, use of the antibodies, reagents, chromogen, counterstaining, and the programming of the ICC. In this "open" system, the individual laboratories have to standardize their own immunostaining. They can work on the newer biomarkers and the laboratory can bypass the proprietary reagents that are relatively costly.

In comparison, the closed system deals with the fixed programming, with proprietary reagents, substrates and counterstaining. The ICC staining technique is fully standardized and controlled. As barcoded slides and reagents are used in this system, the machine provides all the information about the expiry date of the different reagents, volume lot number, etc. Overall, the closed system reduces the time of the technicians and error rate. The staining quality is consistent and comparable for the different laboratories. **Table 1** is highlighting the difference between the open and closed systems of automated immunostainer.

TABLE 1: Comparison of an open and closed systems.

Features	Open	Closed
Programming of immunostaining	Flexible	Rigid
Reagents	The different commercially available reagents can be used	Only proprietary reagents can be used
Tracking	Expiry date, lot number and volume of the reagents cannot be traced	Full details of the reagents are always traceable
Cost	Relatively less costly reagents	More costly
Reproducibility	Variable	More reproducible as more consistent staining
Need	Research laboratory	Routine clinical laboratory

CHOICE OF THE AUTOMATED IMMUNOSTAINING MACHINE

Most of the laboratories need the machine that can fulfil the multiple tasks. The vendors will always push to sell their own instrument by highlighting the most advantageous factor. The laboratory should purchase the machine according to their own need. The critical factors in this regard include:

- Do you need high-speed ICC to do a large batch of slides of limited protocol or flexible protocol with less throughput?
- Do you have enough money to purchase the proprietary reagents?
- Do you wish to do other tests also by this machine?
- Do you have adequate skilled staff to standardize the stain?

The common automated systems available in the market are mentioned and compared in **Table 2**.

Reagent Dispensing System

There are mainly two types of reagents dispensing systems in the machine: Matrix system where the slides and reagents are arranged in rows and columns. Leica Biosystems is an example of the array system. Array stainer mimics the manual staining and is relatively flexible. The robotic arm moves to the slide and dispenses the chemicals on the section. In this system, one can remove the slide when the staining is completed. The other type of arrangement is the rotary system (**Fig. 1**). The Ventana automated stainer is an example of the rotary type. Here the slides and reagents are arranged in a circular rack (**Fig. 2**). The reagents are arranged over the slides in a circular track. This track rotates over the slides and finally dispenses the agents over the particular slide. In this system, the access of the slide is not possible during the ongoing staining procedure.

Reagent dispensing system may vary between different automated systems. In most of the array system, the reagent is dispensed on the section by disposable pipettes. Alternatively, the reagents can also be dispensed by capillary action. In rotary type machine, the reagents are dispensed by print cartridge-like system. Whenever the reagent container reaches above the particular slide, it dispenses the fixed amount of material over the section.

Heat-induced Antigen Retrieval

Heating methods are employed in many of the systems. The heating system is helpful in dewaxing and antigen retrieval.

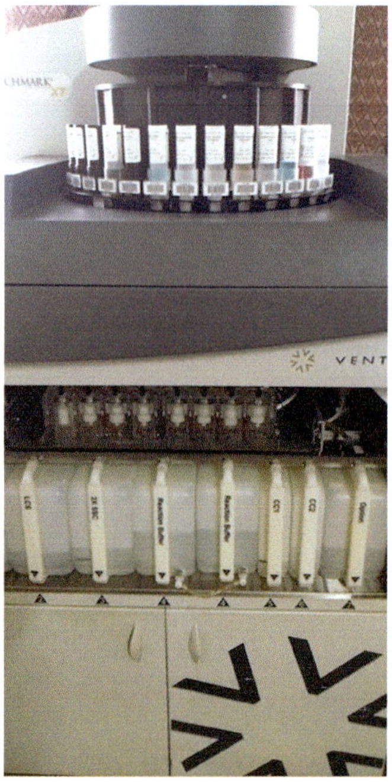

FIG. 1: Automated immunostaining machine by Roche–Ventana.

TABLE 2: Comparison of different commonly used systems.

Features	Leica Biosystems: Bond-Max™ and Bond III™	Biocare medical: IntelliPath™, FLX	Roche-Ventana: Benchmark X™ and Benchmark ULTRA™	DAKO: Autostainer link 48™	Thermo scientific: Autostainer series
Slide baking	Yes	No	Yes	No	No
Slide heating during incubation	Yes	No	Yes	Separate module	Separate module
Counterstaining	Yes	Yes	Yes	Yes	Yes
Coverslipping	No	No	No	No	No
Slide capacity	3 racks, 10 slides each: 30	5 racks, 10 slides each: 50	30 slides	4 racks, 12 slides each: 48	3–7 racks, 12 slides each: 36–84
Robotic movement	Array/Matrix system	Array/Matrix system	Rotary	Array/Matrix system	Array/Matrix system
Reagent application	Probe tips	Probe tips	Liquid coverslip	Probe tips	Probe tips
Attached information technology	Yes	Yes	No	Yes	Yes
Special features	Covertiles™ system protect the tissue during application of the reagent	Chilling chamber for reagent, Simultaneous multiplex staining	Liquid coverslip	Open platform	Open platform

FIG. 2: Carousel of the rotary stainer of the machine.

Head-induced antigen retrieval (HIAR) is the crucial point of immunostaining and it reduces the significant time of the staining. The heating of the slide is the additional feature of many systems as heating the sections during the incubation period reduces overall incubation period.

Slide Capacity and Batch Size

The various automated systems differ markedly regarding the total load of the slides and batch size. Usually, the system contains 10–12 slides in a batch and overall 1–5 batch. So the number of slides in each time varies from 30–60. However, it is essential to remember that total time protocol for the immunostaining is more critical than the batch size. The time protocol of the different system varies widely, and that may reduce the turnaround time.

System Cost

The cost of the automated staining system is an essential factor. When purchasing the machine one should remember that along with the initial cost of the machine, there are other hidden costs such as the cost of the maintenance, extra cost of proprietary reagents, and waste disposal. This cost should be balanced with the saving manpower, high turnaround, and the time need to standardization. Moreover, the reproducibility and comparability of the staining result are also necessary. However, before purchasing the machine, the laboratory personnel or authority should be careful of their need as there is no "best machine" that exists. The price of the machine should always be justified before the purchase.

REFERENCES

1. Grogan TM. Automated immunohistochemical analysis. Am J Clin Pathol. 1992;98(4):S35-8.
2. Prichard JW. Overview of automated immunohistochemistry. Arch Pathol Lab Med. 2014;138(12):1578-82.
3. Le Neel T, Moreau A, Laboisse C, Truchaud A. Comparative evaluation of automated systems in immunohistochemistry. Clin Chim Acta. 1998;278(2):185-192.

2

SECTION

Diagnostic Applications of Immunocytochemistry

CHAPTER 6

Essential Antibodies for Routine Application in Laboratory

INTRODUCTION

The antigen (Ag) is often unique to a particular tissue, and therefore, the corresponding antibody (Ab) is used to identify that tissue. At times, the Ab may cross-react with other tissues, e.g., EMA may be expressed in the synovial sarcoma or CD10 may be positive in renal cell carcinoma. Therefore, a panel of Abs should be used instead of a single Ab to identify that particular tissue. Some Abs may be organ-specific and may help to determine the exact primary origin of the tumor under challenging situations. The current chapter provides an overview of the different tissue and organ-specific antibodies. It will help to understand the rationality of the application of the Ab in the specific situation.

EXPRESSION OF DIFFERENT ANTIBODIES IN ROUTINE LABORATORY

Depending on the distribution of the Ag in the cell, the following four patterns of antigen-antibody reaction are seen.

1. *Membranous pattern*: Here the Ag is distributed mainly on the cell membrane and therefore the Ab is mostly visualized along the cell membrane (**Fig. 1**).
   Examples:
   a. All cluster of differentiation (CD) markers.
   b. Epithelial membrane antigen (EMA).
   c. Cytokeratin (CK)
   d. Hector Battifora mesothelial-1 (HBME-1).
   e. Glucose transporter isoform 1.
2. *Cytoplasmic pattern*: Here, the Ag is present in the cytoplasm and therefore, the predominant pattern of antigen-antibody reaction is distributed within the cytoplasm of the cells. Usually, the Ab against the cytoplasmic materials such as actin, vimentin, melanin, cytoskeletal materials shows cytoplasmic pattern (**Fig. 2**).
   Examples:
   a. Vimentin
   b. Actin
   c. Desmin
   d. Melanin
   e. Calponin

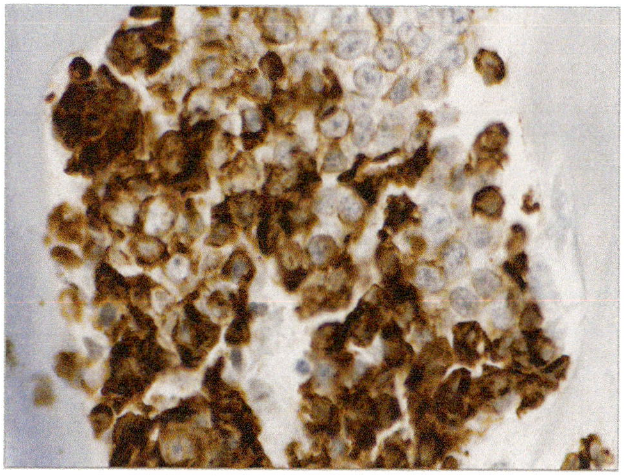

FIG. 1: Membranous positivity of cytokeratin (CK).

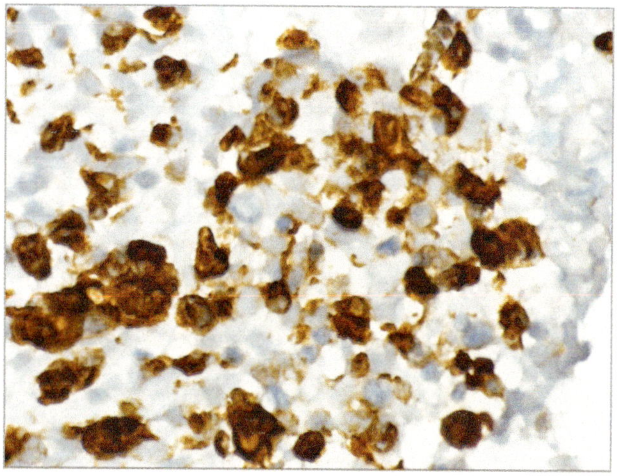

FIG. 2: Cytoplasmic expression of desmin in the case of rhabdomyosarcoma.

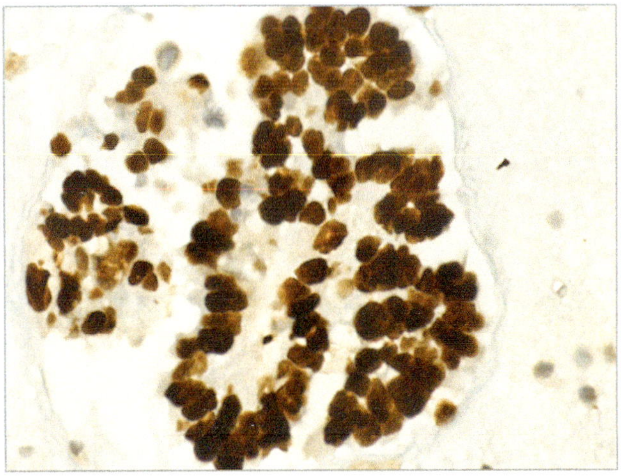

FIG. 3: Nuclear positivity of p53 in ovarian adenocarcinoma in peritoneal effusion.

   f. Chromogranin A
   g. Synaptophysin
   h. Human chorionic gonadotropin (HCG)
3. *Nuclear pattern*: Here, the Ag is present in the nucleus. The transcription factor and hormonal receptors usually show a nuclear pattern (**Fig. 3**).
   Examples:
   a. p53
   b. p16
   c. p63
   d. D2-40
   e. Wilms tumor 1 (WT-1)
   f. Estrogen receptors (ERs)
   g. Progesterone receptors
   h. MyoD1 and myogenin
   i. OCT4
   j. SALL4
   k. Thyroid transcription factor (TTF-1)
   l. CDX2
   m. GATA 3
4. *Extracellular staining pattern*: This pattern is noted when the antigen is present in the extracellular material such as collagenous material.

The different types of antibodies are used to identify the particular types of cells.

EPITHELIAL CELL MARKERS

The commonly used epithelial cell markers are:
- Cytokeratin
- Epithelial membrane antigen
- Carcinoembryonic antigen (CEA)
- BER-EP4

Cytokeratin

Positivity: CK is the most important epithelial marker. According to molecular weight, CK is classified into 20 different types that often are tissue-specific[1-3] (**Table 1**).

TABLE 1: Organ-specific CK positivity.

Tumor type	CK7	CK20	CK5/6
Pancreatobiliary carcinoma	Positive	Positive*	Negative
Transitional cell carcinoma		Positive*	Negative
Colonic adenocarcinoma	Negative	Positive	Negative
Gastric adenocarcinoma	Negative	Positive*	Negative
Merkel cell carcinoma	Negative	Positive	Negative
Lung adenocarcinoma	Positive	Negative	Positive
Ductal carcinoma of breast	Positive	Negative	Positive in basal type of carcinoma
Ovarian serous carcinoma	Positive	Negative	Negative
RCC: clear cell	Negative	Negative	Negative
HCC	Negative	Negative	Negative
Adrenocortical carcinoma	Negative	Negative	Negative
Prostate	Negative	Negative	Negative

*Occasional focal

(HCC: hepatocellular carcinoma; RCC: renal cell carcinoma)

Cytokeratin is mainly positive in the epithelial tumor. However, certain sarcomas are also positive for CK (synovial sarcoma, epithelioid sarcoma and Ewing's sarcoma)

Localization: Membranous positive stain pattern

Organ specificity: The organ-specific CK staining is described in **Table 1**.

Pan-cytokeratin

The mixture of AE1 and AE3 is used as pan-CK. This cocktail is a useful marker of most of the epithelial cell tumors. AE1/AE3 is weakly expressed in hepatocellular carcinoma and prostatic adenocarcinoma. Moreover, this cocktail antibody may cross-react with glial fibrillary acidic protein (**Table 2**).

CK5/6: CK5 is an intermediate filament and is often used to identify squamous cell carcinoma (SqCC). CK5 is expressed mainly in pleural mesothelioma. Therefore, CK5 stain helps to differentiate epithelial mesothelioma from the metastatic carcinoma.CK5 is considered a vital marker to identify the basal-like type of breast carcinoma (**Table 3**).

CK7: CK7 is a type II intermediate filament and is mainly present in the epithelium of duct and gland. It is a marker of adenocarcinoma of nongastrointestinal tract origin. CK7 is a marker of lung, breast, bile duct and ovarian adenocarcinoma. A combination of CK7 and CK20 should be applied. CK7 is also positive in urothelial cell carcinoma (**Table 4**).

CK19: CK19 is an intermediate filament and is present in both simple and complex epithelial lining. It is an acidic (type I) keratin. CK19 is positive in the majority of carcinoma, and therefore it cannot distinguish the different sites of origin of the tumor (**Table 5**). CK19 can be used as a marker to differentiate follicular variant of papillary thyroid carcinomas (positive) from the follicular neoplasm (negative) of the thyroid. The cholangiocarcinoma cases are positive for both CK7 and CK19.

TABLE 2: Pan-cytokeratin expression.

Antibody	Main use	Positive in nonepithelial tumors	Localization	Remarks
AE1 and AE3	Majority of the epithelial tumors	• Mesothelioma • Clear cell sarcoma • Malignant melanoma • Epithelioid sarcoma • Synovial sarcoma	Membranous	• Cross reactive with glial fibrillary acidic protein • Negative in hepatocellular carcinoma, and prostatic adenocarcinoma

TABLE 3: CK5 expression.

Antibody	Main use	Positive in nonepithelial tumors	Localization	Remarks
CK5	• Squamous cell carcinoma • The basal-like phenotype of breast carcinoma	• Mesothelioma • Myoepithelial tumors	Membranous	• CK5 is positive in mesothelial cells, and so it cannot discriminate pleural squamous cell carcinoma and mesothelioma

(CK: cytokeratin)

TABLE 4: CK7 expression.

Antibody	Main use	Positive in nonepithelial tumors	Localization	Remarks
CK7	• Adenocarcinoma of lung, breast, bile duct and ovary • Urothelial cell carcinoma	• Mesothelioma • Papillary carcinoma of the thyroid • Synovial sarcoma	Membranous	• CK7 is expressed in SqCC of esophagus and head-neck • Nonepithelial tumor such as synovial sarcoma is also positive

(CK: cytokeratin; SqCC: squamous cell carcinoma)

TABLE 5: CK19 expression.

Antibody	Main use	Positive in other tumors	Localization	Remarks
CK19	• Adenocarcinoma of lung, bile duct, GIT, and urothelial cell carcinoma	• Papillary thyroid carcinoma • Mesothelioma	Membranous	• CK19 expression cannot distinguish the subtypes of carcinomas • CK19 positivity can help to distinguish cholangiocarcinoma (positive) from hepatocellular carcinoma

(CK: cytokeratin; GIT: gastrointestinal tract)

TABLE 6: CK20 expression.

Antibody	Main use	Positive in other tumors	Localization	Remarks
CK20	• Colorectal adenocarcinoma • Mucinous neoplasm of ovary • Urothelial cell carcinoma	• Merkel cell carcinoma • Adenocarcinoma of lung	Membranous	• CK20 and CK7 combination can identify the possible origin of different tumors.

(CK: cytokeratin)

CK20: CK20 is a low molecular weight type I CK. It is mainly expressed in the colorectal adenocarcinoma, Merkel cell carcinoma, mucinous neoplasm of ovary, and urothelial cell carcinomas. The limited expression of CK20 in the different types of tumor is a helpful factor in identifying the site of origin of the tumors (**Table 6**). Combined use of CK20 with CK7 is helpful to identify the possible source of different tumors.

Epithelial Membrane Antigen

Epithelial membrane antigen is also known as MUC-1. It is a transmembrane glycoprotein. EMA is unrelated with intermediate filament.[4] EMA/MUC-1 is membranous mucin and therefore is distributed along the cell membrane. It is one of the human milk fat globular membrane proteins. Epithelial membrane antigen is a relatively good marker of the normal epithelial cell and epithelial tumors. However, it is also positive in many nonepithelial tumors (**Table 7**).

Localization: Membranous positivity. The pure cytoplasmic positivity of EMA without membranous staining should be neglected as spurious.

Nonepithelial positivity of EMA:[5]

- Mesothelial cells
- Anaplastic large cell lymphoma (ALCL)
- Mesothelioma

TABLE 7: Epithelial membrane antigen (EMA) expression.

Antibody	Main use	Nonepithelial positivity	Localization	Remarks
EMA	Carcinoma of breast, lung, prostate, endometrium, thyroid and skin	• Mesothelial cells • Anaplastic large cell lymphoma • Mesothelioma • Epithelioid sarcoma • Synovial sarcoma • T cell-rich B cell lymphoma	Membranous	• Pure cytoplasmic positivity of EMA should be ignored • It is not a reliable epithelial marker as large number of non- epithelial tumors are positive for EMA

TABLE 8: Carbohydrate antigen 125 (CA125)/MUC-16 expression.

Marker	Positivity in ovarian tumors	Other tumors showing positivity	Localization
CA125 (MUC-16)	• Serous adenocarcinoma of the ovary • Endometrioid carcinoma • Clear cell carcinoma • Yolk sac tumor	• Lung adenocarcinoma • Breast carcinoma • GIT carcinoma • ALCL • Desmoplastic small round cell tumor	Cytoplasmic

(ALCL: anaplastic large cell lymphoma; GIT: gastrointestinal tract)

- Epithelioid sarcoma
- Synovial sarcoma
- T cell-rich B-cell lymphoma

Mucin

Mucins are hyperglycosylated proteins produced by the epithelial cells. There are two types of mucin:
1. *Gel-forming mucin*: MUC-2, MUC-5AC, MUC5B and MUC-6.
2. *Membrane-bound mucin*: MUC-1, MUC-3A, MUC-3B, MUC-4, MUC-12, MUC-13.

MUC-1: Described above as EMA

MUC-2: It is a gel-forming secretory mucin. MUC-2 is produced by the goblet cells of the stomach and small intestine. It is a marker of colonic adenocarcinoma, intestinal-type carcinoma of the duodenum, gastric carcinoma, mucinous carcinoma of ovary and carcinoma of the breast.

MUC-3: MUC-3 is positive in invasive ductal carcinoma of the breast and gastric adenocarcinoma.

MUC-4: It is a membrane-bound mucin and is involved in cellular signaling process. MUC-4 is a marker of adenocarcinoma of lung, stomach, and pancreas.

MUC-5AC: It is expressed in adenocarcinoma of the gastrointestinal tract such as stomach, esophagus, and colon. It is also positive in pancreatic adenocarcinoma and ovarian mucinous carcinoma.

MUC-5B: It is a gel-forming mucin and is expressed in the mucinous gland of the respiratory tract.

MUC-6: It is membrane-bound mucin and is mainly noted in the pyloric glands of the stomach. MUC-6 is expressed in carcinoma of breast and stomach.

MUC-16: MUC-16 I also known as carbohydrate antigen 125 (CA125).

CA125/MUC-16 is expressed in ovarian carcinoma (nonmucinous). It is also positive in various other tumors of lung, breast, and GIT (**Table 8**).

Claudin 5

Claudin 5 is an integral transmembrane glycoprotein related with the tight junction between epithelial and endothelial cells. It is involved in the passage of molecules through the intercellular spaces. Claudin 5 is a marker of endothelial differentiation of the tumor. It is also expressed in many nonendothelial tumors (**Table 9**).

LYMPHOID MARKERS

Leukocyte Common Antigen

Leukocyte common antigen [LCA (CD45)]:

Positivity: CD45 is a common marker of almost all lymphoid cells. However, CD45 is variably positive in plasmacytoid lymphoma and ALCL.

Localization: Membranous positivity

The common markers of different lymphocytes are:

B cell:
- CD19
- CD20
- CD79a
- SIg

T cell:
- CD3
- CD4
- CD8

Natural killer (NK) cell:
- CD3
- CD56

TABLE 9: Claudin 5 expression.

Marker	Expression in vascular tumors	Other tumors expressing this antigen	Localization	Comments
Claudin 5	Vascular tumor	• Adenocarcinoma • Synovial sarcoma • Myxoid chondrosarcoma • Desmoplastic small round cell tumors • Merkel cell carcinoma • Renal cell carcinoma	Cytoplasmic	This is a relatively sensitive but less specific marker for vascular tumor

MESENCHYMAL MARKERS

Vimentin

Positivity: Vimentin is an intermediate filament. It is widely expressed in endothelial cell, fibroblasts, and muscle cells. Vimentin positivity is also noted in mesothelial cells, renal cell carcinoma and even lymphoma. Vimentin positivity can be used as a control of immunocytochemistry (ICC) that indicates that the various other tissue markers are preserved.

Localization: Cytoplasmic positivity.

Desmin

Positivity: Desmin is one of the intermediate filaments. Positive in smooth and striated muscle and the neoplasm of those tissues such as rhabdomyosarcoma and leiomyosarcoma.[6] Desmin positivity is noted in desmoplastic small round cell tumor.

Localization: Cytoplasmic positivity.

Actin

Positivity: Actin is one of the contractile myofilaments of muscle. It is positive in the myoepithelial cells of the salivary gland and breast. It is also positive in rhabdomyosarcoma.

Localization: Cytoplasmic positivity.

Caldesmon

Positivity: Caldesmon is a cytoplasmic protein and is related to smooth muscle differentiation. Heavy (h) caldesmon is a reliable marker of smooth muscle and specific marker of leiomyoma and leiomyosarcomas.

Localization: Cytoplasmic positivity

Calponin

Positivity: It is positive in the cells of smooth muscle differentiation, myoepithelial cells, and myofibroblasts. The leiomyoma and myofibroblastic tumors are positive for calponin.

Localization: Cytoplasmic positivity.

Myoglobin

Positivity: It is an oxygen-transport protein that exclusively found in the striated muscle. It is positive in pleomorphic rhabdomyosarcoma.

Localization: Cytoplasmic positivity. The antibody often produces background staining as the commercially available Ab is polyclonal.

MyoD1 and Myogenin

Positivity: Both these markers are specific for skeletal muscle differentiation. They are difficult to stain in routine paraffin-embedded sections. These two markers are mainly positive in alveolar and embryonal rhabdomyosarcoma.

Localization: Nuclear positivity.

ENDOTHELIAL CELL MARKERS

CD34

Positivity: This is a good marker of the endothelial cell and its neoplasm. It is also a marker of the progenitor cell of the hemopoietic cells. CD34 is positive in angiosarcoma, dermatofibrosarcoma protuberans, solitary fibrous histiocytoma, and gastrointestinal stromal sarcoma.

Localization: Membranous pattern of staining.

CD31

Positivity: CD31 is an endothelial cell marker. It is positive in hemangioma, hemangioendothelioma and angiosarcoma.

Localization: Cytoplasmic and membranous pattern.

Factor VIII-related Antigen (von Willebrand Factor)

Positivity: Factor VIII-related antigen is synthesized in the endothelial cells of the blood vessel and functions as antihemophilic factor. It is used as a marker of the endothelial cells. It is commonly positive in hemangioma and hemangioendothelioma.

Localization: Cytoplasmic stain.

NEUROENDOCRINE MARKERS

Chromogranin

Positivity: Chromogranin is present in the neurosecretory granules. It is specific for neuroendocrine tumor. The sensitivity of this marker is also good.

Localization: Cytoplasmic positivity.

Synaptophysin

Positivity: Synaptophysin is located in the membrane of the synaptic vesicles. It is a sensitive marker of neuroendocrine and neuroectodermal tumors. However, its specificity is lower than chromogranin.

Localization: Cytoplasmic positivity.

CD56

Positivity: It is an intercellular adhesion molecule and is present in NK cells, neurons, astrocytes, and myoblasts. It is used as a marker of neuroendocrine tumors and NK-cell lymphoma.

Localization: Membranous positivity.

MELANOMA MARKERS

HMB45

Positivity: It is expressed in melanocytes. HMB45 is used as marker of melanoma, neural crest derived tumors, angiomyolipoma, perivascular epithelioid cell neoplasms (PEComa), epithelioid cell sarcoma, and clear cell sarcoma.[7]

Localization: Cytoplasmic positivity.

Melan A

Positivity: It is a good sensitivity marker of melanoma, peripheral nerve sheath tumor, angiomyolipoma, and melanocytic nevi and adrenocortical tumor.[8]

Localization: Cytoplasmic positivity.

S-100 Protein

Positivity: S-100 protein is expressed in a large variety of cells such as glial cells, Schwann cells, melanocytes, adipocytes, salivary gland cells, myoepithelial cells, and Langerhans histiocytes. It is used as the marker of Schwannoma, malignant peripheral nerve sheath tumors, melanoma, clear cell sarcoma, extraskeletal myxoid chondrosarcoma, and Langerhans cell histiocytosis.[9,10]

Localization: Cytoplasmic and nuclear positivity.

GERM-CELL MARKERS

Placental Alkaline Phosphatase

Positivity: Placental alkaline phosphatase (PLAP) is an oncofetal antigen. It is expressed in all germ-cell tumors. PLAP is consistently positive in seminoma and embryonal carcinoma. It is very helpful to distinguish undifferentiated carcinoma from the germ cell tumor.[11]

Localization: Cytoplasmic and membranous.

Human Chorionic Gonadotropin

Positivity: Human chorionic gonadotropin is present in the syncytiotrophoblasts of placenta. The antibody against β subunit of HCG is used as a marker of choriocarcinoma.

Localization: Cytoplasmic stain.

OCT4

Positivity: OCT4 (Octamer-binding transcription factor 4) is localized in the pluripotent germ cells. It is a sensitive and specific antibody for the diagnosis of dysgerminoma, embryonal carcinoma, and gonadoblastoma.[12]

Localization: Nuclear stain.

SALL4

Positivity: SALL4 (Sal-like protein 4) is the first-line marker of germ-cell tumors. It is a sensitive and specific marker of endodermal sinus tumors, dysgerminoma, and choriocarcinoma.[13]

Localization: Nuclear stain.

MESOTHELIAL MARKERS

Calretinin

Positivity: Calretinin, a calcium-binding protein, is present in the mesothelial cells. It is a marker of both benign and malignant mesothelial cells. Calretinin is also present in ovarian sex cord tumor and adrenocortical cell tumor.[14]

Localization: Cytoplasmic and nuclear stain.

HBME-1

Positivity: HBME-1 is present in the microvillus of the mesothelial cells. It is also expressed in adenocarcinoma and overall HBME-1 stains is a less preferable stain than calretinin.[15]

Localization: Thick membranous positivity in the mesothelial cells compared to thin membranous stain in adenocarcinoma cells.

Wilms Tumor Gene 1

Positivity: *WT-1*, the tumor suppressor gene, is a good marker for reactive mesothelial cells and mesothelioma.[16] It is also an effective marker of ovarian serous carcinoma and desmoplastic small round cell tumors.

Localization: Nuclear positive.

Podoplanin (D2-40)

Positivity: D2-40 is positive in mesothelial cells. It is a very sensitive and specific marker of mesothelioma.[17] It is also positive in ovarian (50%) and lung carcinoma (30%). So the value of D2-40 in distinguishing mesothelial from the adenocarcinoma in effusion is doubtful.

Localization: Nuclear positive.

Glucose Transporter Isoform (GLUT1)

Positivity: Glucose transporter 1 (GLUT1), a glucose transport protein, is the marker of malignant mesothelial cells. It is useful to differentiate reactive mesothelial cells from mesothelioma.[18]

Localization: Membranous staining pattern and at times cytoplasmic stain is seen.

EPITHELIAL MARKERS IN EFFUSION FLUID

MOC-31

Positivity: MOC-31, a transmembrane glycoprotein, is present on the membrane of the malignant epithelial cells. It detects SqCC, adenocarcinoma and small cell carcinoma in the effusion fluid with 80% sensitivity and 90–100% specificity.[19]

Occasionally mesothelial cells also show weakly positive MOC-31.

Localization: Strong membranous positivity in carcinoma cells.

BER EP4

Positivity: BER EP4 recognizes the surface glycoprotein of the malignant epithelial cells and detects most of the carcinomas in the effusion fluid, except renal cell carcinoma. The sensitivity and specificity of BER EP4 are 80 and 90%, respectively.[20,21]

The mesothelial cells are negative for this stain.

Localization: Strong membranous and cytoplasmic positivity in carcinoma cells.

Carcinoembryonic Antigen

Positivity: Carcinoembryonic antigen (CEA) is positive in both normal and neoplastic tissue of endodermal origin. It is mainly positive in gastrointestinal and lung tumors. Mesothelial cells may often show weak positive CEA.

Localization: Membranous positivity.

SITE-SPECIFIC MARKERS

Lung

Thyroid Transcription Factor

Positivity: Thyroid Transcription Factor (TTF-1) is a transcription factor. It is involved in the formation of thyroperoxidase (TPO) enzymes in thyroid and surfactant in the lung. TTF-1 is the marker of adenocarcinoma of lung and most of the thyroid tumors with 88% sensitivity and 100% specificity.

Localization: Nuclear positivity.

Napsin A

Positivity: Napsin A, a proteinase, is the marker of adenocarcinoma of the lung. It is often considered as a superior marker that TTF-1.[22] However, it may also show positivity in renal cell carcinoma.

Localization: Cytoplasmic positivity.

p63 and p40

Positivity: Both p64 and p40 are the markers of SqCC of lung with very high specificity (nearly 100%). Among these two markers, p40 is more specific than p63.

Localization: Nuclear positivity.

Gastrointestinal Marker

CDX2

Positivity: CDX2, a transcription factor, is the marker of colorectal carcinoma. It is also positive in the mucinous tumor of lung, ovary and occasionally in the endometrioid carcinoma.[23]

Localization: Nuclear positivity.

Breast Markers

Estrogen Receptors

Positivity: Estrogen receptor (ER) is often used as a marker of metastatic breast carcinomas. However, it is not specific for breast carcinomas because other carcinomas such as the cervix, endometrial and ovarian carcinomas are also positive for ER.[24]

Localization: Nuclear positivity.

Mammaglobin

Positivity: Mammaglobin is positive in most of the breast carcinomas (90%).[25]

However, it is also expressed in salivary gland tumors and endometrial carcinomas.[26]

Localization: Cytoplasmic positivity.

Gross Cystic Disease Fluid Protein

Positivity: It is a sensitive (70%) and specific (90%) antibody to detect the carcinoma of the breast. Gross cystic disease fluid protein (GCDFP) is also expressed in salivary, sweat gland and lung carcinomas.[27]

Localization: Cytoplasmic positivity.

GATA 3

Positivity: GATA 3 is expressed in the majority of the breast carcinoma (95%). It is also positive in urothelial cell carcinoma and mammary analog secretory carcinoma.[28]

Localization: Nuclear positivity.

Prostate Marker

Prostate-specific Antigen

Positivity: Prostate-specific antigen (PSA) is positive in normal prostate tissue and prostatic carcinoma (95%). It is also positive in salivary duct carcinoma and breast carcinoma.[29,30]

Localization: Cytoplasmic

Ovarian Markers

PAX8

Positivity: PAX8 is positive in all the tumors of Müllerian origin and is a good marker of nonmucinous ovarian carcinoma.

Localization: Nuclear positivity.

Hepatocyte Nuclear Factor 1 β

Positivity: Hepatocyte nuclear factor 1 β antibody helps to recognize the clear-cell carcinoma of ovary. However, it is also positive in renal cell carcinoma and hepatocellular carcinoma.

Other Important Markers

CD117

Positivity: CD117 (C-kit) is a reliable marker of gastrointestinal stromal tumors (GISTs). It is also positive in peripheral neuroectodermal tumor, seminoma, melanoma, renal cell carcinoma and granulocytic sarcomas.[31,32]

Localization: Strong and diffuse cytoplasmic stain.

CD99

Positivity: CD99, also known as MIC-2, is a marker of primitive neuroectodermal tumors (PNET) and Ewing's sarcomas. It is also positive in neuroectodermal lineage, alveolar rhabdomyosarcomas, lymphoblastic lymphomas, and certain neuroendocrine carcinomas.[33,34]

Localization: Membranous pattern.

Table 10 shows the list of the antibodies, application and the staining pattern.

TABLE 10: The list of the antibodies, application, and the staining pattern.

Antibody	Application	Staining pattern	Comments
Epithelial cell markers			
Cytokeratin (CK)	Epithelial tumor	Membranous	Different types of CK are organ specific
Epithelial membrane antigen (EMA)	Epithelial tumors	Membranous positivity	• The pure cytoplasmic positivity without membranous staining should be ignored as spurious. • Nonepithelial tumors that are EMA positive: Mesothelial cells, ALCL, T Cell-rich B-cell lymphoma, mesothelioma, epithelioid sarcoma, synovial sarcoma
Lymphoid markers			
CD45	All lymphoid tumors	Membranous positivity	
CD19 and CD20	B cell NHL	Membranous positivity	
CD3, CD4, CD8	T cell NHL	Membranous positivity	
CD3, CD56	NK cell	Membranous positivity	CD56 is also a marker of neuroendocrine tumors and small cell carcinoma
Mesenchymal markers			
Vimentin	All connective tissue tumors	Cytoplasmic	• Also positive in mesothelial cells, renal cell carcinoma and even lymphoma • It can be used as the control of the presence of antigen in the tissue
Desmin	Striated muscle: Rhabdomyosarcoma and leiomyosarcoma	Cytoplasmic	• Also positive in desmoplastic small round cell tumor, PEComas, fibrohistiocytic tumors, mesothelioma and Wilms tumor
Actin	• Smooth muscle cells • Breast and salivary gland myoepithelial cells	Cytoplasmic	• Positive in rhabdomyosarcoma • Anti-α smooth muscle actin is not very specific and is positive in many spindle cell sarcomas and sarcomatoid carcinomas
H-Caldesmon	• Smooth muscle differentiation and myoepithelial cells • Positive in leiomyoma and leiomyosarcomas	Cytoplasmic	

Continued

Continued

Antibody	Application	Staining pattern	Comments
Calponin	Smooth muscle differentiation, myoepithelial cells and myofibroblasts	Cytoplasmic	• A reliable marker of the myoepithelial cell
Myoglobin	Positive in striated muscle: Pleomorphic Rhabdomyosarcoma	Cytoplasmic	• Background staining
MyoD1 and myogenin	Specific for skeletal muscle differentiation: Rhabdomyosarcoma	Nuclear	• The cytoplasmic stain should be ignored
Endothelial cell markers			
CD34	• Endothelial cell and its neoplasm • Progenitor cell of the hemopoietic cells	Membranous	• Positive in angiosarcoma, dermatofibrosarcoma protuberans, solitary fibrous histiocytoma and gastrointestinal stromal sarcoma
CD31	Endothelial marker: Positive in hemangioma, hemangio-endothelioma and angiosarcoma	Cytoplasmic and membranous pattern	
Factor VIII-related antigen	Endothelial cells: Positive in hemangioma, and hemangioendothelioma	Cytoplasmic	Sensitivity is low in poorly differentiated vascular tumor
Neuroendocrine markers			
Chromogranin A	Positive in neuroendocrine tumors	Cytoplasmic	A specific and sensitive marker of neuroendocrine tumor Paraganglioma are also positive
Synaptophysin	Positive in neuroendocrine and neuroectodermal tumors	Cytoplasmic	Less specific marker than chromogranin. Paraganglioma and neuroblastomas are also positive
CD56	Positive marker of neuroendocrine tumors and NK cell lymphoma	Membranous	Positive in plasma cell tumor
Melanoma markers			
HMB45	Positive in marker of melanoma, neural crest-derived tumors	Cytoplasmic	Also positive in angiomyolipoma, PEComa, epithelioid cell sarcoma and clear cell sarcoma
Melan A	Positive in melanoma, peripheral nerve sheath tumor, angiomyolipoma and melanocytic nevi and adrenocortical tumor	Cytoplasmic	Sensitive marker of melanoma
S-100 protein	Positive in Schwann cells, melanocytes, adipocytes, salivary gland cells, myoepithelial cells, and Langerhans histiocytes	Cytoplasmic and nuclear	A marker of marker of Schwannoma, malignant peripheral nerve sheath tumors, melanoma, clear cell sarcoma, extraskeletal myxoid chondrosarcoma and Langerhans cell histiocytosis
Germ-cell markers			
Placental alkaline phosphatase	Marker of seminoma and embryonal carcinoma	Cytoplasmic and membranous	
Human chorionic gonadotropin	A marker of choriocarcinoma	Cytoplasmic	
OCT4	Specific marker of dysgerminoma, embryonal carcinoma, and gonadoblastoma	Nuclear	Present in the pluripotent germ cells
SALL4	A marker of endodermal sinus tumors, dysgerminoma, and choriocarcinoma	Nuclear	A highly sensitive and specific marker of various germ-cell tumors

Continued

Continued

Antibody	Application	Staining pattern	Comments
Mesothelial markers			
Calretinin	Mesothelial cells: Benign and malignant	Cytoplasmic and nuclear stain	Also positive in ovarian sex cord tumor and adrenocortical cell tumor
HBME-1	Mesothelial cells	Thick membranous positivity	Not a useful marker for mesothelial cells as adenocarcinoma cells are also positive
Wilms tumor gene 1	A good marker for reactive mesothelial cells and mesothelioma	Nuclear	An effective marker of ovarian serous carcinoma and desmoplastic small round cell tumors
D2-40	Mesothelioma	Nuclear	Also positive in ovarian and lung carcinoma
Glucose transporter isoform 1	A marker of malignant mesothelial cells	Membranous staining pattern and at times cytoplasmic pattern	
Lung-specific markers			
TTF 1	Marker of adenocarcinoma of lung and most of the thyroid tumors	Nuclear	High sensitivitiy and specificity for lung adenocarcinoma
Napsin A	Adenocarcinoma of lung	Cytoplasmic	Positive in renal cell carcinoma
p63 and p40	Squamous cell carcinoma of lung	Nuclear	p40 is more specific than p63
Gastrointestinal marker			
CDX2	Colorectal carcinoma	Nuclear	Also positive in occasional cases of mucinous tumor of lung, ovary
Breast markers			
Estrogen receptors (ERs)	Metastatic breast carcinomas	Nuclear	Carcinomas such as cervix, endometrial and ovarian carcinomas are also positive for ER
Mammoglobin	Positive in most of the breast carcinomas	Cytoplasmic	Also positive in salivary gland malignancies, and endometrial adenocarcinomas
Gross cystic disease fluid protein	Carcinoma of breast	Cytoplasmic	Also positive in salivary, sweat gland and some lung carcinomas
GATA 3	Carcinoma of breast	Nuclear	Also positive in urothelial carcinoma and mammary analog secretory carcinoma
Ovarian markers			
PAX8	Positive in all the tumors of Müllerian origin	Nuclear	
Hepatocyte nuclear factor 1 β	Clear cell carcinoma of ovary		
Other important markers			
CD117	Marker of gastrointestinal stromal tumors	cytoplasmic	Also positive in peripheral neuroectodermal tumor, seminoma, melanoma, renal cell carcinoma and granulocytic sarcomas
CD99	A marker of PNET and Ewing's sarcomas	Membranous	Occasionally positive in uterine tumor resembling ovarian sex cord tumors

(ALCL: anaplastic large cell lymphoma; HBME: Hector Battifora mesothelial-1; NHL: non-Hodgkin lymphoma; NK: natural killer; OCT 4: Octamer-binding transcription factor 4; PEComas: perivascular epithelioid cell neoplasms; PNET: peripheral neuroectodermal tumor; SALL4: sal-like protein 4; TTF 1: thyroid transcription factor)

REFERENCES

1. Moll R. Cytokeratins as markers of differentiation in the diagnosis of epithelial tumors. Subcell Biochem. 1998;31:205-62.
2. Mortimer G, Jones DN, Assaf H, al-Ahmadi T. Patterns of cytokeratin expression by neoplastic and non-neoplastic epithelium. Ir J Med Sci. 1993;162(3):77-81.
3. Chu P, Wu E, Weiss LM. Cytokeratin 7 and cytokeratin 20 expression in epithelial neoplasms: a survey of 435 cases. Mod Pathol. 2000;13(9):962-72.
4. Pinkus GS, Kurtin PJ. Epithelial membrane antigen—a diagnostic discriminant in surgical pathology: immunohistochemical profile in epithelial, mesenchymal, and hematopoietic neoplasms using paraffin sections and monoclonal antibodies. Hum Pathol. 1985;16(9):929-40.
5. Schnitt SJ, Vogel H. Meningiomas. Diagnostic value of immunoperoxidase staining for epithelial membrane antigen. Am J Surg Pathol. 1986;10(9):640-9.
6. Rangdaeng S, Truong LD. Comparative immunohistochemical staining for desmin and muscle-specific actin. A study of 576 cases. Am J Clin Pathol. 1991;96(1):32-45.
7. Friedman HD, Tatum AH. HMB-45-positive malignant lymphoma. A case report with literature review of aberrant HMB-45 reactivity. Arch Pathol Lab Med. 1991;115(8):826-30.
8. Busam KJ, Iversen K, Coplan KA, Old LJ, Stockert E, Chen YT, et al. Immunoreactivity for A103, an antibody to melan-A (Mart-1), in adrenocortical and other steroid tumors. Am J Surg Pathol. 1998;22(1):57-63.
9. Loeffel SC, Gillespie GY, Mirmiran SA, Miller EW, Golden P, Askin FB, et al. Cellular immunolocalization of S100 protein within fixed tissue sections by monoclonal antibodies. Arch Pathol Lab Med. 1985;109(2):117-22.
10. Hisaoka M, Hashimoto H. Extraskeletal myxoid chondrosarcoma: updated clinicopathological and molecular genetic characteristics. Pathol Int. 2005;55(8):453-63.
11. Wick MR, Swanson PE, Manivel JC. Placental-like alkaline phosphatase reactivity in human tumors: an immunohistochemical study of 520 cases. Hum Pathol. 1987;18(9):946-54.
12. Hattab EM, Tu PH, Wilson JD, Cheng L. OCT4 immunohistochemistry is superior to placental alkaline phosphatase (PLAP) in the diagnosis of central nervous system germinoma. Am J Surg Pathol. 2005;29(3):368-71.
13. Rabban JT, Zaloudek CJ. A practical approach to immunohistochemical diagnosis of ovarian germ cell tumours and sex cord-stromal tumours. Histopathology. 2013;62(1):71-88.
14. Doglioni C, Dei Tos AP, Laurino L, Iuzzolino P, Chiarelli C, Celio MR, et al. Calretinin: a novel immunocytochemical marker for mesothelioma. Am J Surg Pathol. 1996;20(9):1037-46.
15. Politi E, Kandaraki C, Apostolopoulou C, Kyritsi T, Koutselini H. Immunocytochemical panel for distinguishing between carcinoma and reactive mesothelial cells in body cavity fluids. Diagn Cytopathol. 2005;32(3):151-5.
16. Hecht JL, Lee BH, Pinkus JL, Pinkus GS. He value of Wilms tumor susceptibility gene 1 in cytologic preparations as a marker for malignant mesothelioma. Cancer. 2002;96(2):105-9.
17. Saad RS, Lindner JL, Lin X, Liu YL, Silverman JF. The diagnostic utility of D2-40 for malignant mesothelioma versus pulmonary carcinoma with pleural involvement. Diagn Cytopathol. 2006;34(12):801-6.
18. Afify A, Zhou H, Howell L, Paulino AF. Diagnostic utility of GLUT-1 expression in the cytologic evaluation of serous fluids. Acta Cytol. 2005;49(6):621-6.
19. Kundu UR, Krishnamurthy S.Use of the monoclonal antibody MOC-31 as an immunomarker for detecting metastatic adenocarcinoma in effusion cytology. Cancer Cytopathol. 2011;119(4):272-8.
20. Bailey ME, Brown RW, Mody DR, Cagle P, Ramzy I. Ber-EP4 for differentiating adenocarcinoma from reactive and neoplastic mesothelial cells in serous effusions. Comparison with carcinoembryonic antigen, B72.3 and Leu-M1. Acta Cytol. 1996;40(6):1212-6.
21. Latza U, Niedobitek G, Schwarting R, Nekarda H, Stein H. Ber-EP4: new monoclonal antibody which distinguishes epithelia from mesothelial.J Clin Pathol. 1990;43(3):213-9.
22. Mukhopadhyay S, Katzenstein AL. Subclassification of non-small cell lung carcinomas lacking morphologic differentiation on biopsy specimens: Utility of an immunohistochemical panel containing TTF-1, napsin A, p63, and CK5/6. Am J Surg Pathol. 2011;35(1):15-25.
23. Jaffee IM, Rahmani M, Singhal MG, Younes M. Expression of the intestinal transcription factor CDX2 in carcinoid tumors is a marker of midgut origin. Arch Pathol Lab Med. 2006;130(10):1522-6.
24. Recine MA, Deavers MT, Middleton LP, Silva EG, Malpica A. Serous carcinoma of the ovary and peritoneum with metastases to the breast and axillary lymph nodes: a potential pitfall. Am J Surg Pathol. 2004;28(12):1646-51.
25. Wang Z, Spaulding B, Sienko A, Liang Y, Li H, Nielsen G, et al. Mammaglobin, a valuable diagnostic marker for metastatic breast carcinoma. Int J Clin Exp Pathol. 2009;2(4):384-9.
26. Bhargava R, Beriwal S, Dabbs DJ. Mammaglobin vs GCDFP-15: an immunohistologic validation survey for sensitivity and specificity. Am J Clin Pathol. 2007;127(1):103-13.
27. Striebel JM, Dacic S, Yousem SA. Gross cystic disease fluid protein-(GCDFP-15): expression in primary lung adenocarcinoma. Am J Surg Pathol. 2008;32(3):426-32.
28. Schwartz LE, Begum S, Westra WH, Bishop JA. GATA3 immunohistochemical expression in salivary gland neoplasms. Head Neck Pathol. 2013;7(4):311-5.
29. Bodey B, Bodey B, Jr, Kaiser HE. Immunocytochemical detection of prostate specific antigen expression in human breast carcinoma cells. Anticancer Res. 1997;17(4A):2577-81.
30. Fan CY, Wang J, Barnes EL. Expression of androgen receptor and prostatic specific markers in salivary duct carcinoma: an immunohistochemical analysis of 13 cases and review of the literature. Am J Surg Pathol. 2000;24(4):579-86.
31. Tsuura Y, Hiraki H, Watanabe K, Igarashi S, Shimamura K, Fukuda T,et al. Preferential localization of c-kit product in tissue mast cells, basal cells of skin, epithelial cells of breast, small cell lung carcinoma and seminoma/dysgerminoma in human: immunohistochemical study on formalin-fixed, paraffin-embedded tissues. Virchows Arch. 1994;424(2):135-41.
32. Natali PG, Nicotra MR, Sures I, Santoro E, Bigotti A, Ullrich A. Expression of c-kit receptor in normal and transformed human nonlymphoid tissues. Cancer Res. 1992;52(22):6139-43.
33. Lumadue JA, Askin FB, Perlman EJ. MIC2 analysis of small cell carcinoma. Am J Clin Pathol. 1994;102(5):692-4.
34. Riopel M, Dickman PS, Link MP, Perlman EJ. MIC2 analysis in pediatric lymphomas and leukemias. Hum Pathol. 1994;25(4):396-9.

CHAPTER 7

Immunocytochemistry of Effusion

INTRODUCTION

Diagnosis of malignancy in body cavity effusion fluid is a challenging area due to several reasons. The reactive mesothelial cells often look like adenocarcinoma and vice versa. Therefore, the diagnosis of carcinoma in an effusion sample on the basis of cytomorphology alone is often difficult. The selective use of an antibody in the effusion fluid solves the following issues:

- Identification of carcinoma cells among the reactive mesothelial cells.
- The source of primary malignancy.
- Differentiating malignant from the benign mesothelial cells.

MARKERS OF MESOTHELIAL CELLS[1]

- *Calretinin:* Nuclear and cytoplasmic positivity
- *Wilms' tumor (WT1):* Nuclear positivity
- *Podoplanin (D2-40):* Nuclear positivity
- *Cytokeratin (CK5/6):* Membrane positivity

Out of all the above markers, calretinin is the most widely used and reliable marker of the mesothelial cells (**Figs. 1** and **2**). These markers should be used along with the other markers for adenocarcinoma.

Calretinin

Calretinin is present in cells: Mesothelial cells, steroid-producing cells, renal convoluted tubules, and neurons. It is a specific marker of mesothelial cells (**Table 1**).

Calretinin-positive tumors are:

- Mesothelioma
- Desmoplastic small round cell tumors
- Adrenal cortical tumors
- Granular cell tumors
- Thymic carcinoma and also thymoma
- Synovial sarcoma

Wilms' Tumor (WT1)

Wilms' tumor (WT1), a transcriptional regulator protein, is encoded from chromosome 11p13. WT1 is associated with the development of genitourinary tract, mesothelial cells, and ovary. It is a helpful antibody to detect serous epithelial tumors of the ovary, endometrioid carcinoma, and clear cell carcinoma. However, many other tumors are also positive for WT1 and careful interpretation is need (**Table 2**).

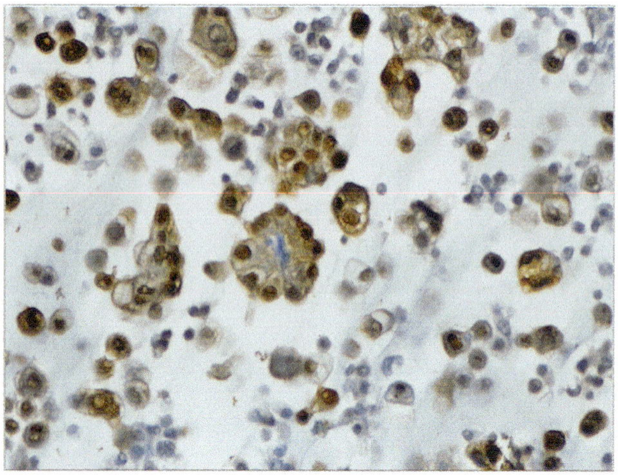

FIG. 1: Calretinin: The nuclear positivity in the mesothelial cells.

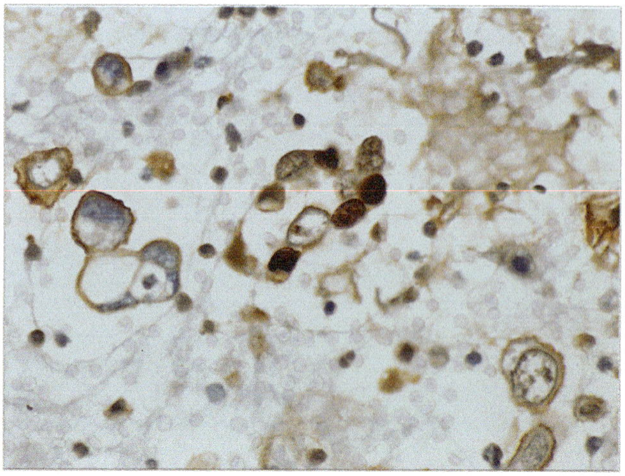

FIG. 2: Nuclear-positive WT1 in mesothelial cells.

TABLE 1: Calretinin expression.

Marker	Main use in different tumors	Other tumors showing calretinin positivity	Localization	Remarks
Calretinin	• Malignant mesothelioma • Ovarian sex cord stromal tumor • Adrenal cortical tumors	• Desmoplastic small round cell tumors • Granular cell tumors • PEComa • Thymic carcinoma and also thymoma • Synovial sarcoma	Cytoplasmic and nuclear stain	Calretinin should be used along with other markers of mesothelial cells as it is widely expressed in different tumors

(PEComa: perivascular epithelioid cell tumor)

TABLE 2: Wilms' tumor (WT1) expression.

Marker	Main use in different tumors	Other tumors showing WT1 positivity	Localization
WT1	• Serous adenocarcinoma of the ovary • Malignant mesothelioma • Tubal serous carcinoma	• Wilms' tumor • Peritoneal serous carcinoma • Desmoplastic small round cell tumor • Endometrial stromal sarcoma • Neuroblastoma • Rhabdomyosarcoma • Sex cord stromal tumor	Nuclear

TABLE 3: Podoplanin (D2-40) expression.

Marker	Main use in different tumors	Other tumors expressing this antigen	Localization	Comments
Podoplanin (D2-40)	• Lymphatic tumor • Mesothelioma	• Germ cell tumors • Meningioma • GIST • Synovial sarcoma • MPNST • Kaposi's sarcoma	Membranous and cytoplasmic	Wide variety of tumors express this marker and so it should be used along with a panel of other markers

(GIST: gastrointestinal stromal tumor; MPNST: malignant peripheral nerve sheath tumor)

Podoplanin

Podoplanin (D2-40), a membranous glycoprotein, is seen in the lymphatic endothelium and mesothelial cells. In soft-tissue tumor, podoplanin is mainly used in the demonstration of lymphatic differentiation of the vascular tumor (**Table 3**).

The markers of adenocarcinomas are:

- *MOC31*: Membranous stain, positive for most of the adenocarcinomas in the effusion fluid.
- *BER-EP4*: Membranous stain
- *Carcinoembryonic antigen (CEA):* Cytoplasmic staining
- *B72.3*: Cytoplasmic staining

The distinguishing features of reactive mesothelial cells and adenocarcinoma are mentioned in **Table 4**.

The approach to diagnosis of malignancy in effusion fluid is described in **Box 1**.

TABLE 4: Distinguishing characteristics of reactive mesothelial cells and adenocarcinoma.

Antibody	Reactive mesothelial cells	Metastatic adenocarcinoma
Calretinin	+	–
WT1	+	–
D2-40	+	–
H-Caldesmon	+	–
BER-EP4	–	+
MOC31	–	+

+ = Positive, – = Negative

BOX 1 Diagnostic key points of malignancy in effusion.

- Always correlate the immunocytochemistry with the cytological features and clinical history
- Past history and radiological features may be helpful
- A panel of markers:
  - Mesothelial markers: Calretinin and any other markers (WT1/D2-40/H-Caldesmon)
  - Epithelial markers: BER-EP4, MOC31

IMMUNOSTAINING PATTERN OF DIFFERENT MALIGNANCIES

It is often necessary to detect the primary source of malignancy in the effusion cytology. The clinical, biochemical, and radiological findings are necessary to detect the panel of antibodies for the identification of the primary site (**Box 2**).

Metastasis from Primary Lung Carcinoma

Lung Adenocarcinoma

The three commonly used markers include the following:[2]

- *Thyroid transcription factor* 1 *(TTF-1):* Nuclear positivity
- *Napsin A:* Cytoplasmic staining
- *CK7:* Membranous stain

BOX 2 Detection of the primary site of tumor.

- Salient clinical features
- Age: Certain tumors may occur in lower or higher age groups
- Sex: Ovarian carcinomas in female
- Location:
  - Peritoneal: Ovarian and intestinal carcinomas
  - Pleural: Lung and breast carcinomas
- Cytology:
  - Squamous cell carcinoma: Lung, cervix
  - Adenocarcinoma: Colorectal, ovary, lung
  - Clear cell carcinoma: Kidney, ovary
- Panel for the detection of the location of primary:
  - Colorectal: CDX2, CK20
  - Ovary: PAX8, CK7
  - Lung: TTF-1, napsin, CK7
  - Prostate: PSA
  - Breast: GATA3, mammaglobin, GCDFP
  - Renal cell carcinoma: PAX8, CD10, vimentin, claudin
  - Neuroendocrine: Synaptophysin, neuron-specific enolase (NSE), Chromogranin
  - Thyroid: TTF-1, thyroglobulin
  - Pancreatobiliary: CK7, CK19, MUC1

The positivity of these markers in a selected case usually indicates the primary lung origin of the malignant cells (**Figs. 3A** to **F**).

Small Cell Carcinoma

The small cell carcinoma (SCC) of lung shows (**Figs. 4** to **9**):

- *Chromogranin:* Cytoplasmic positivity
- *Synaptophysin:* Cytoplasmic positivity
- *CD56*: Membranous positivity
- *TTF-1*: Occasionally positive
- *CK7*: Occasionally positive

Squamous Cell Carcinoma of the Lung

The markers of the metastatic squamous cell carcinoma are:[3]

- *p40/p63*: Strong nuclear positivity
- *TTF-1*: Negative

It is preferable to use both p40/p63 and TTF-1 immunostaining in the effusion cytology specimen in the suspected metastatic squamous cell carcinoma case (**Figs. 10** and **11**).

Breast Carcinoma

The metastatic carcinoma from the breast in the effusion cytology (**Figs. 12** and **13**) is positive for:[4]

- *GATA3*: Nuclear positivity
- *ER*: Nuclear positivity
- *PR*: Nuclear positivity
- *Gross cystic disease fluid protein (GCDFP):* Nuclear positivity
- *Mammaglobin:* Cytoplasmic positivity
- *CK7*: Membranous positive

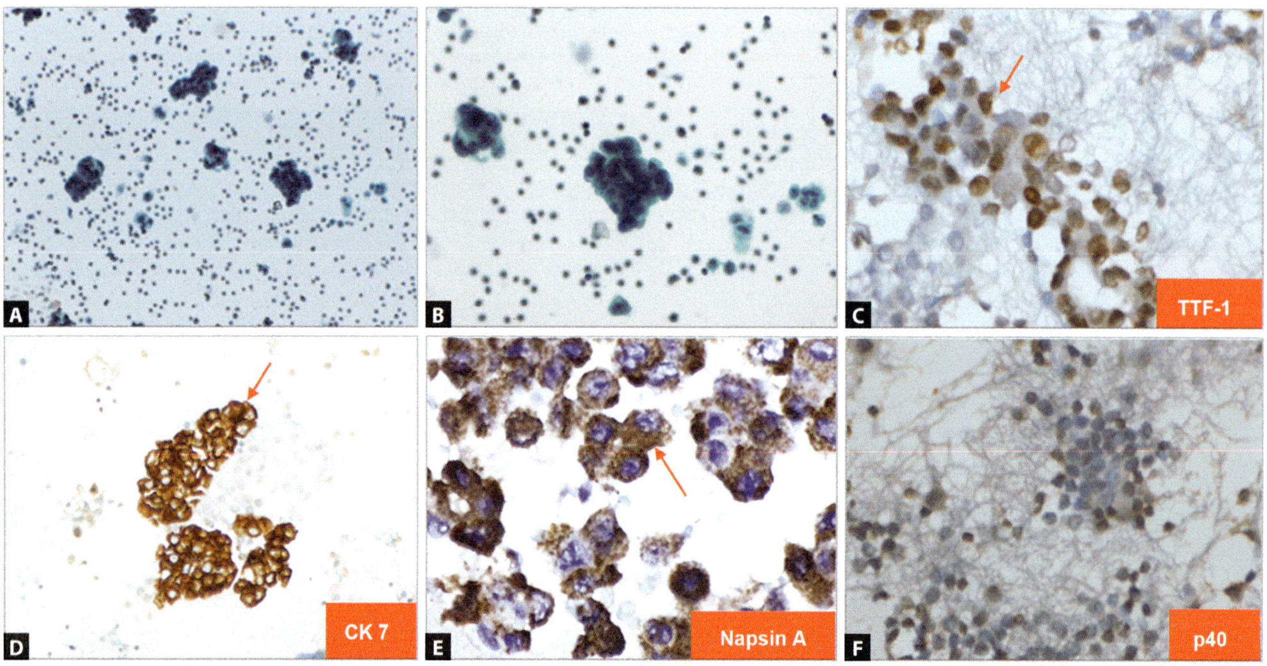

FIGS. 3A TO F: Metastatic lung adenocarcinoma in pleural effusion. (A) Multiple clusters of tumor cells. (B) Three-dimensional cohesive clusters of tumor cells. (C) Nuclear positivity of TTF-1 (see arrow). (D) Tumor cells showing membranous positivity of CK7 (see arrow). (E) Tumor cells showing cytoplasmic positivity of napsin (see arrow). (F) Tumor cells are negative for p40 immunostaining.

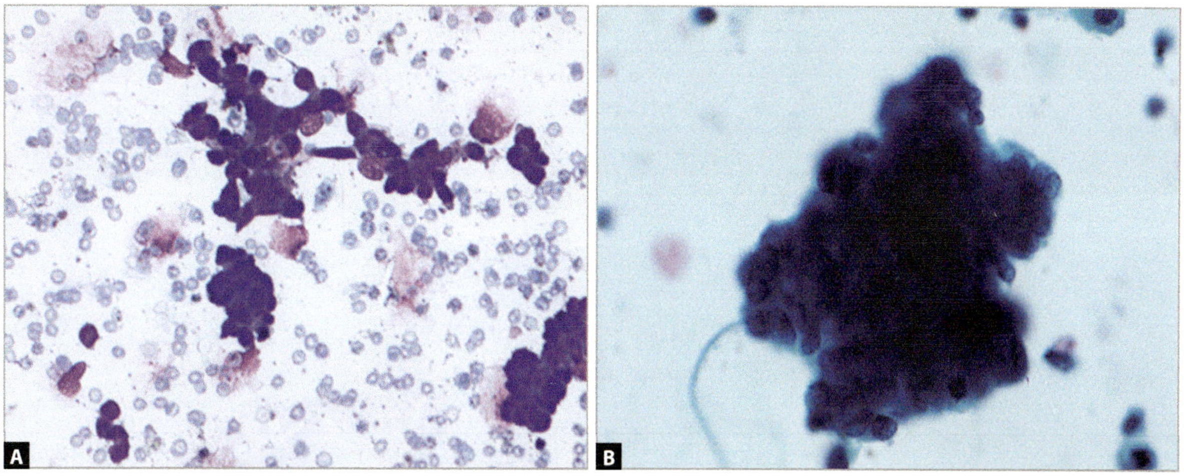

FIGS. 4A AND B: Metastatic small cell carcinoma in pleural fluid. (A) Small round cells in cohesive clusters. (B) Papanicolaou-stained smear showing small cells with scanty cytoplasm.

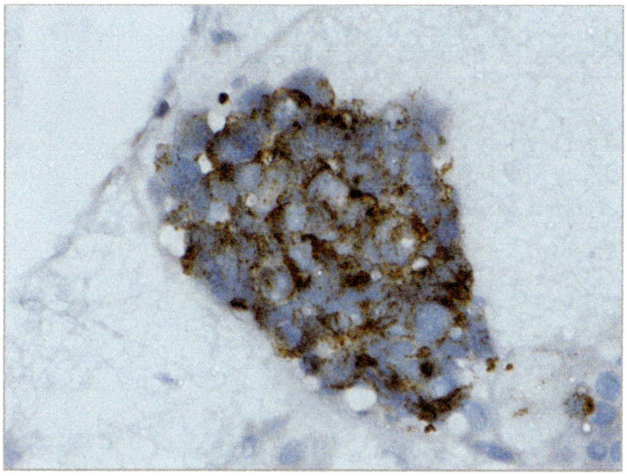

FIG. 5: Metastatic small cell carcinoma in pleural fluid: Chromogranin-positive cells.

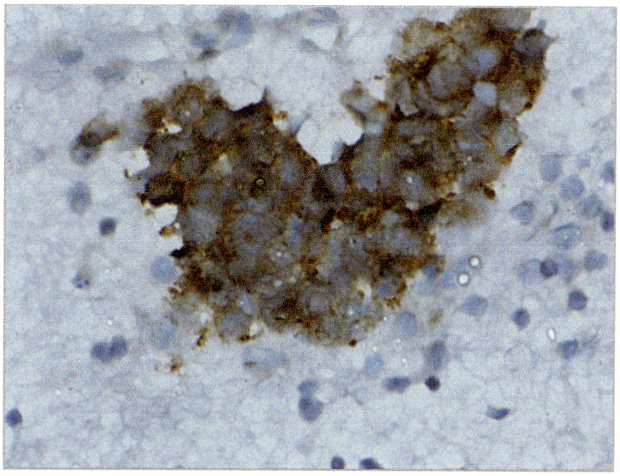

FIG. 6: Metastatic small cell carcinoma in pleural fluid: Synaptophysin-positive cells.

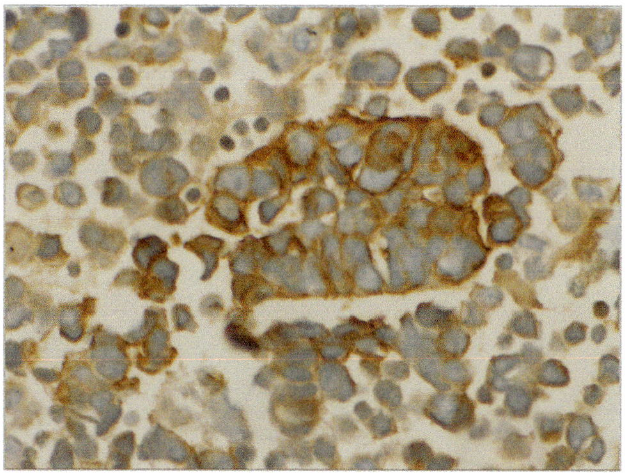

FIG. 7: Metastatic small cell carcinoma in pleural fluid: Membranous positivity of CD56 immunostaining.

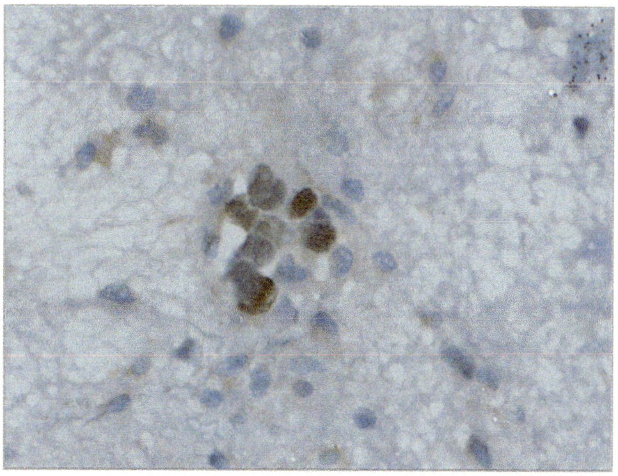

FIG. 8: Metastatic small cell carcinoma in pleural fluid: Occasional cells showing nuclear positivity of TTF-1 immunostaining.

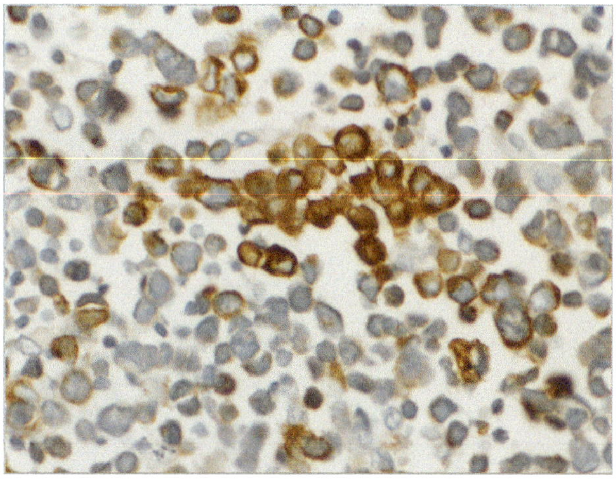

FIG. 9: Metastatic small cell carcinoma in pleural fluid: Membranous positivity of CK7 immunostaining.

In a suspected case of metastatic carcinoma of the breast, it is preferable to use GATA3/GCDFP/mammaglobin staining.

The differentiating features of lung carcinoma versus breast carcinoma metastasis in the effusion fluid are highlighted in **Table 5**.

Metastatic Ovarian Carcinoma

Ovarian serous carcinoma in the effusion fluid shows:[5]

- PAX8: Positive
- BER-EP4: Positive
- MOC31: Positive
- WT1: Positive
- p53: Positive
- CA125: Positive

The combination of PAX8, p53, and WT1 positivity in the effusion fluid usually indicates high-grade ovarian serous adenocarcinoma (**Figs. 14** to **20**).

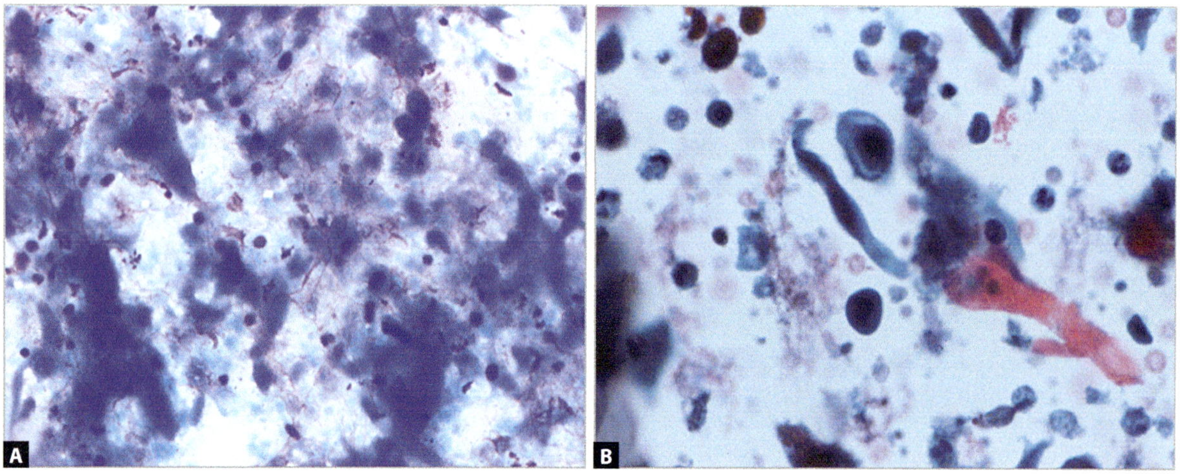

FIGS. 10A AND B: Pleural effusion in a metastatic squamous cell carcinoma of the lung. (A) Discrete polyhedral cells with hyperchromatic nuclei. (B) Papanicolaou-stained smear showing fiber-like cells.

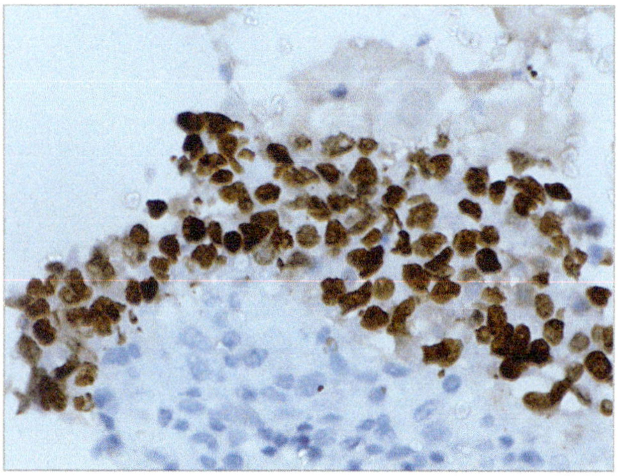

FIG. 11: Pleural effusion in a metastatic squamous cell carcinoma of the lung showing nuclear positivity of p40.

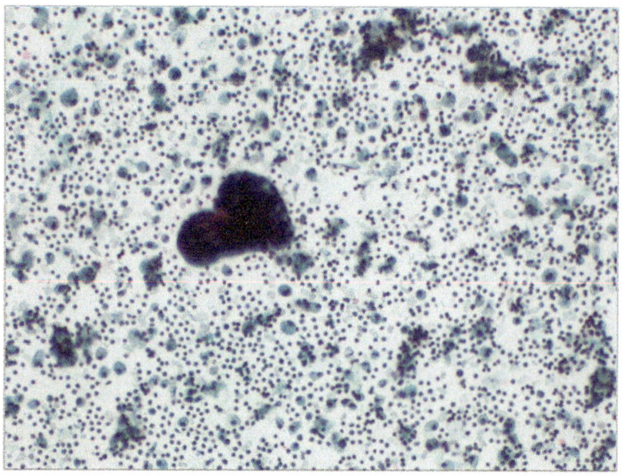

FIG. 12: Metastatic carcinoma from the breast: Smear showing three-dimensional cluster of malignant cells.

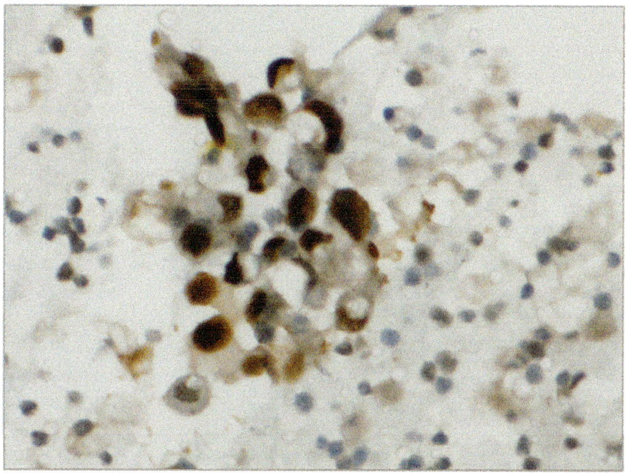

FIG. 13: Metastatic carcinoma from the breast: GATA3-positive cells.

Metastatic Colonic Carcinoma in Effusion

The metastatic adenocarcinoma from colon shows the following pattern of immunostaining (**Figs. 21** to **24**):

TABLE 5: Distinguishing characteristics of immunocytochemistry in breast and lung carcinomas in effusion.

Antibody	Breast carcinoma	Lung adenocarcinoma
GATA3	+	–
Mammaglobin	+	–
GCDFP	+	–
TTF-1	–	+
Napsin A	–	+

+ = Positive, – = Negative

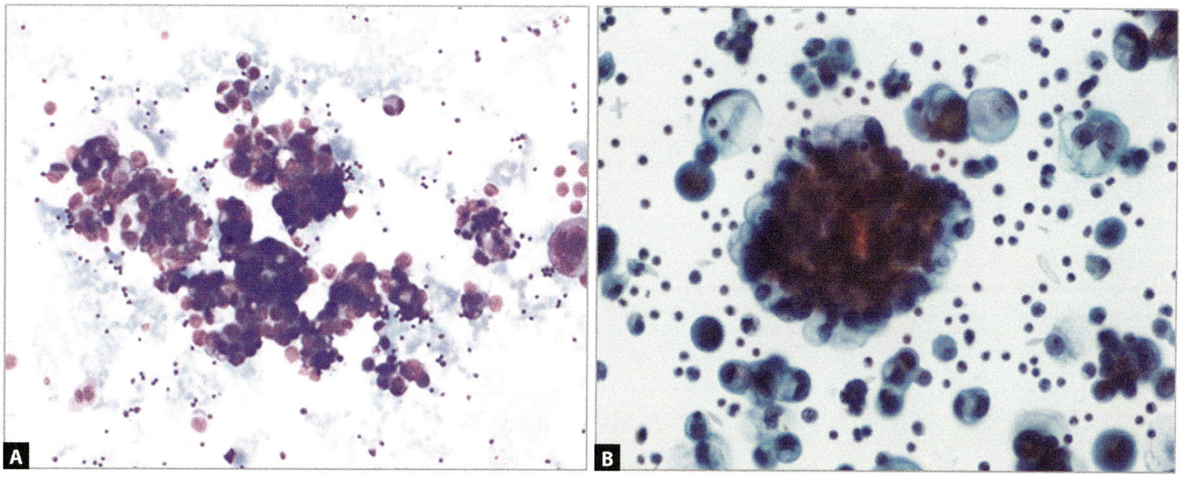

FIGS. 14A AND B: (A) Metastatic adenocarcinoma of the ovary in effusion fluid showing multiple clusters of tumor cells. (B) Metastatic adenocarcinoma of the ovary in effusion fluid showing three-dimensional ball-like clusters in Papanicolaou-stained smear.

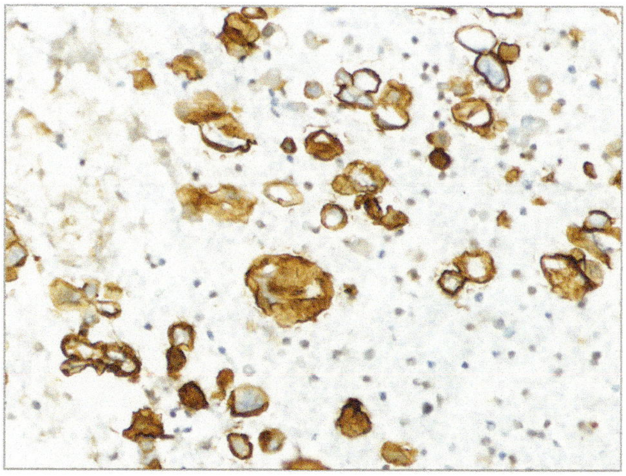

FIG. 15: Metastatic adenocarcinoma of the ovary in effusion fluid showing membranous positivity of CA125 immunostaining.

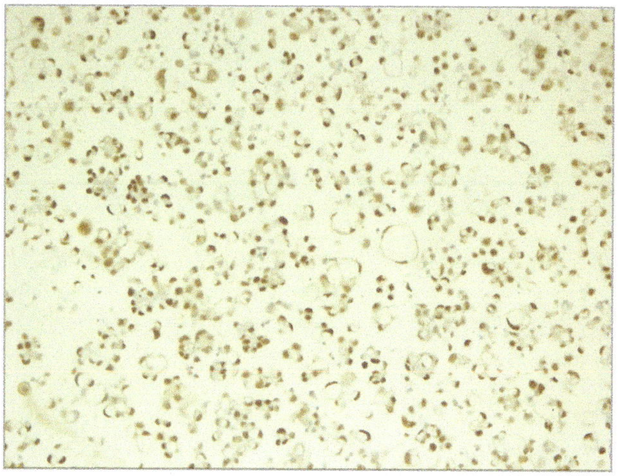

FIG. 16: Metastatic adenocarcinoma of the ovary in effusion fluid showing nuclear positivity of PAX8 immunostaining.

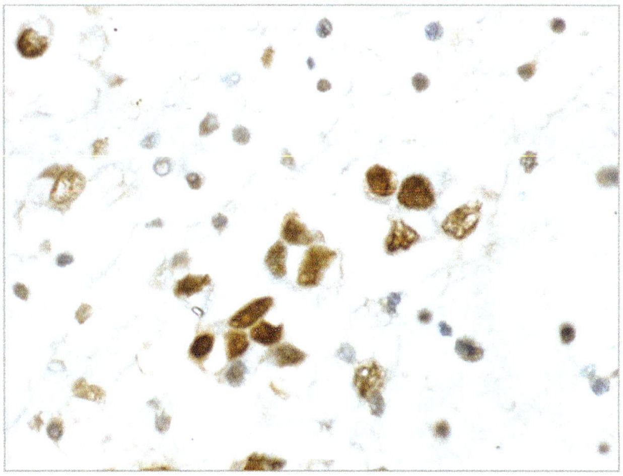

FIG. 17: Metastatic adenocarcinoma of the ovary in effusion fluid showing nuclear positivity of p53 immunostaining.

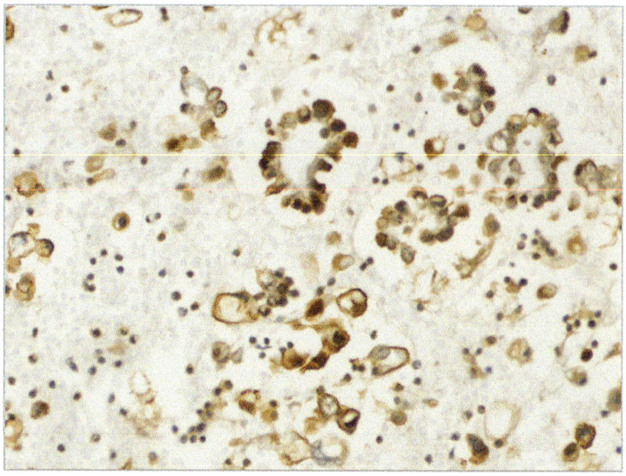

FIG. 18: Metastatic adenocarcinoma of the ovary in effusion fluid showing nuclear positivity of WT1 immunostaining.

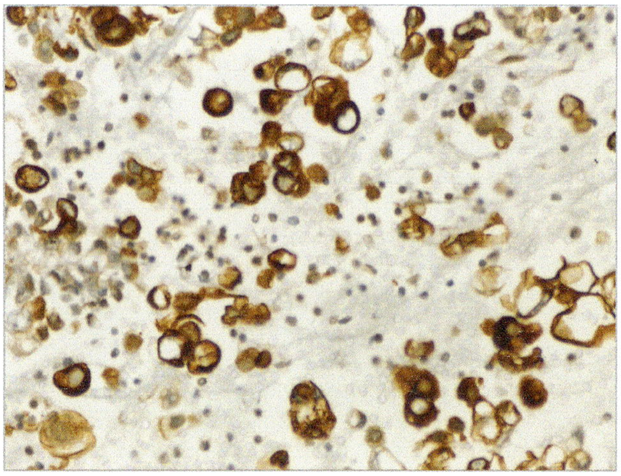

FIG. 19: Metastatic adenocarcinoma of the ovary in effusion fluid showing strong membranous CK7 immunostaining.

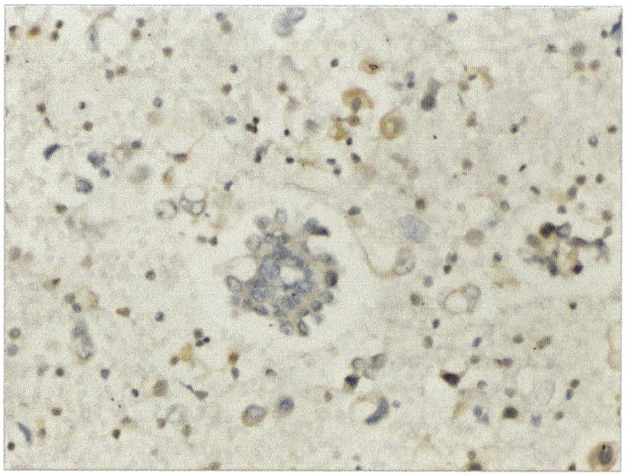

FIG. 20: Metastatic adenocarcinoma of the ovary in effusion fluid showing negative CK20 immunostaining.

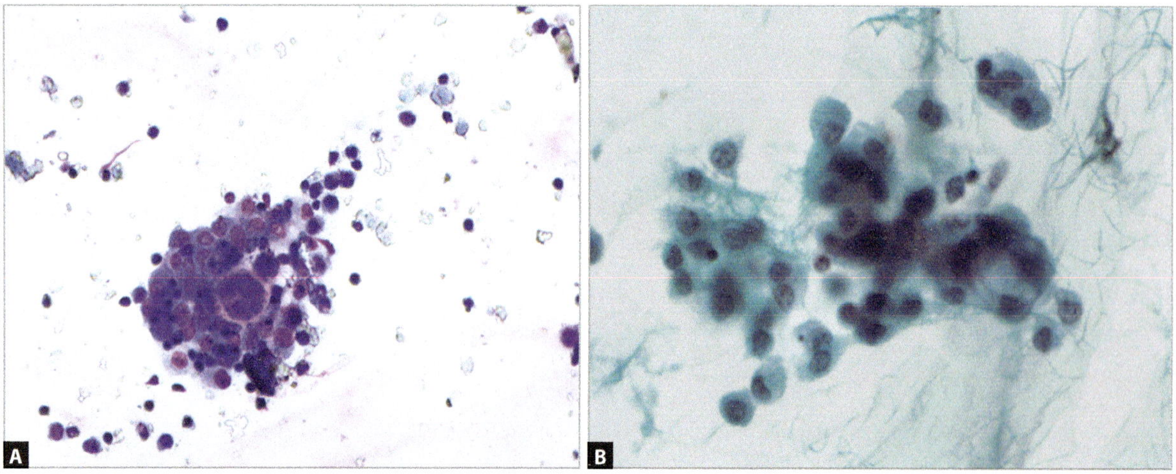

FIGS. 21A AND B: Metastatic colonic carcinoma in peritoneal effusion. (A) Clusters of tumor cells. (B) Papanicolaou-stained smear showing cells with vacuolated cytoplasm having enlarged nuclei.

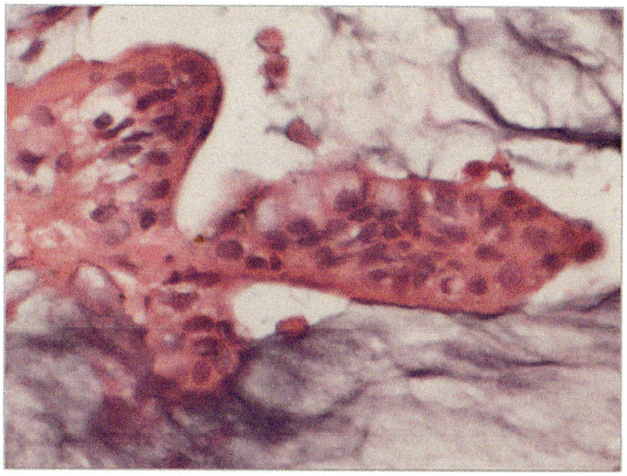

FIG. 22: Metastatic colonic carcinoma in peritoneal effusion: Hematoxylin and eosin-stained cell block section showing sheets of tumor cells in a mucinous background.

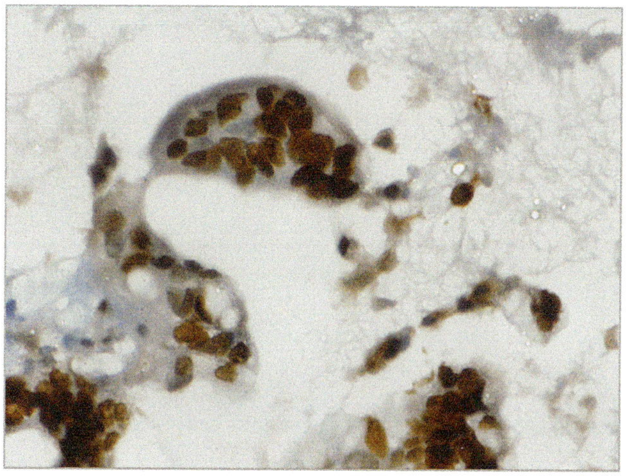

FIG. 23: Metastatic colonic carcinoma in peritoneal effusion: CDX2 immunostaining showing strong nuclear-positive tumor cells.

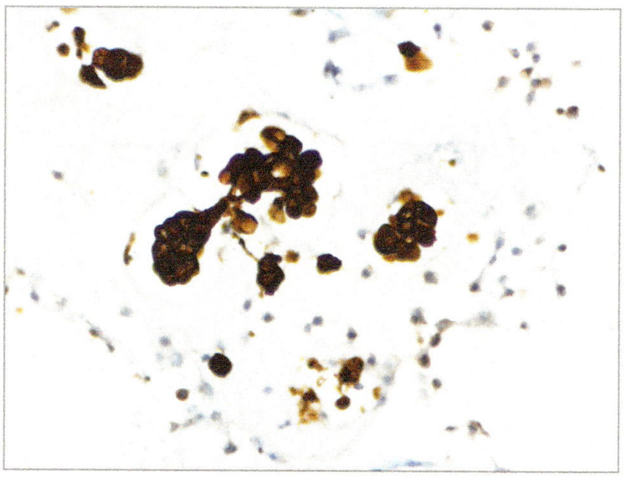

FIG. 24: Metastatic colonic carcinoma in peritoneal effusion: CK20 immunostaining showing strong membranous-positive tumor cells.

TABLE 6: Distinguishing characteristics of ovarian from colonic carcinoma in effusion fluid.

Antibody	Colonic carcinoma	Serous adenocarcinoma	Mucinous adenocarcinoma
CDX2	+	–	–
CK20	+	–	–
CK7	–	+	+
PAX8	–	+	+
MUC5AC	–	–	+
CA125	–	+	+
Beta-catenin	+	–	–

+ = Positive, – = Negative

- *CDX2*: Nuclear positivity. It is a relatively specific marker of colonic carcinoma.
- *CK20*: Membranous positivity

The above two markers should be used along with the other markers to differentiate colonic carcinoma from the metastatic carcinoma of ovary[6] (**Table 6**; **Figs. 25A** to **E**).

Pancreatic and Biliary Carcinoma

The pancreaticobiliary carcinomas show following immunocytochemistry (ICC) (**Figs. 26** to **30**):

- CK7: Positive
- CK19: Positive
- CK20: Positive
- MUC1: Positive
- CA19.9: Positive

A panel of CK20, MUC1, and CDX2 correctly identifies the pancreatobiliary carcinomas where the CDX2 is negative and CK20 and MUC1 are positive.[7]

Metastatic Renal Cell Carcinoma

The metastatic renal cell carcinoma (RCC) shows the following immunocytochemical features:

- PAX8: Positive
- CD10: Positive
- Vimentin: Positive
- Claudin-4: Positive
- CK7: Negative
- CK20: Negative

PAX8 is also a marker of gynecological tumors. Similarly, CD10 is positive in urothelial and prostatic carcinomas. Therefore, it is preferable to have a panel of marker to confirm a clinically suspicious case of renal cell carcinoma in the effusion fluid[8,9] (**Figs. 31A** to **D**).

Metastatic Urothelial Cell Carcinoma

Urothelial cell carcinoma (UCC) in effusion fluid shows the following pattern of ICC:[10]

- GATA3: Positive
- CK7: Positive
- CK20: Positive

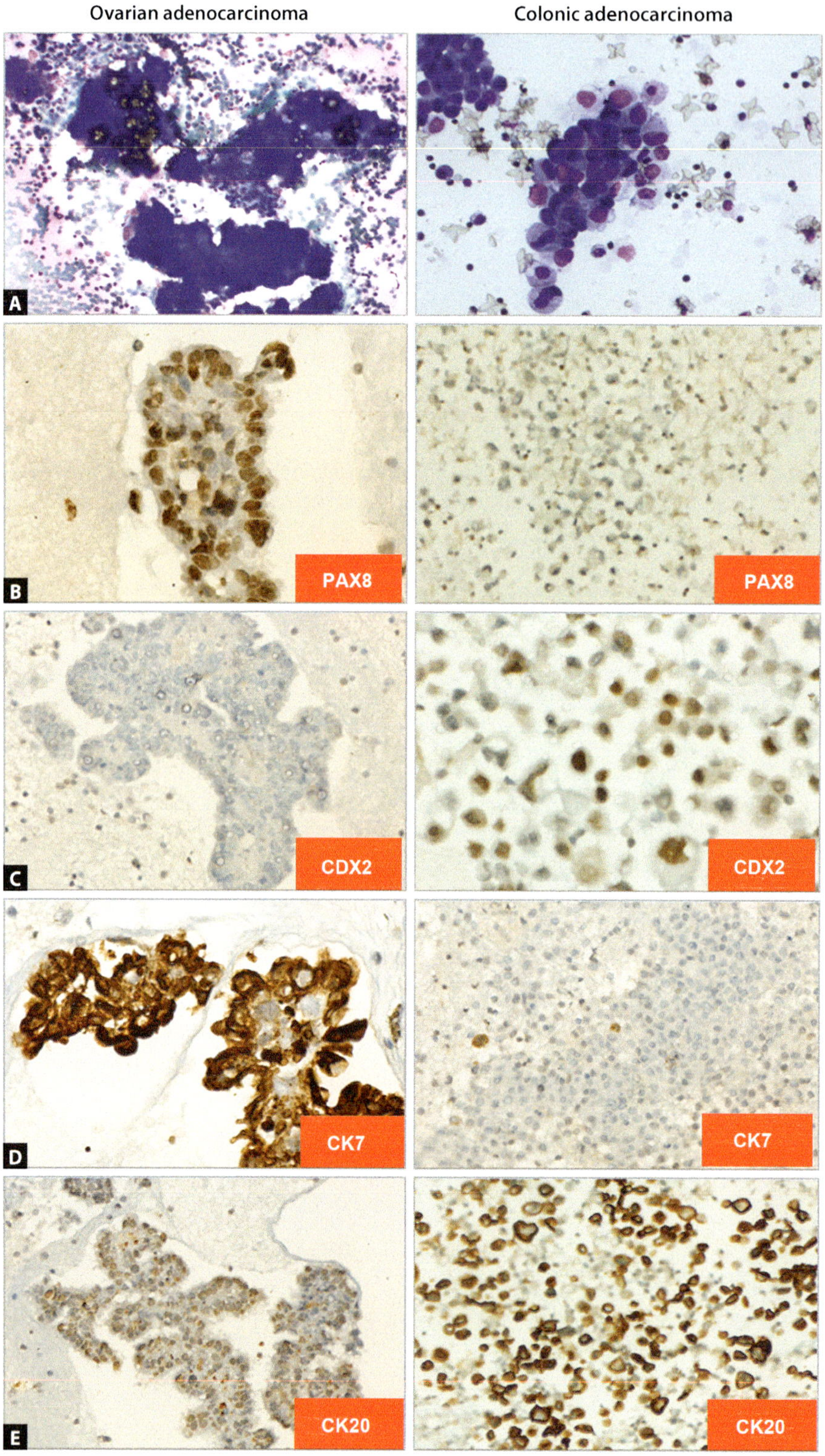

FIGS. 25A TO E: Distinguishing characteristics of immunocytochemistry of colonic versus ovarian serous adenocarcinoma in effusion cytology.

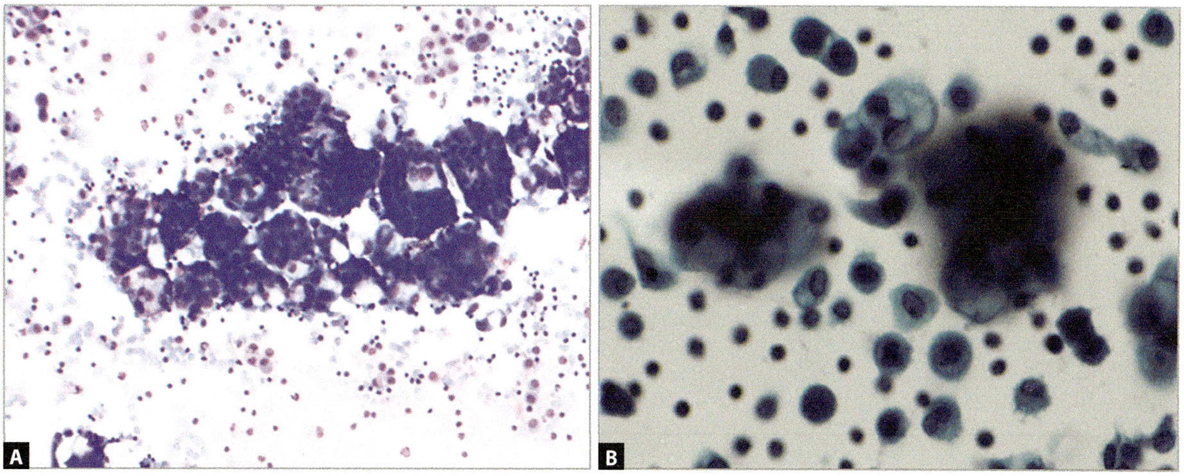

FIGS. 26A AND B: Pancreatobiliary carcinoma in peritoneal effusion. (A) Multiple three-dimensional clusters of tumor cells. (B) Cells with vacuolated cytoplasm having enlarged pleomorphic nuclei.

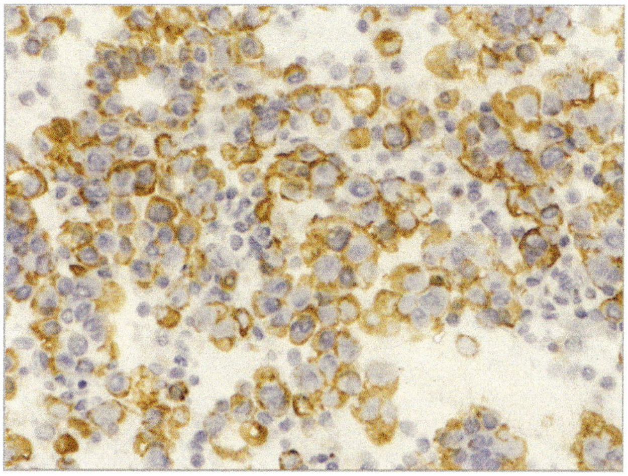

FIG. 27: Pancreatobiliary carcinoma in peritoneal effusion showing membranous positivity of CK7.

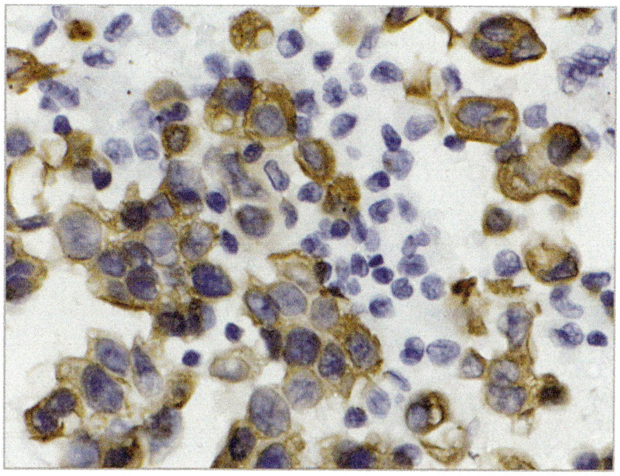

FIG. 28: Pancreatobiliary carcinoma in peritoneal effusion showing membranous positivity of CK19.

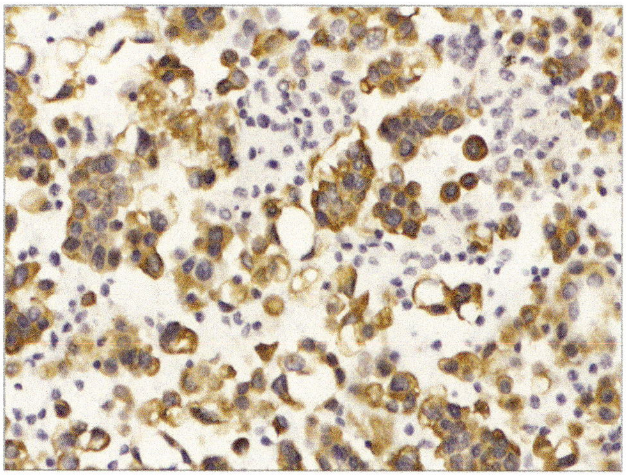

FIG. 29: Pancreatobiliary carcinoma in peritoneal effusion showing membranous positivity of CK20.

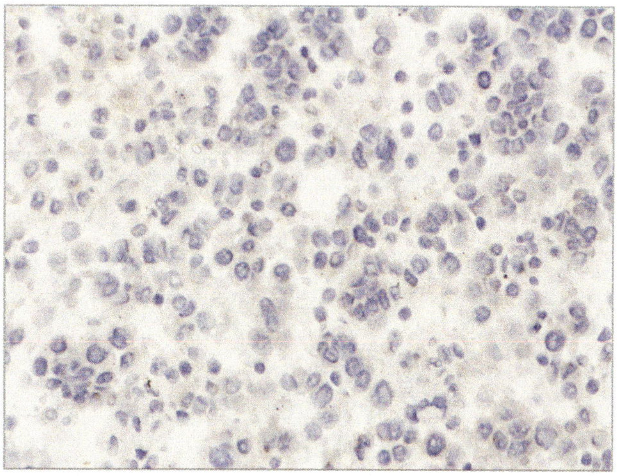

FIG. 30: Pancreatobiliary carcinoma in peritoneal effusion showing negative CDX2 immunostaining.

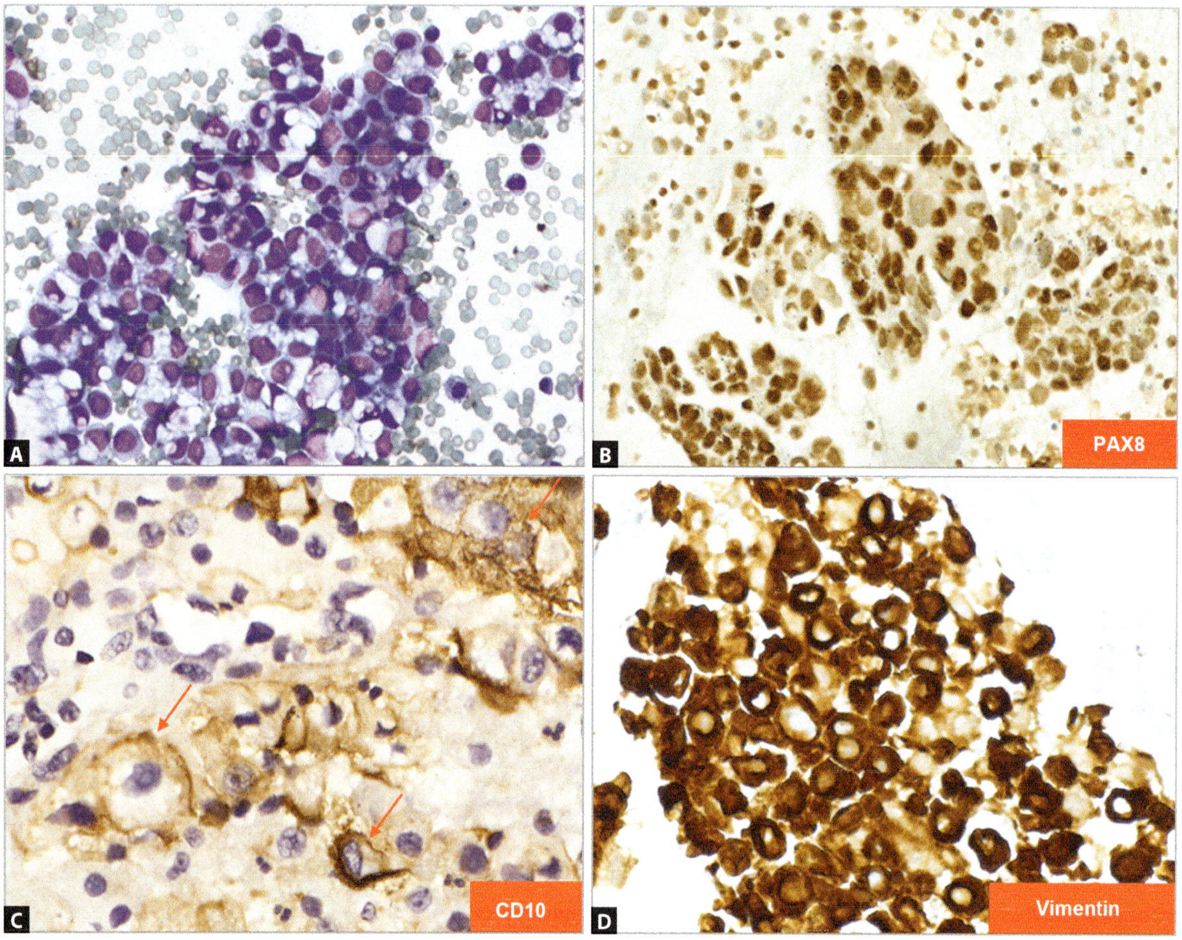

FIGS. 31A TO D: Renal cell carcinoma in peritoneal effusion. (A) Sheets of cell with clear cytoplasm and moderately pleomorphic nuclei. (B) PAX8-positive cells. (C) Membranous CD10-positive cells (see red arrows). (D) cytoplasmic positivity of vimentin.

- Thrombomodulin: Positive
- Uroplakin III: Positive

Table 7 highlights the summary of the different immunostaining patterns in various malignancies in effusion cytology.

Mesothelioma

Mesothelioma in the effusion fluid is predominantly confirmed by the ICC pattern along with cell morphology. The mesothelioma cells show the following ICC pattern (**Figs. 32** to **35**):

- Calretinin: Positive
- WT1: Positive
- D2-40: Positive
- CK5/6: Positive

It is essential to differentiate the reactive mesothelial cells and the mesothelioma. The characteristic ICC pattern is mentioned in **Table 8**.

It is preferable to use a panel of marker to differentiate between mesothelioma and reactive mesothelial cells.[1]

Malignant Round Cell Tumor

A panel of antibody is required to identify the individual malignant round cell tumors[11] (**Table 9**).

Non-Hodgkin Lymphoma

A full panel of antibody is necessary for the diagnosis of non-Hodgkin lymphoma (NHL) in the effusion fluid.[12]

The B-cell NHL (BHNL) shows:

- *Light chain restriction:* Either kappa or lambda light chain is expressed
- *CD20*: Positive
- *CD19*: Positive

Depending on the type of NHL, the ICC markers are used (**Table 10**).

TABLE 7: Immunostaining pattern in different malignancies in effusion cytology.

Primary site	Positive immunostaining	Negative immunostaining	Comments
Lung adenocarcinoma	TTF-1, napsin, CK7	p63, p40, CK5/6	TTF-1 and napsin positivity is diagnostic of lung adenocarcinoma
Small cell carcinoma of lung	Chromogranin, synaptophysin, CD56, TTF-1,* CK7*	p63, p40, CK5/6	Primary site is not possible to predict
Squamous cell carcinoma of lung	p40/p63, CK5/6	TTF-1, napsin, CK7	From the immunocytochemistry pattern, it may not be possible to indicate lung origin
Breast carcinoma	GATA3, ER, PR, GCDFP, mammaglobin, CK7		A panel consisting of GATA3/GCDFP/mammaglobin staining may be helpful to indicate breast origin
Ovarian serous carcinoma	PAX8, CK7, p53, WT1	CDX2, CK20	A panel of antibody consisting of PAX8, CK7, CK20, and CDX2 is helpful
Colonic adenocarcinoma	CDX2, CK20, beta-catenin	PAX8, CK7	A panel of antibody consisting of PAX8, CK7, CK20, and CDX2 is helpful
Pancreatic and biliary carcinoma	CK7, CK19, CK20, MUC1, CA19.9		A panel of antibody consisting of CK7, CK19, and MUC1 is helpful
Renal cell carcinoma	PAX8, CD10, vimentin, claudin	CK7, CK20	PAX8, CD10, EMA and vimentin are helpful panels
Urothelial cell carcinoma	GATA3, CK7, CK20, thrombomodulin, uroplakin III		A panel of antibody consisting of GATA3, CK7, and CK20 is helpful

*Occasionally positive.

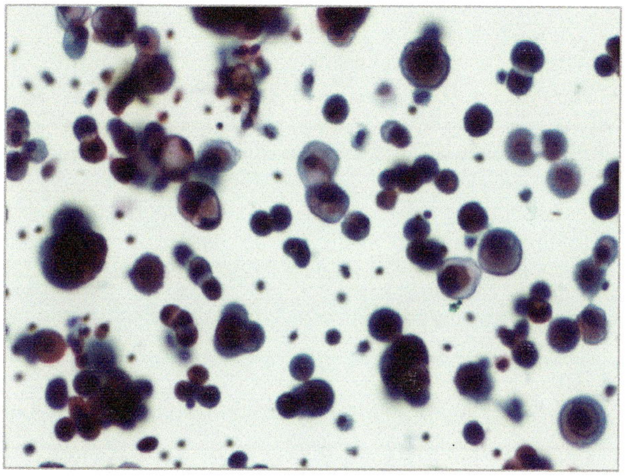

FIG. 32: Mesothelioma: Abundant mesothelial cells.

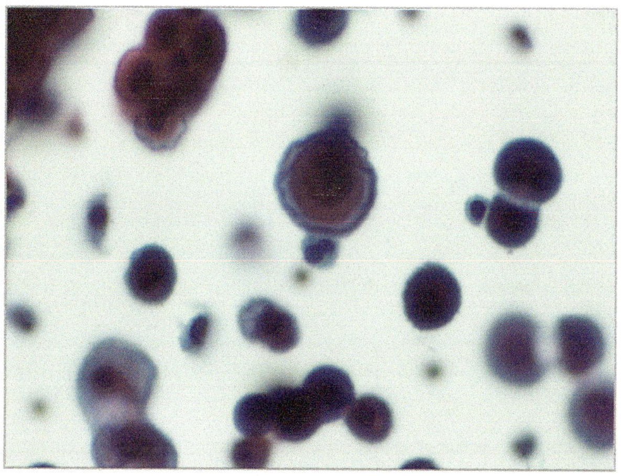

FIG. 33: Mesothelioma: Mesothelial cells showing enlarged and pleomorphic nuclei.

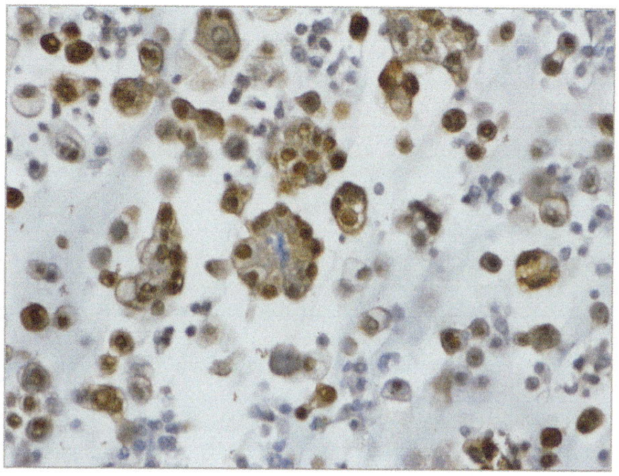

FIG. 34: Mesothelioma: Malignant mesothelial cells showing calretinin positivity.

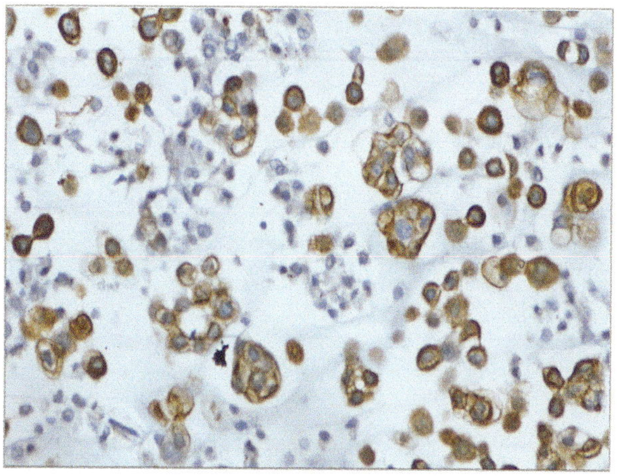

FIG. 35: Mesothelioma: Malignant mesothelial cells showing CK5/6 positivity.

TABLE 8: Distinguishing characteristics of reactive mesothelial cells and mesothelioma.

Antibody	Reactive mesothelial cells	Mesothelioma
GLUT-1	–	+
IMP3	–	+
p53	–	+

+ = Positive, – = Negative

TABLE 9: Distinguishing characteristics of different malignant round cell tumors in effusion.

Antibody	Lymphoma	RMS	NB	ES	DSRCT
CD45	+	–	–	–	–
CD99	–	–	–	+	May be positive
Desmin	–	+	–	–	+
CAM5.2	–	–	–	–	+
CD56	–	–	+	–	–

+ = Positive, – = Negative

(DSRCT: desmoplastic small round cell tumors; ES: Ewing's sarcoma; NB: neuroblastoma; RMS: rhabdomyosarcoma)

TABLE 10: Markers of different lymphomas in effusion cytology.

Antibody	B-SLL	MCL	PL	FL	BL	B-ALL	DLBCL	ALCL	PEL
CD19	+	+	+	+	+	+	+	–	–
CD20	+	–	+	–	+	+	+	–	–
CD3	–	–	–	–	–	–	–	+	–
CD5	+	+	–	–	–	–	–	–	–
CD23	+	–	–	–	–	–	–	–	–
CD10	–	–	–	+	–	–	±	–	–
CD38	–	–	+	–	–	–	–	–	–
TdT	–	–	–	–	–	+	–	–	–
ALK-1	–	–	–	–	–	–	–	+	–
CD30	–	–	–	–	–	–	–	+	+

+ = positive; – = negative.

(ALCL: anaplastic large cell lymphoma; B-ALL: B acute lymphoblastic lymphoma; B-SLL: B-small lymphocytic lymphoma; DLBCL: diffuse large B-cell lymphoma; FL: follicular lymphoma; MCL: mantle cell lymphoma; PEL: primary effusion lymphoma; PL: plasmacytoid lymphoma)

REFERENCES

1. Husain AN, Colby TV, Ordóñez NG, Allen TC, Attanoos RL, Beasley MB, et al. Guidelines for Pathologic Diagnosis of Malignant Mesothelioma 2017 Update of the Consensus Statement From the International Mesothelioma Interest Group. Arch Pathol Lab Med. 2018;142(1):89-108.
2. Jagirdar J. Application of immunohistochemistry to the diagnosis of primary and metastatic carcinoma to the lung. Arch Pathol Lab Med. 2008;132(3):384-96.
3. Kaur G, Nijhawan R, Gupta N, Singh N, Rajwanshi A. Pleural fluid cytology samples in cases of suspected lung cancer: An experience from a tertiary care centre. Diagn Cytopathol. 2017;45(3):195-201.
4. Francis IM, Alath P, George SS, Jaragh M, Al Jassar A, Kapila K. Metastatic breast carcinoma in pleural fluid: Correlation of receptor and HER2 status with the primary carcinoma-a pilot study. Diagn Cytopathol. 2016;44(12):980-6.
5. Mittal K, Soslow R, McCluggage WG. Application of immunohistochemistry to gynecologic pathology. Arch Pathol Lab Med. 2008;132(3):402-23.
6. Groisman GM, Meir A, Sabo E. The value of Cdx2 immunostaining in differentiating primary ovarian carcinomas from colonic carcinomas metastatic to the ovaries. Int J Gynecol Pathol. 2004;23(1):52-7.
7. Chang DK, Jamieson NB, Johns AL, Scarlett CJ, Pajic M, Chou A, et al. Histomolecular phenotypes and outcome in adenocarcinoma of the ampulla of Vater. J Clin Oncol. 2013;31(10):1348-56.
8. Pan Z, Grizzle W, Hameed O. Significant variation of immunohistochemical marker expression in paired primary and metastatic clear cell renal cell carcinomas. Am J Clin Pathol. 2013;140(3):410-8.
9. Tong GX, Yu WM, Beaubier NT, Weeden EM, Hamele-Bena D, Mansukhani MM et al. Expression of PAX8 in normal and neoplastic renal tissues: an immunohistochemical study. Mod Pathol. 2009;22(9):1218-27.
10. Liu H, Shi J, Wilkerson ML, Lin F. Immunohistochemical evaluation of GATA3 expression in tumors and normal tissues: a useful immunomarker for breast and urothelial carcinomas. Am J Clin Pathol. 2012;138(1):57-64.
11. Zhang PJ, Barcos M, Stewart CC, Block AW, Sait S, Brooks JJ. Immunoreactivity of MIC2 (CD99) in acute myelogenous leukemia and related diseases. Mod Pathol. 2000;13(4):452-8.
12. Das DK. Serous effusions in malignant lymphomas: a review. Diagn Cytopathol. 2006;34(5):335-47.

CHAPTER 8

Immunocytochemistry in Undifferentiated Malignancy and Tumor of the Unknown Primary Site

INTRODUCTION

In the cases of undifferentiated malignancies, the major goals of the cytopathologists are:

- To identify the exact type of the undifferentiated malignant tumor.
- To identify the probable primary site of the malignancy. A stepwise approach is essential in this aspect (**Box 1**).

STEP I: IDENTIFICATION OF THE EXACT TYPE OF MALIGNANCY[1-3]

- *Carcinoma:* Usually AE1/3 and CAM5.2 combination can detect the cells of the epithelial origin.
- *Lymphoma:* Leukocyte common antigen (LCA) (CD45) is a reliable marker for non-Hodgkin lymphoma (NHL). Once LCA is positive, one can proceed to have other lineage-specific B or T markers.
- *Sarcoma:* Vimentin as a marker of soft-tissue tumor is unreliable. Once the tumor is positive for vimentin, one can proceed for other markers such as desmin, myogenin, and MyoD for the muscle and S-100 for neural markers. Most of the sarcomas are negative for epithelial markers except synovial sarcoma and epithelioid sarcoma.

BOX 1 Stepwise approach to identify the type and site of malignancy.

Step I: Identification of the exact type of the malignancy

- Carcinoma: AE1/3 and CAM5.2 combination
- Lymphoma: LCA (CD45)
- Sarcoma: Vimentin
- Melanoma: HMB-45
- Others
- Germ cell: PLAP, OCT4, SALL4, and CD117
- Mesothelial cells: Calretinin

Step II: Identify the possible primary sites of carcinoma on the basis of cytokeratin profile: CK7 and CK20 combination

Step III: Further work-up to identify the exact primary site on the basis of CK7 and CK20 staining pattern

Step IV: The markers based on cell product and differentiation

- *Melanoma:* The melanomas are positive for Hmb-45, melan-A, and SOX-10. Whenever a tumor is positive for S-100 protein and CK negative, one can raise the suspicion of melanoma and the melanoma-specific markers should be done.
- *Others (e.g., germ cell, mesothelial cells):* The markers of germ cell tumors are PLAP, OCT4, SALL4, and CD117. These markers can detect most of the germ cell tumors.

A panel of antibody to solve the issue of an undifferentiated tumor.

In a particular situation, one should first evaluate the following factors:

- *Clinical presentation:*
  - Localizing symptoms
  - Age, sex
  - *Site of the present tumor:* Supraclavicular lymph node means to find out the primary in the upper gastrointestinal tract (GIT).
- *History:* Any known malignancy in the past.
- *Family history:* Any significant history of carcinoma.
- *Radiological findings:* Any other space-occupying lesions.
- *Serological information:* Serum tumor markers such as alpha fetoprotein or prostate-specific antigen (PSA) level

If the tissue is adequate, then a panel of combined antibody helps a lot. The panel should consist of:

- Epithelial markers: Two broad-spectrum antibodies AE1/1E3 and CAM5.2
- Neural/melanoma marker: S-100
- Mesenchymal marker: Vimentin
- Lymphoid marker: LCA (CD45)

Once the result of this panel is available (**Flowchart 1**), then we can proceed for further immunostaining (**Flowchart 2**).

STEP II: IDENTIFY THE POSSIBLE PRIMARY SITES OF CARCINOMA BASED ON THE CYTOKERATIN PROFILE

Once the diagnosis of carcinoma is confirmed, the next step should be to make a tentative list of the likely primary sites. The cytokeratins are intermediate filaments, and the different

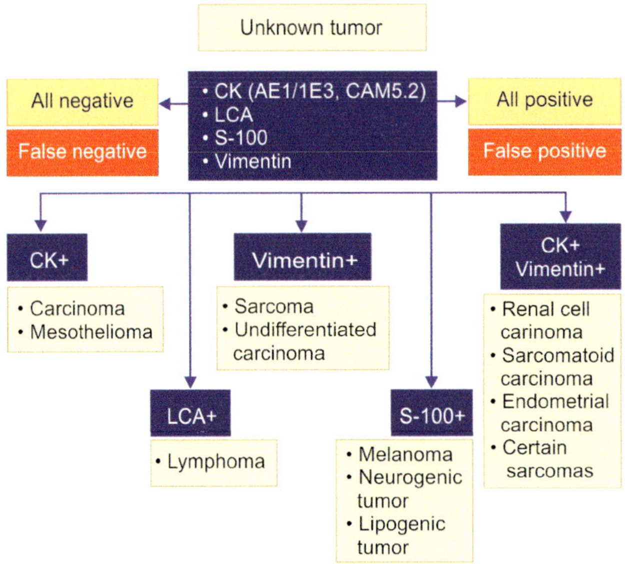

FLOWCHART 1: Diagram showing the interpretation of a combined panel.

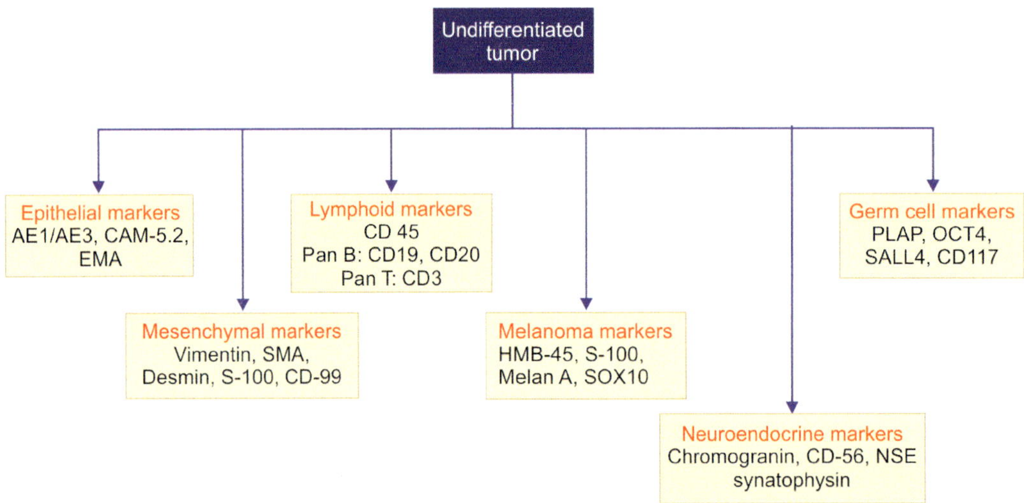

FLOWCHART 2: The essential markers to differentiate undifferentiated tumors in cytology.

subtypes of cytokeratin are organ-specific. A combination of cytokeratin CK7 and CK20 can help to make a shortlist of different organ-specific carcinomas[3-6] (**Table 1**).

STEP III: FURTHER WORK-UP TO IDENTIFY THE SPECIFIC PRIMARY SITE BASED ON CK7 AND CK20 STAINING PATTERN

Once the result of CK7 and CK20 is obtained, further immunocytochemistry (ICC) can be done to indicate the more specific site (**Box 2**).

CK7 Positive and CK20 Negative Tumors

The tumors with CK7 positive and CK20 negative pattern are mainly of ovarian, lung, breast, and gastrointestinal origin.[1,6] Now depending on the history, and radiological findings, one should proceed to do further ICC (**Flowchart 3**). The following ICC pattern is seen in these tumors (**Figs. 1** to **4**):

- *Breast*
    - GATA3: Positive
    - GCDFP: Positive
    - ER/PR: Positive
- *Lung*
    - TTF-1: Positive
    - Napsin: Positive
- *Serous carcinoma of the ovary*
    - PAX8: Positive
    - WT1: Positive
- *Clear cell carcinoma of the ovary*
    - Napsin
    - HNF-1β

TABLE 1: Primary origin of different carcinomas based on CK7 and CK20 expression.

CK7+ CK20–	CK7+ CK20+	CK7– CK20+	CK7– CK20–
• Lung adenocarcinoma • Breast carcinoma • Endometrial adenocarcinoma • Ovarian serous carcinoma • Small cell carcinoma of lung • Cholangiocarcinoma • Squamous cell carcinoma of cervix • Pancreatic duct carcinoma • Gastric adenocarcinoma • Thymoma • Thymic carcinoma	• Urothelial carcinoma • Mucinous carcinoma of ovary • Pancreatic and biliary carcinoma • Gastric adenocarcinoma	• Colonic adenocarcinoma • Merkel cell carcinoma	• Prostatic adenocarcinoma • Squamous cell carcinoma of lung • Hepatocellular carcinoma • Adrenocortical carcinoma • Renal cell carcinoma

BOX 2 CK7 positive and CK20 negative tumors.

Common tumors: Lung adenocarcinoma, breast carcinoma, endometrial adenocarcinoma, ovarian serous carcinoma, small cell carcinoma of lung, cholangiocarcinoma, squamous cell carcinoma of cervix, pancreatic duct carcinoma, and gastric adenocarcinoma

Immunocytochemical markers

- Breast: GATA3, mammaglobin, GCDFP
- Lung adenocarcinoma: TTF1, napsin
- Serous carcinoma of ovary: PAX8, p53, WT1
- Follicular carcinoma of thyroid: Thyroglobulin, TTF-1
- Stomach: CDX2, MUC1, CEA
- Pancreatic duct adenocarcinoma: CK7, CK19

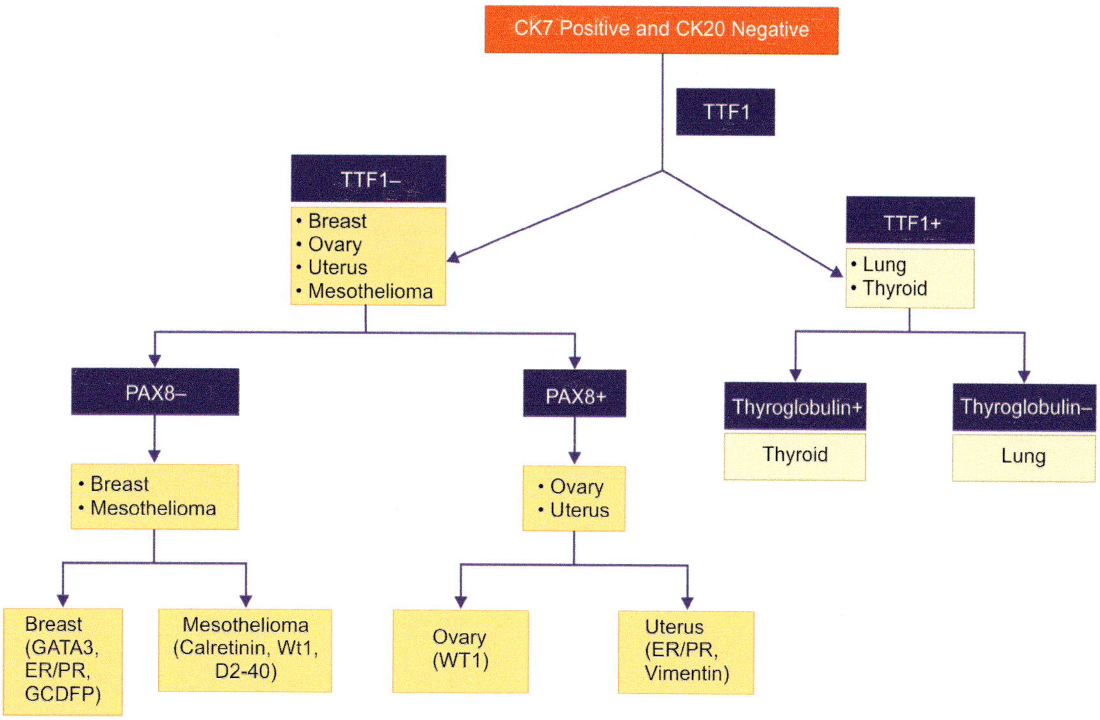

FLOWCHART 3: Algorithmic approach in CK7 positive and CK20 negative tumors in a female patient.

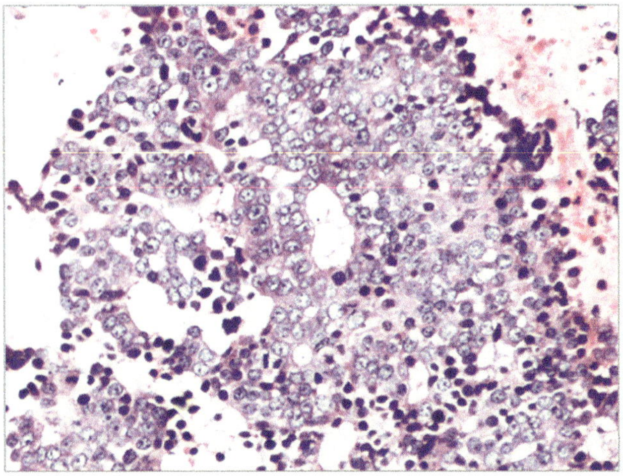

FIG. 1: Cell block section from an omental mass.

- *Endometrial carcinoma*
  - Pax8: Positive
  - ER: Positive
  - Vimentin: Positive
- *Follicular carcinoma of the thyroid*
  - Thyroglobulin: Positive
  - TTF-1: Positive
- *Medullary carcinoma of the thyroid*
  - Calcitonin: Positive
  - TTF-1: Positive
- *Stomach*
  - Carcinoembryonic antigen (CEA): Positive
  - CDX2: Occasional positive
  - MUC1: Occasional positive
- *Urinary bladder*
  - GATA3: Positive
  - p63: Positive
  - CK5/6: Positive

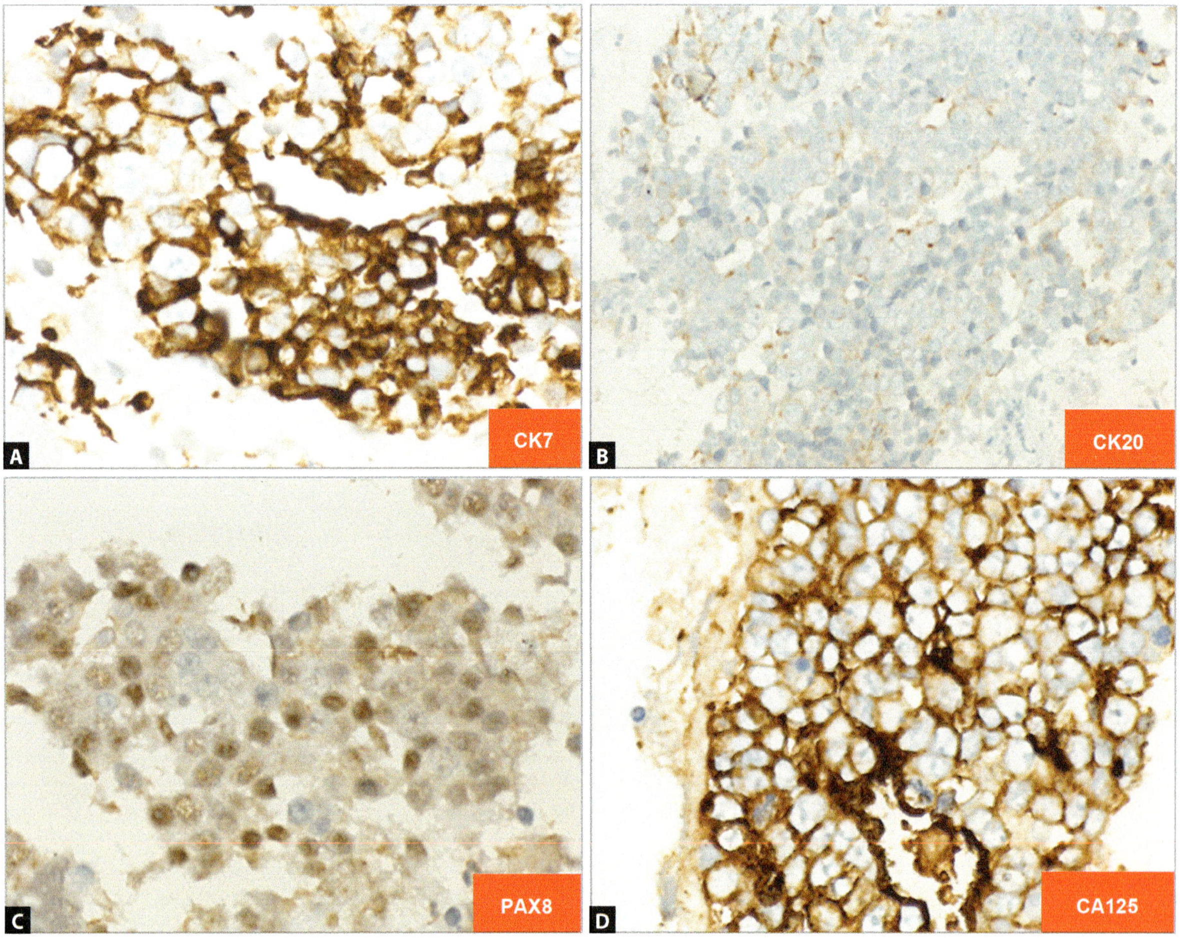

FIGS. 2A TO D: (A) Omental mass: CK7 positive. (B) Omental mass: CK20 negative. (C) Omental mass: PAX8 positive. (D) Omental mass: CA125 positive. The overall features are suggestive of a primary in the ovary.

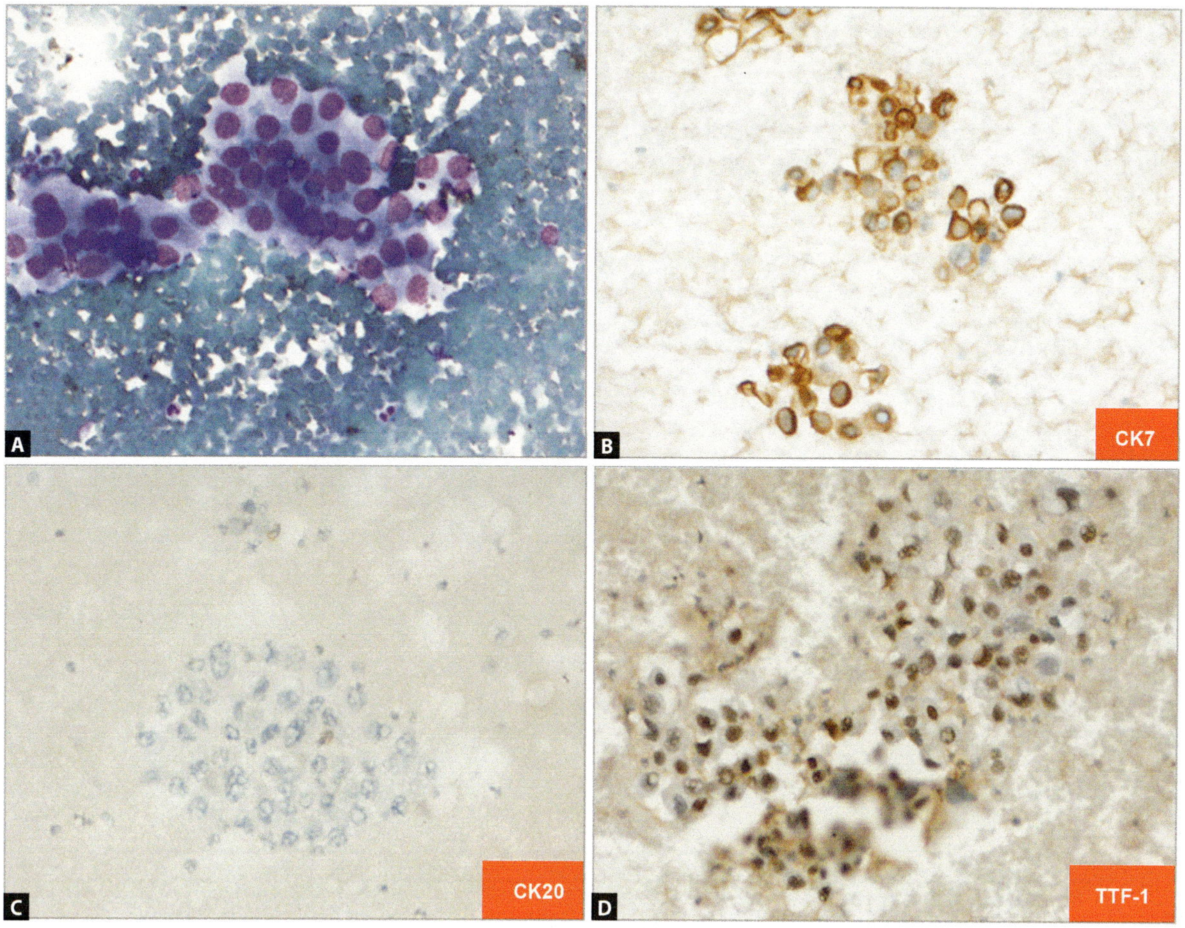

FIGS. 3A TO D: A 55-year-old male with metastasis in the first lumber vertebra. (A) Fine needle aspiration cytology smears showing clusters of malignant cells. (B) Membranous positivity of CK7. (C) CK20 negative tumor cells. (D) TTF-1 positive cells. The overall features are suggestive of primary lung adenocarcinoma.

- *Thymus*
  - CD5: Occasional positive
  - GLUT1: Occasional positive
  - CD117: Occasional positive
- *Pancreatic duct adenocarcinoma*
  - CK7: Positive
  - CK19: Positive
- *Mesothelioma*
  - Calretinin: Positive
  - WT1: Positive
  - D2-40: Positive
  - CK5/6: Positive

Once the result of CK7 positive and CK20 negative tumor is available, we should proceed further to identify the primary site of the tumor by using a battery of tests.

CK7 Positive and CK20 Positive Tumors

Both CK7 positive and CK20 positive tumors are mainly urothelial carcinoma of the urinary bladder, pancreatic and biliary carcinoma, gastric adenocarcinoma, and mucinous carcinoma of the ovary[1-9] (**Box 3**).

The ICC patterns in these tumors are (**Flowchart 4**):

- *Urinary bladder (**Figs. 5A** to **F**)*
  - GATA3: Positive
  - p63: Positive
  - CK5/6: Positive
- *Stomach*
  - CEA: Positive
  - CDX2: Occasional positive
  - MUC1: Occasional positive
- *Mucinous carcinoma of the ovary*
  - DPC4: Positive
  - CA125: Positive

CK7 Negative and CK20 Positive Tumors

The CK7 positive and CK20 positive tumors are mainly colonic adenocarcinoma and Merkel cell carcinoma[1-9] (**Box 4**).

The ICC patterns of these tumors are:

- *Colonic adenocarcinoma (**Figs. 6** and **7**)*
  - CDX2: Positive
  - MOC-31: Positive
  - BER-EP4: Positive

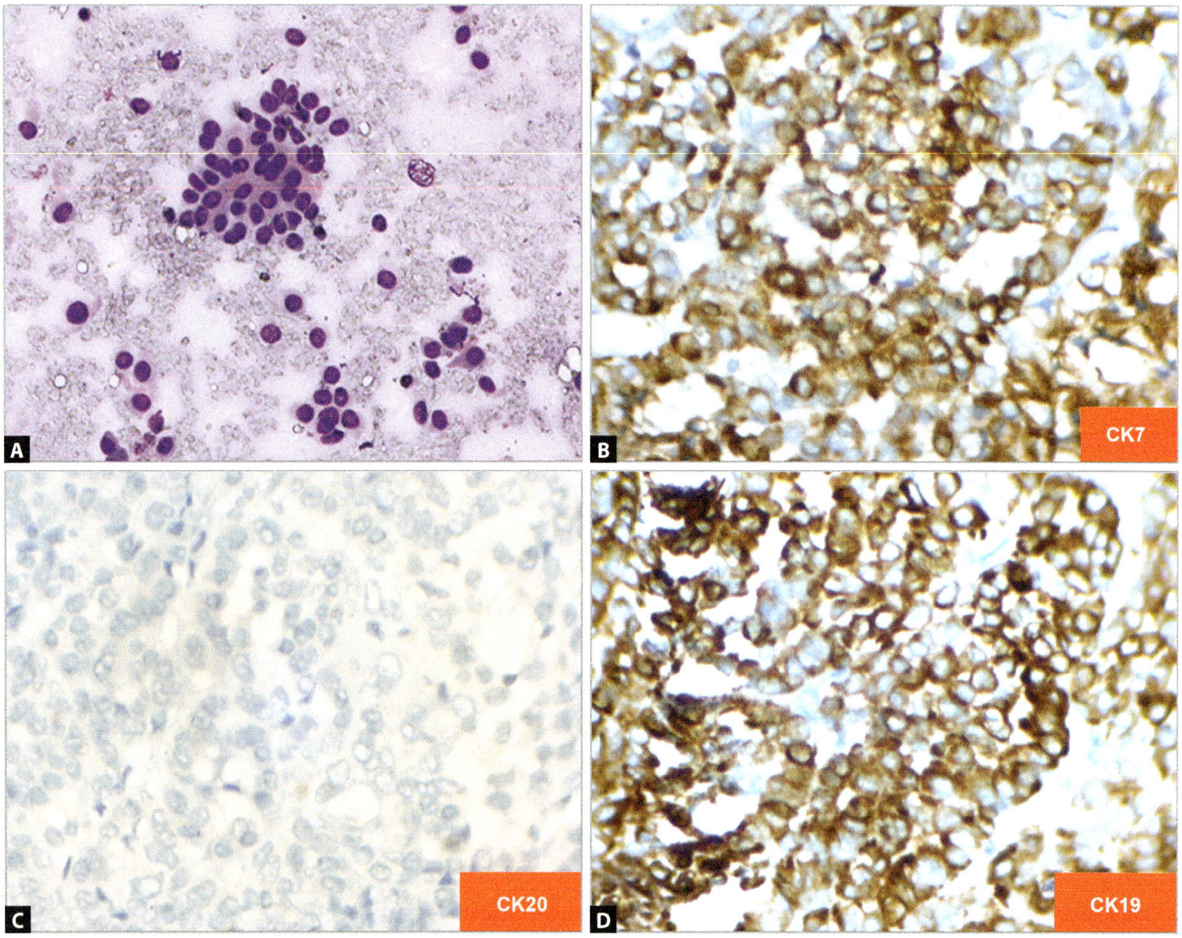

FIGS. 4A TO D: Metastatic carcinoma in skin from a pancreaticobiliary primary site. (A) Fine needle aspiration cytology smear showing multiple clusters of tumor cells. (B) Tumor cells are positive for CK7. (C) Negative for CK20. (D) Positive for CK19.

BOX 3 CK7 positive and CK20 positive tumors.

Common tumors: Urothelial carcinoma, mucinous carcinoma of ovary, pancreatic and biliary carcinoma, gastric adenocarcinoma

Immunocytochemical markers

- Urinary bladder: GATA3, p63
- Mucinous carcinoma of ovary: DPC4, CA125
- Pancreas: Maspin, IMP3

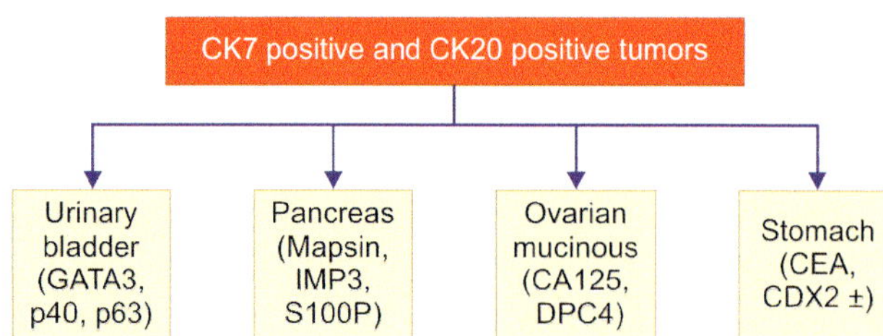

FLOWCHART 4: Immunocytochemistry of CK7 positive and CK20 positive tumors.

- *Merkel cell carcinoma*
    - Chromogranin: Positive
    - Synaptophysin: Positive
    - CD56: Positive

CK7 Negative and CK20 Negative Tumors

Both CK7 and CK20 negative tumors are prostatic adenocarcinoma, squamous cell carcinoma of the lung, hepatocellular carcinoma, adrenocortical carcinoma, and renal cell carcinoma[1-9] (**Box 5**).

The following ICC pattern is noted in these tumors:

- *Prostatic adenocarcinoma*
    - PSA: Positive
    - P501S: Positive
- *Squamous cell carcinoma of the lung*
    - p63
    - p40
- *Renal cell carcinoma*
    - PAX8: Positive
    - CD10: Positive
    - Vimentin: Positive

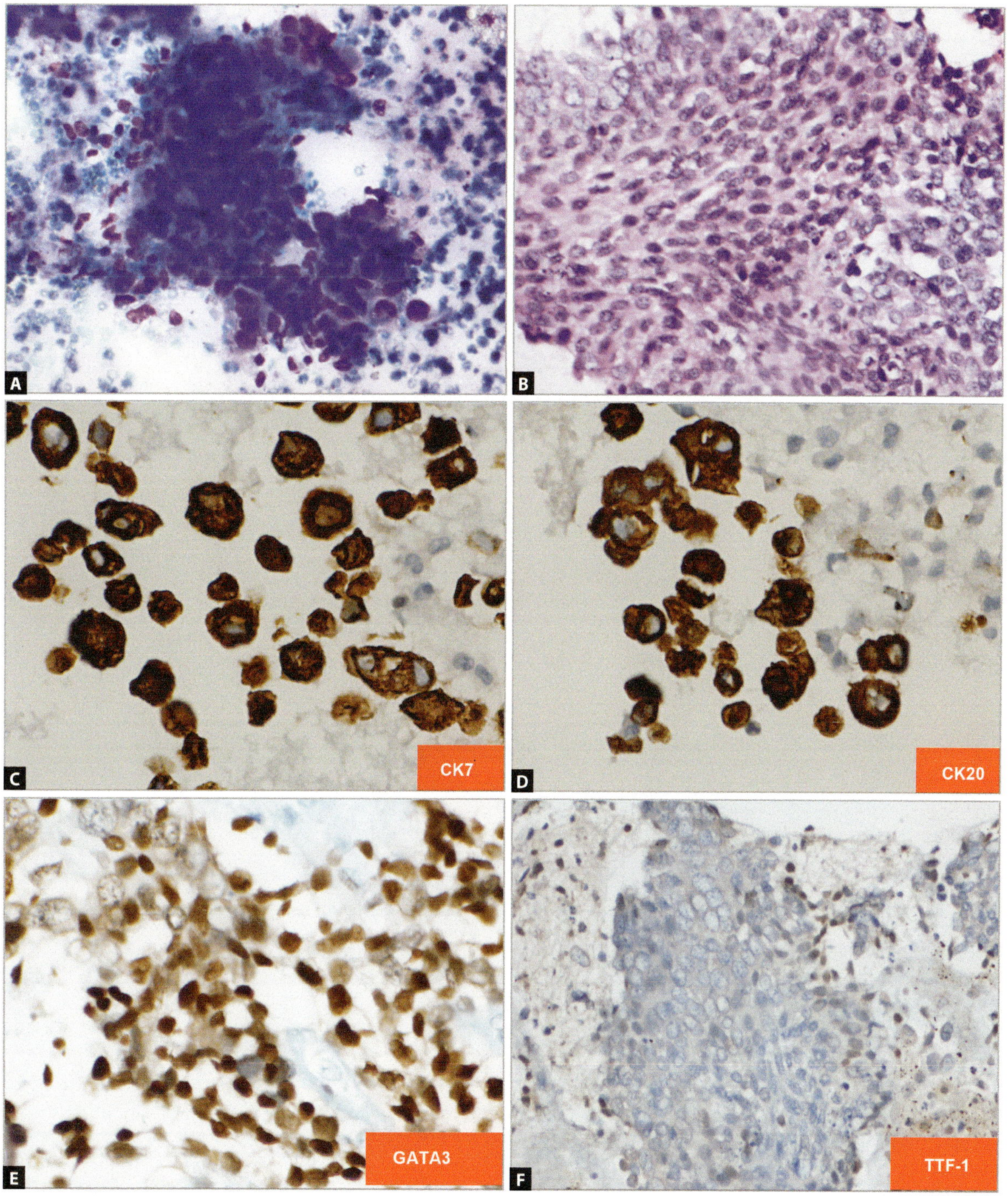

FIGS. 5A TO F: (A) Cytology smear in a space-occupying lesion of the upper lobe of the right lung in a 56-year-old male. (B) Cell block section of the same. (C) CK7 is positive in the tumor cells. (D) CK20 is positive. (E) Tumor cells are positive for GATA3. (F) TTF-1 negative tumor cells. Overall immunocytochemistry favors the primary as urothelial carcinoma of the urinary bladder.

- *Hepatocellular carcinoma*
  - HepPar1: Positive
  - Glypican-3: Positive
  - Alpha fetoprotein: Positive
- *Germ cell tumors*
  - OCT4: Positive

BOX 4 **CK7 negative and CK20 positive tumors.**

Common tumors: Colonic adenocarcinoma, Merkel cell carcinoma

Immunocytochemical markers

- Colonic adenocarcinoma: CDX2, CK20, MOC-31, BER-EP4
- Merkel cell carcinoma: Chromogranin, Synaptophysin, CD56

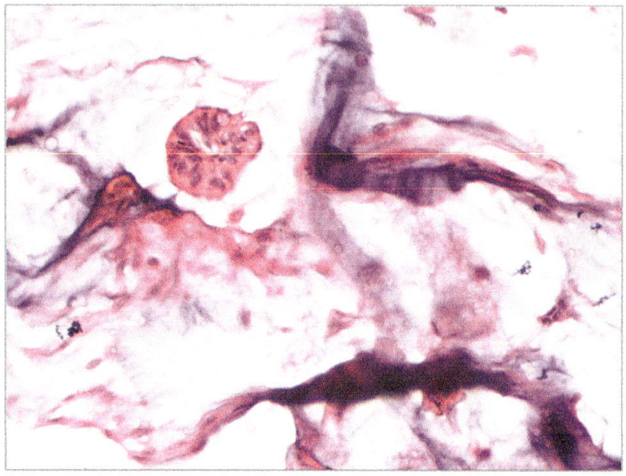

FIG. 6: Cell block section in an omental mass of a 55-year-old male patient showing malignant cells in a glandular arrangement.

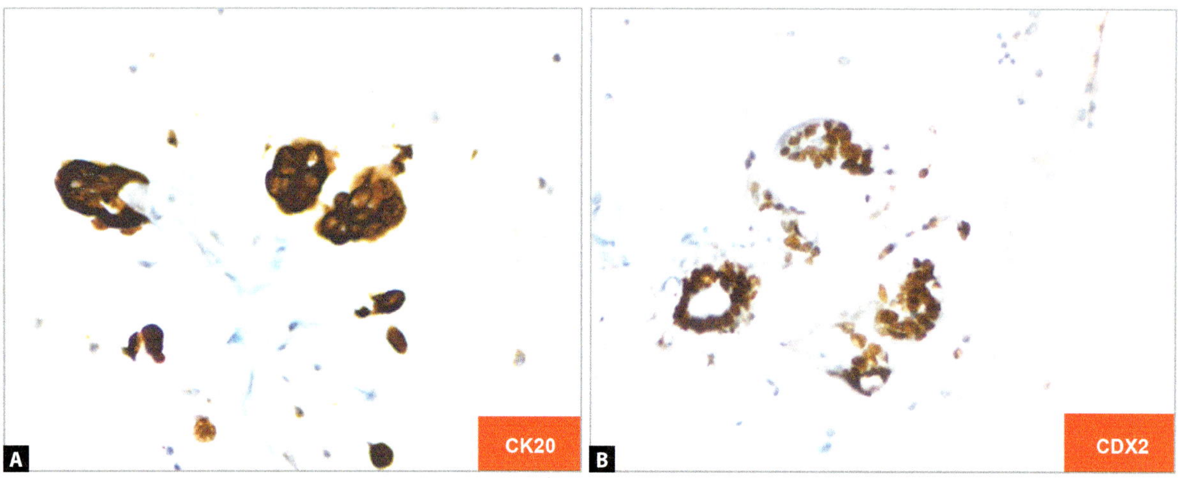

FIGS. 7A AND B: (A) The tumor cells showing CK20 positive cells. These cells were negative for CK7. (B) Strong nuclear positive CDX2 immunostaining. The overall immunostaining pattern indicates the diagnosis of primary colonic carcinoma.

BOX 5 CK7 negative and CK20 negative tumors.

Common tumors: Prostatic adenocarcinoma, squamous cell carcinoma of lung, renal cell carcinoma, hepatocellular carcinoma, germ cell tumors

Immunocytochemical markers

- Prostatic adenocarcinoma: PSA, P501S
- Squamous cell carcinoma of lung: p63, p40
- Renal cell carcinoma: PAX8, CD10, Vimentin
- Hepatocellular carcinoma: HepPar1, Glypican-3, Alpha fetoprotein
- Germ cell tumors: OCT4, CD117, PLAP, SALL4, NANOG

    - CD117: Positive
    - PLAP: Positive
    - SALL4: Positive
    - NANOG: Positive

Cytokeratin 5/6 Expression

Cytokeratin 5/6 is expressed in these following tumors:

- Squamous cell carcinoma of the lung.
- Epithelial mesothelioma
- Basal-like phenotype of breast cancer.
- Most of the cases of urothelial carcinoma (>50%).

Strongly positive CK5/6 helps to differentiate metastatic carcinoma from epithelial mesothelioma. CK5/6 positivity, along with p63 positivity, helps to confirm the poorly differentiated squamous cell carcinoma of the lung (**Figs. 8A** to **D**). CK5/6, particularly CK5 immunostaining, is helpful to identify the molecular phenotype of basal-like breast carcinoma.[10]

Nonepithelial Tumors that Express Cytokeratin

Various nonepithelial tumors without any evidence of epithelial differentiation may also express keratin (**Box 6**). These tumors include:[1]

- *Synovial sarcoma:* Expresses CAM5.2 and AE1/AE3.
- *Desmoplastic round cell tumor:* Expresses CAM5.2 and AE1/AE3.
- *Epithelioid sarcoma:* Expresses CAM5.2 and AE1/AE3.
- *Malignant rhabdoid tumor:* Expresses CAM5.2 and AE1/AE3.

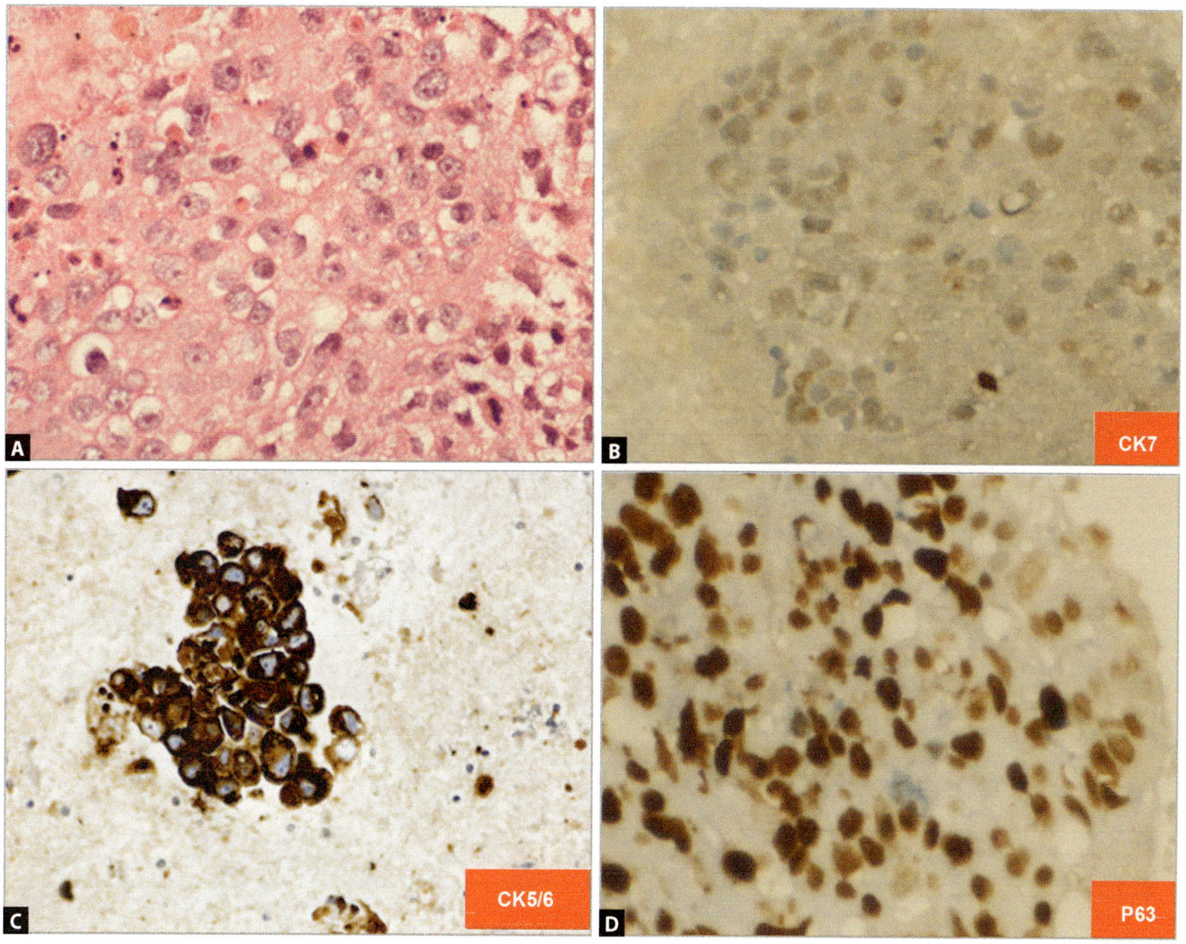

FIGS. 8A TO D: A case of metastatic squamous cell carcinoma in liver showing (A) Cell block section; (B) CK7 negative cells; (C) CK5/6 positive tumor cells; (D) Nuclear positive p63 tumor cells.

BOX 6 Nonepithelial tumors that express cytokeratin.

- Tumors
    - Synovial sarcoma
    - Desmoplastic round cell tumor
    - Epithelioid sarcoma
    - Malignant rhabdoid tumor
    - Anaplastic large cell lymphoma
    - Plasmacytoma
    - Malignant melanoma
- Cytomorphology: The details of cytomorphological features have great importance
- Other markers: Vimentin, desmin, CD45, CD38, HMB45

- *Anaplastic large cell lymphoma:* Expresses epithelial membrane antigen (EMA).
- *Plasmacytoma:* Expresses EMA
- *Malignant melanoma:* Expresses CK8 and CK18

Carcinomas that Lose Keratin Expression

It is important to remember that certain carcinomas may have reduced or total loss of expression of keratin.[1-9] These tumors are (**Box 7**):

BOX 7 Carcinomas that lose keratin expression.

- Tumors
    - Small cell carcinoma
    - Renal cell carcinoma
    - Anaplastic carcinoma of thyroid
    - Adrenocortical carcinoma
- Cytomorphology: The characteristic cytological features are helpful clues
- Other markers: Vimentin, calretinin, PAX8, CD10, chromogranin

- *Small cell carcinoma:* May lose the expression of CAM5.2 and AE1/AE3.
- *Renal cell carcinoma:* May lose the expression of both CK7 and CK20.
- *Anaplastic carcinoma of the thyroid:* Negative for CKs except for CAM5.2.
- *Adrenocortical carcinoma:* Negative for both CK7 and CK20.

Carcinomas that Express Both Cytokeratin and Vimentin

Vimentin is expressed in almost all cases of mesenchymal tumors. In cell block section, if the blood vessel wall does not show any vimentin expression, then it indicates loss of antigenicity in the tissue. Vimentin is expressed in all cases of metastatic adenocarcinoma in the effusion, and this finding has no diagnostic importance.

Vimentin is an intermediate filament, and unfortunately, the expression of vimentin is quite non-specific as various carcinomas also express this marker. Certain types of carcinomas consistently express vimentin and therefore, the combined pattern of cytokeratin and vimentin expression is helpful in the diagnosis of such carcinomas.[1]

The carcinomas that commonly express vimentin include:
- Renal cell carcinoma (**Figs. 9A** to **F**)
- Thyroid follicular carcinoma
- Endometrial adenocarcinoma
- Salivary gland carcinoma

The carcinomas that rarely express vimentin include:
- Small cell carcinoma of lung
- Adenocarcinoma of prostate
- Lobular carcinoma of breast
- Colorectal adenocarcinoma

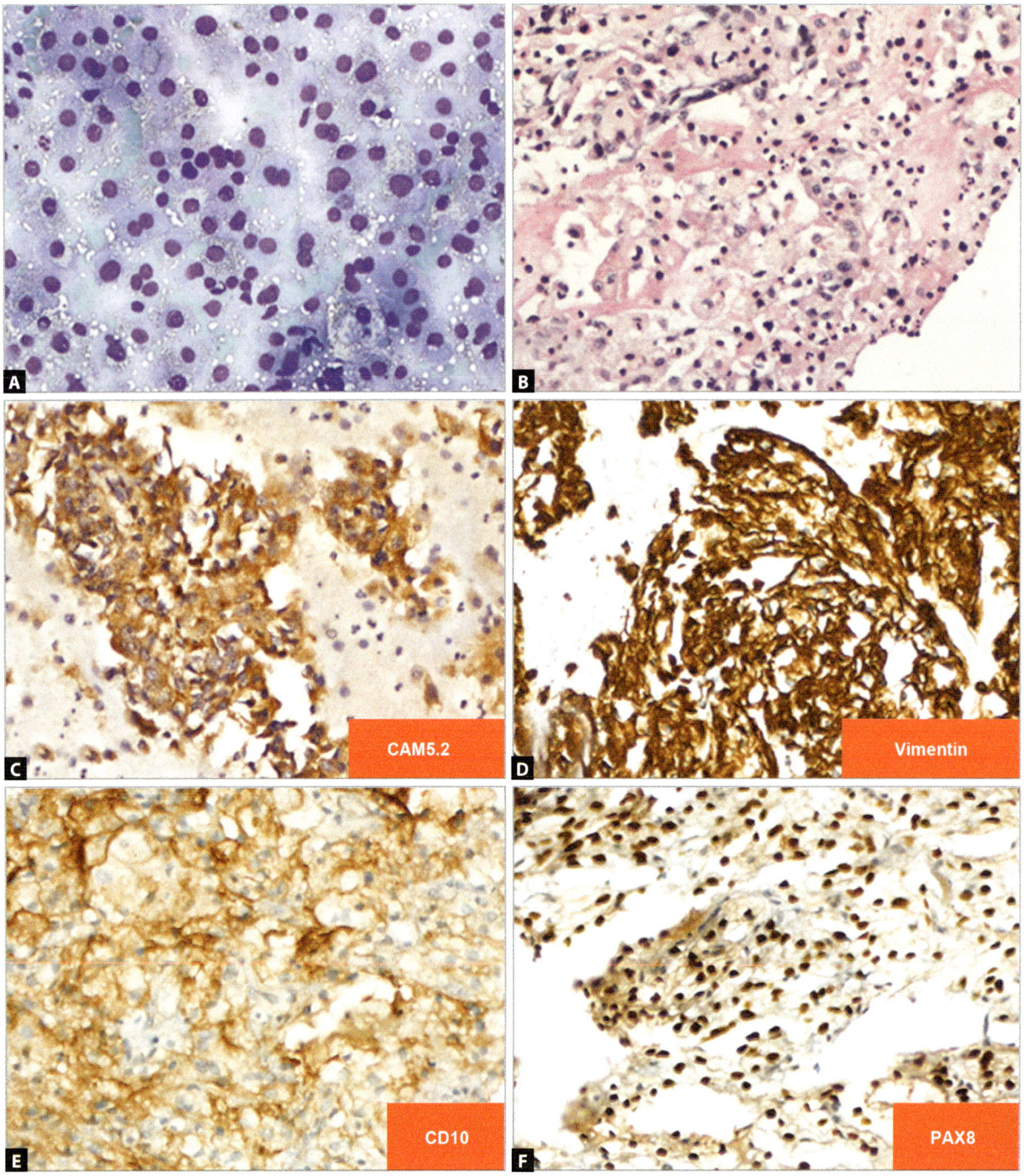

FIGS. 9A TO F: A case of metastatic clear cell carcinoma of kidney in the lumber fifth vertebra. (A) Fine needle aspiration cytology smear. (B) Cell block section. (C) Tumor cells are positive for CAM5.2. (D) Cells are strongly positive for vimentin. (E) CD10 positive cells. (F) Nuclear positive PAX8.

BOX 8 The markers of epithelial differentiation other than cytokeratin.

Markers: EMA, CEA, BER-EP4, and MOC-31

EMA
- EMA/MUC1 is a membranous mucin
- Localization: Membranous positivity
- A good marker of the normal epithelial cell and epithelial tumors
- Other nonepithelial tumors showing positivity: Anaplastic large cell lymphoma, mesothelioma , epithelioid sarcoma, synovial sarcoma, T-cell rich B-cell lymphoma

CEA
- A glycoprotein
- Localization: Cytoplasmic positivity
- Positive: Adenocarcinoma of lung, colon, breast

BER-EP4
- Glycoprotein located on the glandular epithelial cells
- Localization: Membranous positivity
- Majority of epithelial malignancy except renal cell carcinoma

MOC31
- Recognizes the epitope of the glycoprotein of the epithelial cells
- Localization: Membranous positivity
- Majority of epithelial malignancy

EMA, BER-EP4, and MOC-31 antibodies are unable to localize the primary site of malignancy

(CEA: carcinoembryonic antigen; EMA: epithelial membrane antigen)

Surrogate Markers of Epithelial Differentiation other than Cytokeratin

Various other markers are also available for the tumors with epithelial differentiation.[1-9] These markers include CEA, EMA BER-EP4, and MOC-31 (**Box 8**).

Epithelial Membrane Antigen

Epithelial membrane antigen (EMA) has been described in Chapter 6.

Carcinoembryonic Antigen

Carcinoembryonic antigen (CEA) is a glycoprotein, and various antibodies are available against the different antigenic determinant sites of CEA. The polyclonal CEA cross-reacts with bile canalicular cells in hepatocellular carcinoma. CEA is expressed in the adenocarcinoma of both colon and lung. Tumor with both CEA and CK7 positivity is seen in lung adenocarcinomas whereas CEA positive and CK20 positive tumor is seen in colonic adenocarcinoma.

The following adenocarcinomas show CEA positivity:
- Lung
- Colon
- Breast
- Salivary gland
- Stomach
- Pancreas

BER-EP4

This antibody identifies adenocarcinoma. BER-EP4 antibody is against the glycoprotein located on the glandular epithelial cells. The sensitivity of this antibody is very high, and it can be used along with the other markers to identify the metastatic carcinoma in effusion fluid.

TABLE 2: Markers of different primary sites.

Primary sites or broad origin	Markers
Thyroid	Thyroglobulin, Thyroid transcription factor (TTF-1)
Gastrointestinal tract	CDX2, SATB2Q
Neuroendocrine	Chromogranin, Synaptophysin, CD56, NSE
Gynecological tumor	PAX8
Mesothelial	Calretinin, WT1, D2-40
Breast	Mammaglobin, GCDFP, GATA3
Liver	HepPar-1, Glypican-3, Arginase
Prostate	PSA, prostatic acid phosphatase
Bladder	Uroplakin III, Uroplakin II, Thrombomodulin, GATA3
Germ cell markers	OCT4, PLAP, SALL4

MOC-31

MOC-31 is also another epithelial marker that is against the epitope of the glycoprotein of the epithelial cells. It is also an excellent marker of the adenocarcinoma in effusion fluid.

STEP IV: THE MARKERS BASED ON CELL PRODUCT AND DIFFERENTIATION

The use of cytokeratin helps to make a shortlist of the possible primary sites. After narrowing down the possible primary sites of the tumor, it is essential to apply cell-specific differentiation of the tumor. **Table 2** shows the outline of such markers.[1,11]

Thyroid

Thyroglobulin

Thyroglobulin is a glycosylated protein that is expressed only in thyroid follicular epithelial cells. It is rarely expressed in leukemic blasts.
- *Thyroglobulin positive thyroid tumors:*
  - Follicular carcinoma
  - Papillary carcinoma
  - Poorly differentiated carcinoma of the thyroid: Less expressed
- *Thyroglobulin negative thyroid tumors:*
  - Medullary carcinoma
  - Anaplastic carcinoma of the thyroid

Thyroid Transcription Factor-1

Thyroid transcription factor-1 (TTF-1) is a nuclear transcription factor and is predominantly expressed in thyroid and lung adenocarcinomas.

TTF-1 is very helpful in the differential diagnosis of primary versus metastatic lung adenocarcinomas and identification of lung carcinomas in the effusion fluid.

TTF-1 Positive Tumor

- *Strongly positive*
    - Adenocarcinoma of lung
    - *Thyroid carcinoma:* Papillary carcinoma, follicular carcinoma, insular carcinoma, medullary carcinoma.
    - Small cell carcinoma of GIT, bladder, and cervix
- *Focally positive*
    - Endocervical and endometrial carcinoma
    - Lobular carcinoma of breast
    - Adenocarcinoma of colon

Gastrointestinal Tract

CDX2

CDX2, a transcription protein, is related to the normal development of epithelial cells of the intestine. The adenocarcinomas developed in this region usually express CDX2.[12]

- *CDX2 positive tumors*
    - Adenocarcinoma of the colon and rectum
    - Mucinous tumors of the ovary
    - Adenocarcinoma of the small intestine
    - Neuroendocrine tumors in the GIT
    - Pancreatic and biliary carcinoma
- *Focally positive CDX2*
    - Thyroid carcinoma
    - Prostate carcinoma

Diffused strongly positive CDX2 immunostaining along with CK20+/CK7− combination strongly favors colorectal adenocarcinoma.

SATB2

The special AT-rich sequence binding protein 2 (SATB2) antibody binds with the nuclear matrix attachment region. It is responsible for chromatin modeling and gene transcription. SATB2 is expressed in a large variety of tumors, and its expression helps in the identification of carcinomas of the colorectal region, appendix, and bone tumors.[13,14]

SATB2 positive tumors are:

- Colonic carcinoma
- Rectal carcinoma
- Appendicular carcinoma
- Merkel cell carcinoma
- Osteosarcoma

Neuroendocrine Tumors

The commonly used neuroendocrine markers are chromogranin, synaptophysin, CD56, and neuron-specific enolase (NSE).

Chromogranin

Chromogranin (CHG), the acidic glycoprotein, is the component of neurosecretory granules. It is present in the neuroendocrine cells and endocrine cells. The intensity of the CHG stain increases with the increased number of the neurosecretory granules.[7] The staining intensity of CHG decreases with the decreasing differentiation of the neuroendocrine tumors.

Chromogranin positive tumors are:

- Neuroendocrine tumors
- Desmoplastic small round cell tumor
- Hepatoblastoma fetal type
- Parathyroid neoplasm
- Medullary carcinoma of the thyroid
- Small cell carcinoma

Synaptophysin

Synaptophysin (SNP), a glycoprotein, is the constituent of the membrane of the neurosecretory granules. It has a high sensitivity; however, its specificity is lower than CHG. Therefore, the positive SNP is not the sole criterion of a neuroendocrine tumor, and it should be supported by other markers or characteristic morphology.

Synaptophysin positive tumors are (**Figs. 10A** and **B**):[1-9]

- Neuroendocrine tumors
- Large cell undifferentiated neuroendocrine carcinoma of the lung.

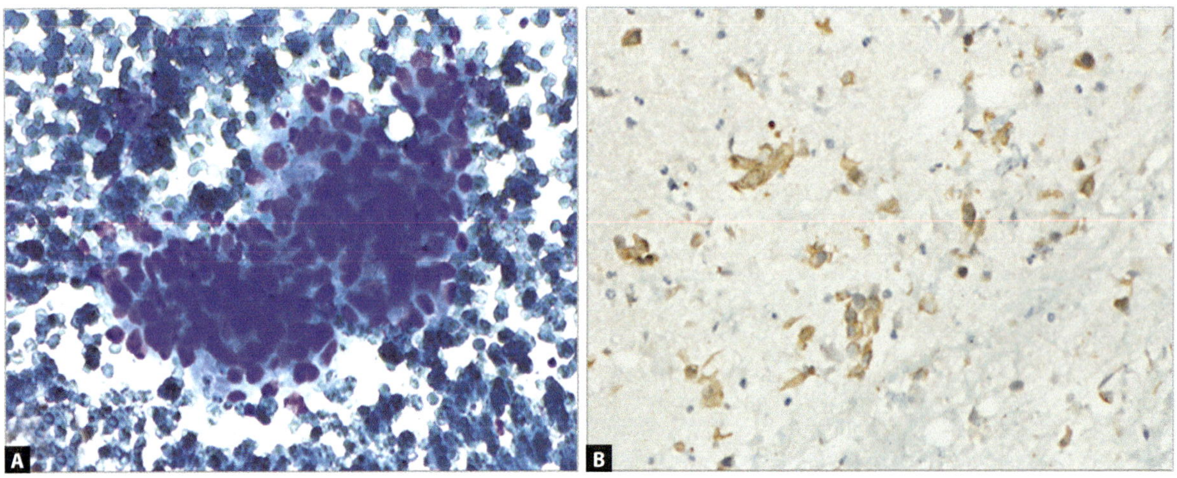

FIGS. 10A AND B: Metastatic neuroendocrine tumor in lymph node. (A) Fine needle aspiration cytology smear showing clusters of cells. (B) Synaptophysin positive tumor cells.

- Adrenocortical adenoma and carcinoma
- Neuroblastoma
- Pheochromocytoma
- Desmoplastic small round cell tumor

CD56

CD56, a glycoprotein, is related to cell-to-cell adhesion. It is also a marker of NK cells and NK lymphoma. CD56 is expressed in various neuroendocrine tumors. CD56 is a less specific neuroendocrine marker and it should be used along with other markers of neuroendocrine tumors.
CD56 positive tumors are:
- Neuroendocrine carcinomas
- Thyroid carcinoma
- Merkel cell carcinoma
- Pheochromocytoma
- Pancreatic neuroendocrine tumor
- Small cell carcinomas of cervix
- NK cell lymphoma

Neuron-specific Enolase

Neuron-specific enolase (NSE) is an enolase enzyme and is noted in neurons and neuroendocrine cells. The term "specific" is not correct and NSE is found in various non-neuronal cells also.

It is quite an unreliable marker for neuroendocrine tumors, and NSE should be used along with other neuroendocrine markers.
Neuron-specific enolase positive tumors are:
- Neuroendocrine tumor
- Melanoma
- Tumor with the neuroectodermal origin

Gynecological Tumors

Paired Box Gene 8 (PAX8)

PAX8 gene expression is needed to develop the Müllerian system, normal renal and thyroid follicular cells.[15]
PAX8 positive tumors are:
- The tumors developed from the Müllerian system: Serous ovarian carcinoma, endometrial adenocarcinoma
- Thyroid carcinomas: Follicular carcinoma, papillary carcinoma
- Renal cell carcinoma
- Thymic tumors
- Seminoma, thymic carcinoma
- Neuroendocrine tumors of duodenum and pancreas

Mesothelial Cell Markers

Calretinin

Calretinin has been discussed in Chapter 7.

Wilms' Tumor

Wilms' tumor (WT1) has been discussed in Chapter 7.

Podoplanin

Podoplanin (D2-40) has been discussed in Chapter 7.

Breast

Mammaglobin

Mammaglobin is an epithelial secretory protein mainly related to breast. It is highly sensitive and specific for the detection of breast carcinoma.[16] However, a few nonbreast carcinomas such as endometrioid carcinoma are also positive for mammaglobin.
Mammaglobin positive tumors are:
- Breast carcinoma
- Mammary analogue secretory carcinoma of the salivary gland
- Endometrial carcinoma
- Sweat gland carcinoma

Gross Cystic Disease Fluid Protein 15

Gross cystic disease fluid protein 15 (GCDFP-15) is a prolactin-inducing protein and is found in the cystic fluid of breast. It is a relatively specific marker (99%) for the breast carcinoma and can be used for the identification of the metastatic breast carcinoma in difficult situations.[17]
GCDFP positive tumors are:
- Breast carcinoma
- Salivary gland carcinoma
- Occasional cases of lung adenocarcinoma

GATA Binding Protein 3

GATA binding protein 3 (GATA3) is a zinc-finger binding protein. It is related to the differentiation of the epithelium of breast[18] (**Figs. 11A to D**), urothelial cells, and trophoblasts.
GATA3 positive tumors are:
- Breast carcinoma: Ductal and lobular
- Urothelial carcinoma
- Paraganglioma
- Choriocarcinoma
- Yolk sac tumor
- Pheochromocytoma
- Salivary duct carcinoma
- Mammary analogue secretory carcinoma

Liver

Hepatocyte Paraffin 1

Hepatocyte paraffin 1 (HepPar1) identifies the mitochondrial antigen of the hepatocytes. It is a sensitive and specific marker of hepatocytes. HepPar1 positivity may help to differentiate primary hepatocellular carcinoma from the metastatic adenocarcinoma in liver.[19]
HepPar1 positive tumors are:
- Hepatocellular carcinoma
- Signet carcinoma of the stomach
- Ovarian carcinoma with hepatoid differentiation
- Intestinal adenocarcinoma

Glypican-3

Glypican-3 is a heparan sulfate proteoglycan located in the cell membrane of the fetal hepatocytes. It is a highly sensitive marker of hepatocellular carcinoma and can be

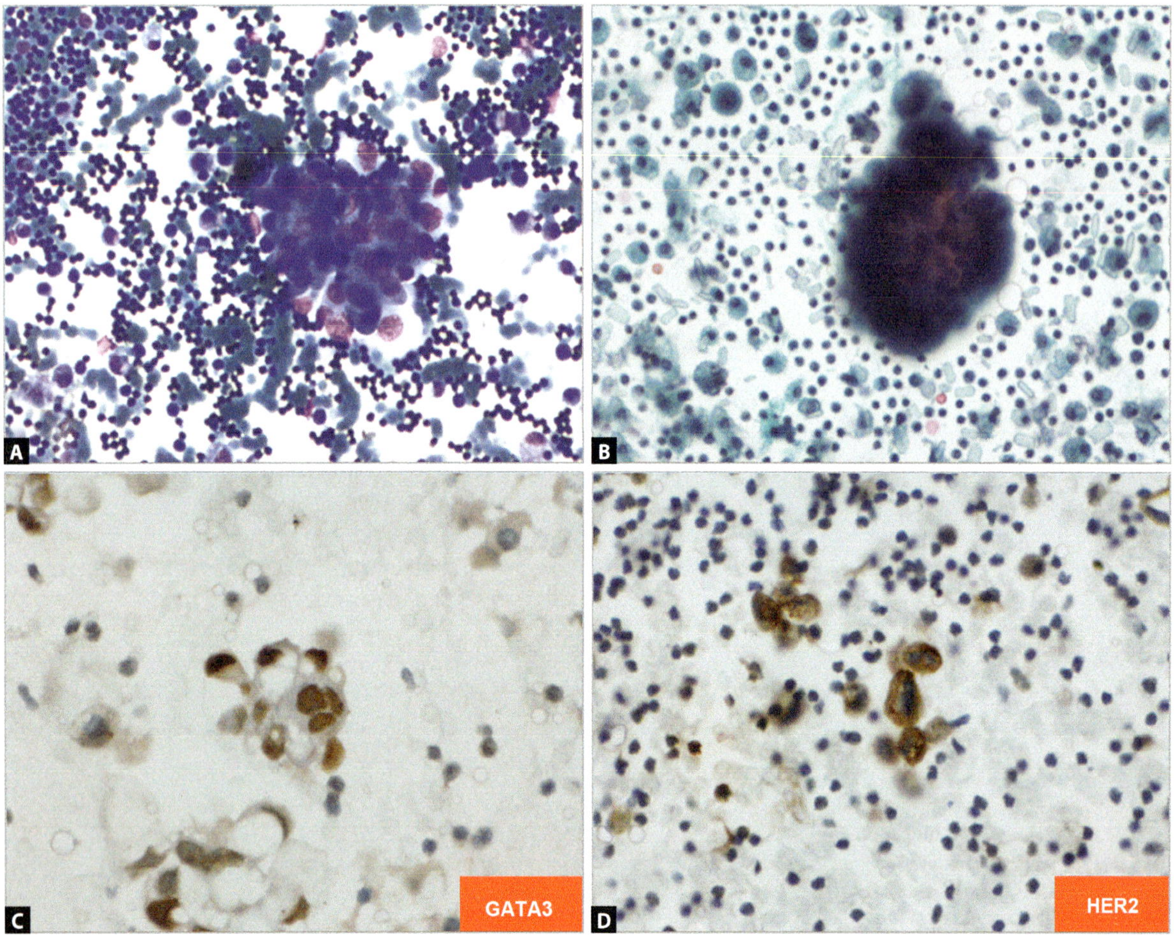

FIGS. 11A TO D: Metastatic carcinoma of breast in pleural effusion. (A) Cytology smear showing large clusters of tumor cells. (B) Papanicolaou-stained smear of the same showing malignant cells in ball-like clusters. (C) GATA3 positive cell. (D) HER-2 positive tumor cells.

used to differentiate benign hepatocytes from hepatocellular carcinoma.[20]

Glypican-3 positive tumors are:

- Hepatocellular carcinoma
- Yolk sac tumor of the ovary

Arginase-1

Arginase-1 is a urea cycle enzyme that catalyzes hydrolysis of arginine to ornithine. It is a specific marker of hepatocellular carcinoma.[21] The nuclear and cytoplasmic positivity of Arginase-1 is accepted as positive result. This is probably the most specific marker of hepatocellular carcinoma.

Prostate

Prostate-specific Antigen

Prostate-specific antigen (PSA) is a serine protease that is highly specific for prostatic epithelial tissue. However, PSA has also been expressed in tumors other than prostate.

Prostate-specific antigen positive tumors are:

- Prostatic adenocarcinoma
- Salivary duct carcinoma
- Sweat gland tumors
- Cystitis glandularis
- Melanoma

Prostatic Acid Phosphatase

Prostatic acid phosphatase is also a specific marker of the prostate. It is:

Prostatic acid phosphatase positive tumors are:

- Prostatic adenocarcinomas
- Bladder adenocarcinomas
- Rectal carcinomas
- Rectal carcinoids

Bladder

Uroplakin III

Uroplakin III is a transmembrane protein. It is specific for the urothelial tumor but only 50% sensitive. In comparison, uroplakin II is more sensitive than uroplakin III.[22]

Thrombomodulin

Thrombomodulin is a glycoprotein of the endothelial cells. It is a more sensitive marker than uroplakin III. However, it lacks specificity.[23]

Thrombomodulin positive tumors are:
- Urothelial carcinoma of the bladder
- Mesothelioma
- Squamous cell carcinoma

Germ Cell Markers

Placental-like Alkaline Phosphatase

Placental-like alkaline phosphatase (PLAP) is an alkaline phosphatase enzyme. It is a highly sensitive marker (100%) for germ cell tumor, especially seminoma/dysgerminoma.[24] It shows intense membranous and cytoplasmic immunostaining in classical seminoma. However, other germ cell tumors (choriocarcinoma, yolk sac tumor, embryonal carcinoma) are only focally positive for PLAP.
Placental-like alkaline phosphatase positive tumors are:
- *Germ cell tumors*
  - Seminoma
  - Focally positive: Choriocarcinoma, yolk sac tumor, embryonal carcinoma
- *Non-germ cell tumors*
  - Renal cell carcinoma
  - Breast carcinoma
  - Lung carcinoma

These non-germ cell tumors are EMA positive whereas seminoma is EMA negative.

Octamer-binding Transcription Factor 3/4

Octamer-binding transcription factor 3/4 (OCT3/4) is a transcription factor that is coded by *POU5F1* gene. OCT3/4 plays a vital role in the self-renewal of the pluripotent germ cells. It takes nuclear stain in the seminoma/dysgerminoma and embryonal carcinoma. It is a very sensitive and specific stain. However, OCT3/4 stain is negative in yolk sac tumor.[25]
OCT3/4 positive tumors are:
- Seminoma/dysgerminoma
- Embryonal carcinoma
- Gonadoblastoma

SALL4

SALL4 is a zinc-finger transcriptional factor and is necessary for the initiation and maintenance of pluripotent germ cells. It is a highly sensitive and specific marker and covers all the varieties of germ cell tumors.[25]
SALL4 positive tumors are:
- *Germ cell tumors*
  - Seminoma/dysgerminoma
  - Embryonal carcinoma
  - Choriocarcinoma
  - Yolk sac tumor
- *Non-germ cell tumors (rarely positive)*
  - Adenocarcinoma of colon
  - Adenocarcinoma of esophagus
  - Adenocarcinoma of endocervix
  - Urothelial carcinoma

Malignant Round Cell Tumors

Malignant round cell tumors (MRCTs) are a group of miscellaneous tumors that consists of cells with a round shape. Under an ordinary light microscope, there may be considerable difficulties to recognize the individual tumors. However, these tumors are different in their origin, genetic constituent and ICC pattern. The biological behavior and management of MRCTs are different, and therefore it is mandatory to identify the individual members of MRCT. The MRCTs include the following neoplasms: NHL, neuroblastoma (NB), Ewing's sarcoma (ES)/peripheral neuroectodermal tumor (PNET), WT1, rhabdomyosarcoma, and DSRCT. Each tumor can be successfully identified with the help of ICC (**Figs. 12** to **30**).[26-28]

The ICC pattern of the different MRCTs is highlighted in **Table 3**.

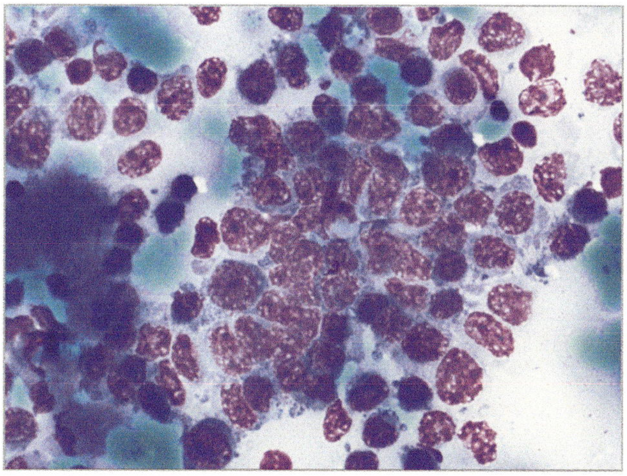

FIG. 12: Neuroblastoma: Posterior mediastinal swelling in a 15-year-old child. Fine needle aspiration cytology smear showing discrete cells with round monomorphic nuclei.

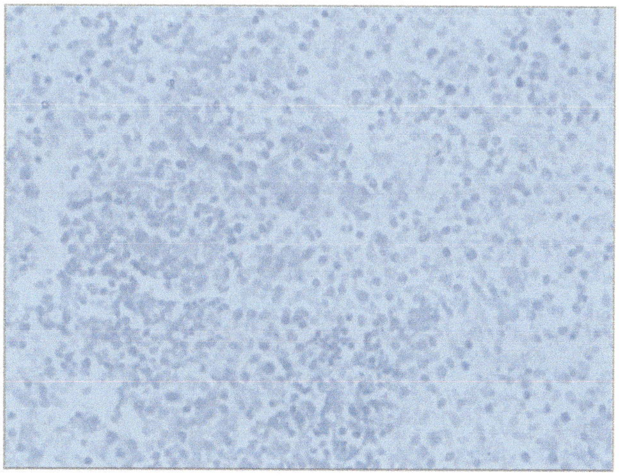

FIG. 13: Neuroblastoma: Cell block section showing sheets of monomorphic cells.

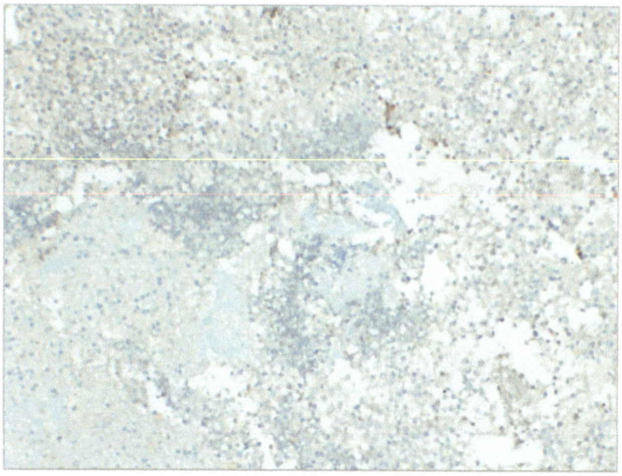

FIG. 14: Neuroblastoma: The tumor cells are negative for Pan CK.

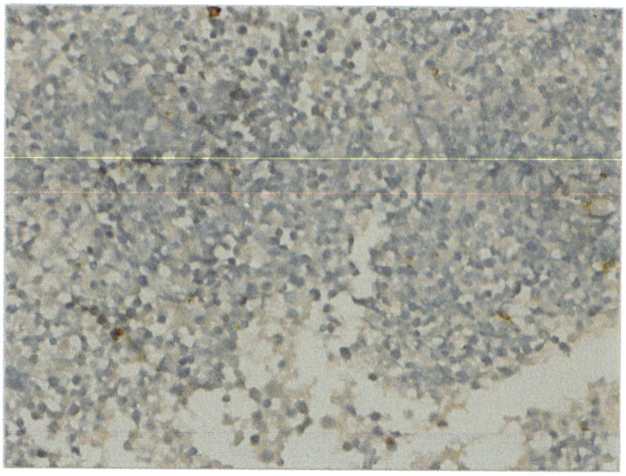

FIG. 15: Neuroblastoma: The tumor cells are negative for desmin.

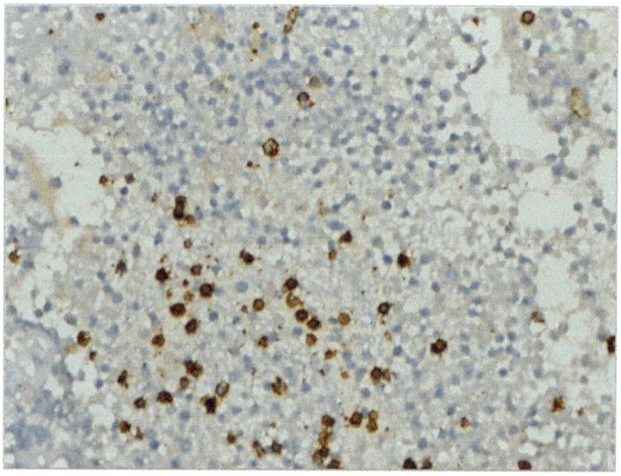

FIG. 16: Neuroblastoma: The tumor cells are negative for CD45. Only occasional cells showing CD45 staining.

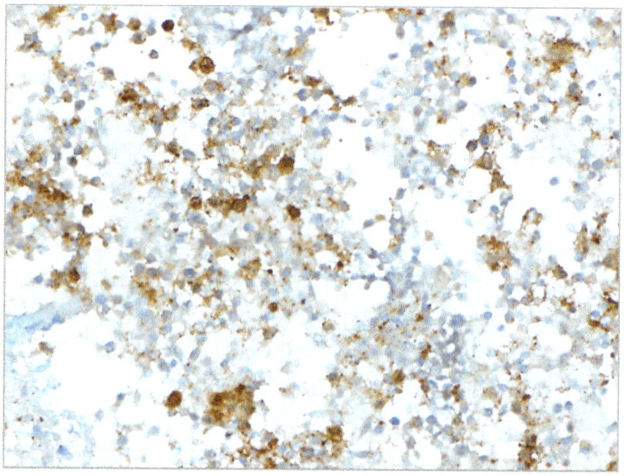

FIG. 17: Neuroblastoma: The tumor cells are positive for NB84.

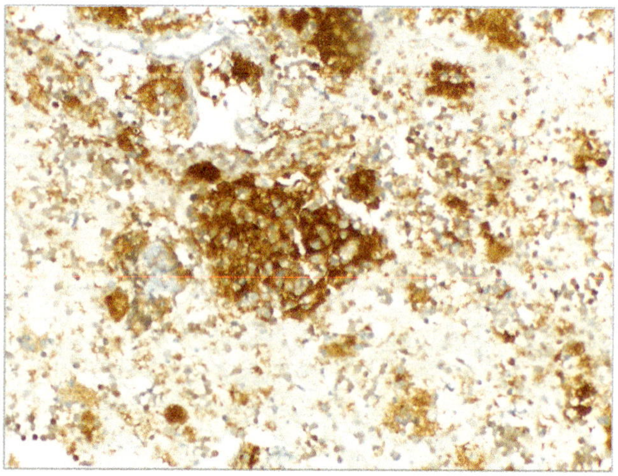

FIG. 18: Neuroblastoma: The tumor cells are positive for neuron-specific enolase (NSE).

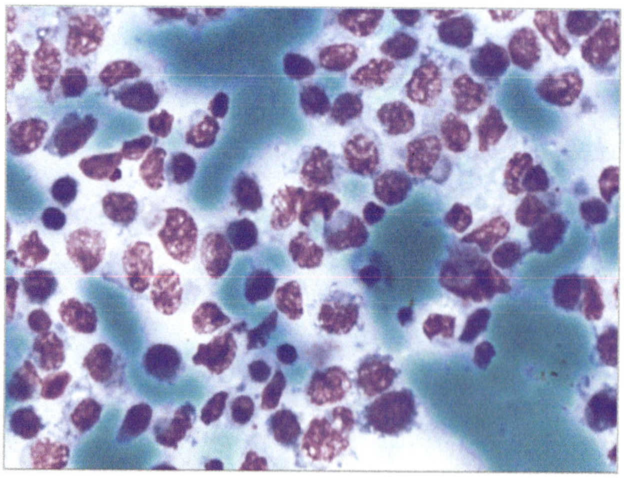

FIG. 19: Peripheral neuroectodermal tumor/soft-tissue Ewing's tumor: Fine needle aspiration cytology smear of a cheek swelling in a 12-year-old female showing discrete monomorphic round cells.

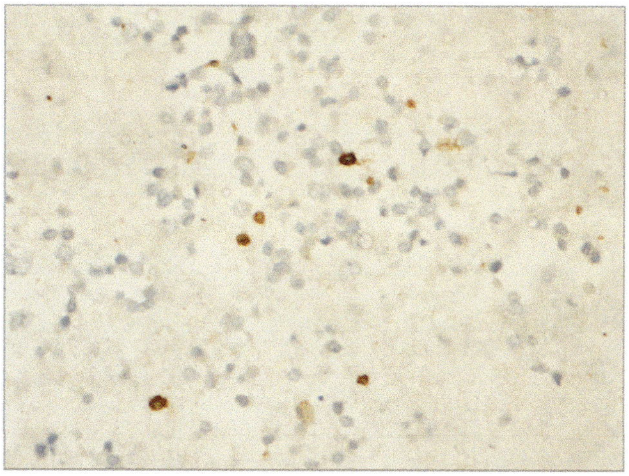

FIG. 20: Peripheral neuroectodermal tumor/soft-tissue Ewing's tumor: The cells are overall negative for leukocyte common antigen (LCA).

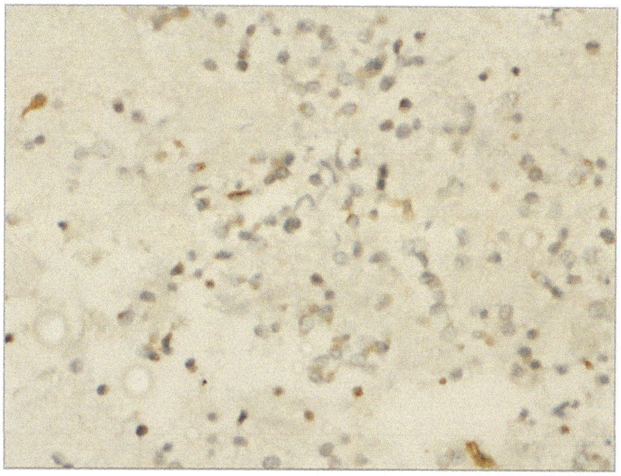

FIG. 21: Peripheral neuroectodermal tumor/soft-tissue Ewing's tumor: The cells are overall negative for desmin.

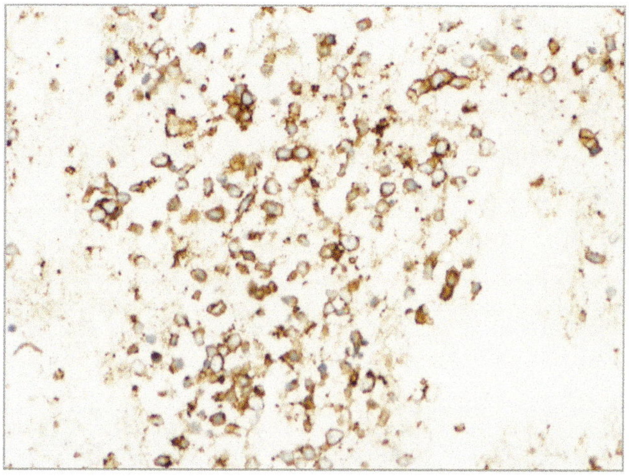

FIG. 22: Peripheral neuroectodermal tumor/soft-tissue Ewing's tumor: The cells are positive for CD99.

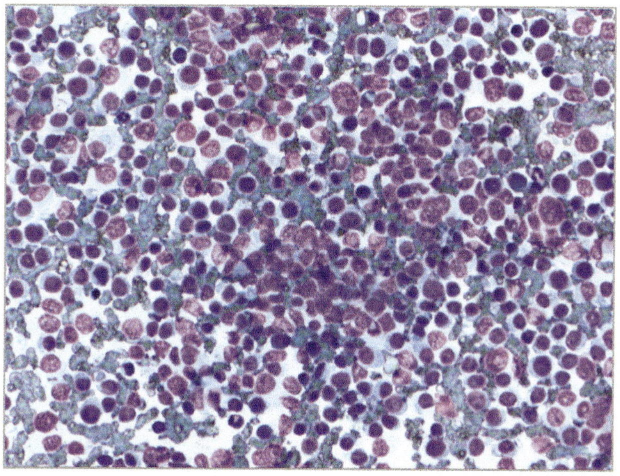

FIG. 23: Rhabdomyosarcoma: Fine needle aspiration cytology smear of an intraorbital swelling in a 10-year-old boy. Smear showing discrete round cells.

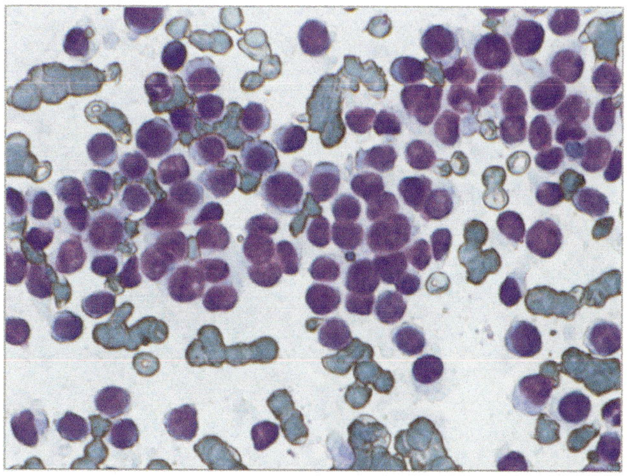

FIG. 24: Rhabdomyosarcoma: Fine needle aspiration cytology smear showing cells with round nuclei having fine chromatin.

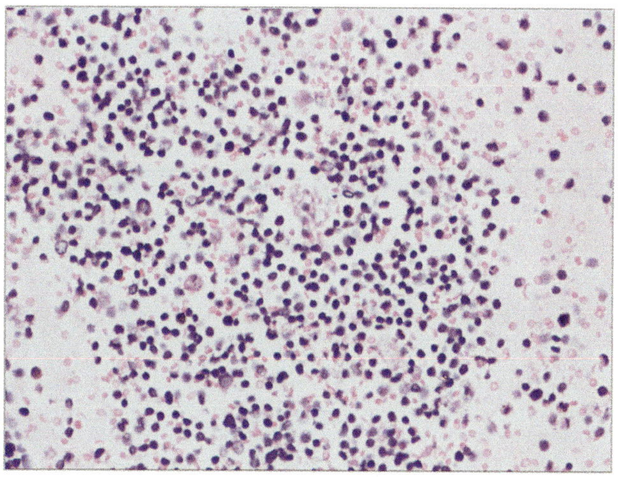

FIG. 25: Rhabdomyosarcoma: Cell block section showing sheets of monomorphic round cells.

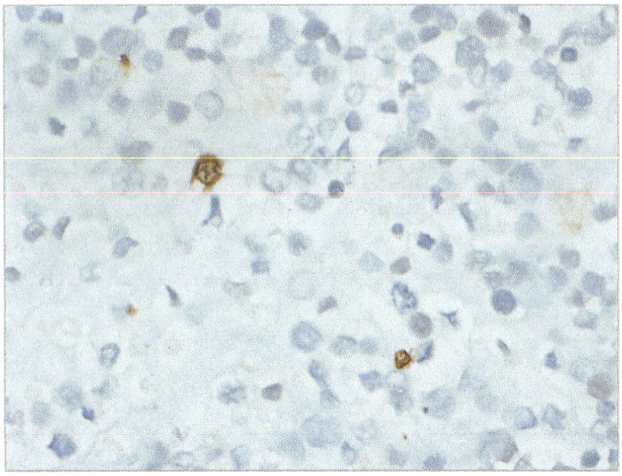

FIG. 26: Rhabdomyosarcoma: The cells are negative for CD45 (leukocyte common antigen).

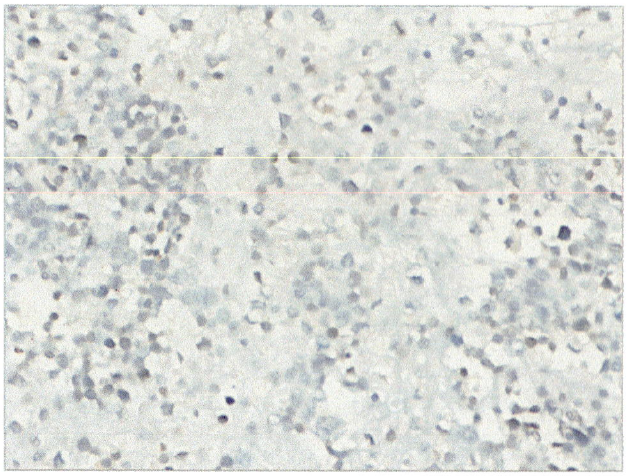

FIG. 27: Rhabdomyosarcoma: The cells are negative for chromogranin.

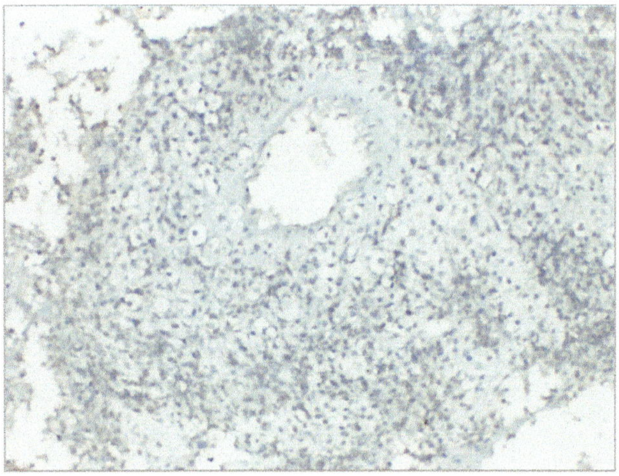

FIG. 28: Rhabdomyosarcoma: The cells are negative for neuron-specific enolase (NSE).

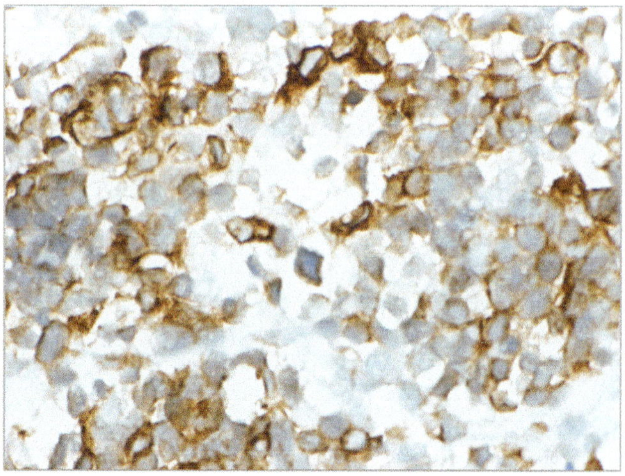

FIG. 29: Rhabdomyosarcoma: The cells are positive for desmin.

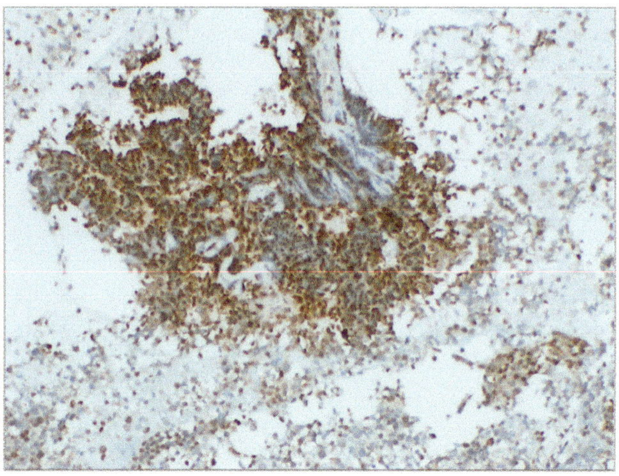

FIG. 30: Rhabdomyosarcoma: The cells are positive for myogenin.

TABLE 3: Immunocytochemistry of small round cell tumor.

	Immunocytochemistry							
	LCA	NB-84	Myo D1	CK	Desmin	CD99 (Mic-2)	NSE	FLI-1
NHL	+	–	–	–	–	–	–	–
NB	–	+	–	–	–	+	+	–
EWS/PNET	–	–	–	–	–	+	–	+
Wilms' tumor	–	+	–	+	+	–	–	–
RMS	–	–	+		+	–	–	–
DSRCT	–	–	–	+	+		+	–
SCC	–	–	–	+	–	–	+	–

– = Negative, + = Positive.

(DSRCT: desmoplastic small round cell tumor; EWS: Ewing's sarcoma; NB: neuroblastoma; NHL: non-Hodgkin lymphoma; PNET: primitive neuro-ectodermal tumor; RHS: rhabdomyosarcoma; SCC: squamous cell carcinoma)

REFERENCES

1. Lin F, Liu H. Immunohistochemistry in undifferentiated neoplasm/tumor of uncertain origin. Arch Pathol Lab Med. 2014;138(12):1583-610.
2. Bahrami A, Truong LD, Ro JY. Undifferentiated tumor: true identity by immunohistochemistry. Arch Pathol Lab Med. 2008;132(3):326-48.
3. Wick MR. Immunohistochemical approaches to the diagnosis of undifferentiated malignant tumors. Ann Diagn Pathol. 2008;12(1):72-84.
4. Park SY, Kim BH, Kim JH, Lee S, Kang GH. Panels of immunohistochemical markers help determine primary sites of metastatic adenocarcinoma. Arch Pathol Lab Med. 2007;131(10):1561-7.
5. Kummar S, Fogarasi M, Canova A, Mota A, Ciesielski T. Cytokeratin 7 and 20 staining for the diagnosis of lung and colorectal adenocarcinoma. Br J Cancer. 2002;86(12):1884-7.
6. Dey P, Togra J, Mitra S. Intermediate filament: structure, function, and applications in cytology. Diagn Cytopathol. 2014;42(7):628-35.
7. Dey P. Special stains and immunocytochemistry. In: Diagnostic Cytology. New Delhi: Jaypee Medical Publishers; 2016. Chapter 17.
8. Krishna M. Diagnosis of metastatic neoplasms: an immunohistochemical approach. Arch Pathol Lab Med. 2010;134(2):207-15.
9. Varadhachary GR, Abbruzzese JL, Lenzi R. Diagnostic strategies for unknown primary cancer. Cancer. 2004;100(9):1776-85.
10. Bhargava R, Beriwal S, McManus K, Dabbs DJ. CK5 is more sensitive than CK5/6 in identifying the "basal-like" phenotype of breast carcinoma. Am J Clin Pathol. 2008;130(5):724-30.
11. Kandalaft PL, Gown AM. Practical applications in immunohistochemistry: Carcinomas of unknown primary site. Arch Pathol Lab Med. 2016;140(6):508-23.
12. Werling RW, Yaziji H, Bacchi CE, Gown AM. CDX2, a highly sensitive and specific marker of adenocarcinomas of intestinal origin: an immunohistochemical survey of 476 primary and metastatic carcinomas. Am J Surg Pathol. 2003;27(3):303-10.
13. Magnusson K, de Wit M, Brennan DJ, Johnson LB, McGee SF, Lundberg E, et al. SATB2 in combination with cytokeratin 20 identifies over 95% of all colorectal carcinomas. Am J Surg Pathol. 2011;35(7):937-48.
14. Dragomir A, de Wit M, Johansson C, Uhlen M, Pontén F. The role of SATB2 as a diagnostic marker for tumors of colorectal origin: Results of a pathology-based clinical prospective study. Am J Clin Pathol. 2014;141(5):630-8.
15. Ordóñez NG. Value of PAX2 immunostaining in tumor diagnosis: a review and update. Adv Anat Pathol. 2012;19(6):401-9.
16. Hou Y, Shen R, Chaudhary S, Tonkovich D, Li Z. Utility of different immunostains for diagnosis of metastatic breast carcinomas in both surgical and cytological specimens. Ann Diagn Pathol. 2017;30:21-7.
17. Luo MH, Huang YH, Ni YB, Tsang JY, Chan SK, Shao MM, Tse GM. Expression of mammaglobin and gross cystic disease fluid protein-15 in breast carcinomas. Hum Pathol. 2013;44(7):1241-50.
18. Braxton DR, Cohen C, Siddiqui MT. Utility of GATA3 immunohistochemistry for diagnosis of metastatic breast carcinoma in cytology specimens. Diagn Cytopathol. 2015;43(4):271-7.
19. Ibrahim TR, Abdel-Raouf SM. Immunohistochemical study of glypican-3 and HepPar-1 in differentiating hepatocellular carcinoma from metastatic carcinomas in FNA of the liver. Pathol Oncol Res. 2015;21(2):379-87.
20. Zhou F, Shang W, Yu X, Tian J. Glypican-3: A promising biomarker for hepatocellular carcinoma diagnosis and treatment. Med Res Rev. 2018;38(2):741-67.
21. Chandan VS, Shah SS, Torbenson MS, Wu TT. Arginase-1 is frequently positive in hepatoid adenocarcinomas. Hum Pathol. 2016;55:11-6.
22. Li W, Liang Y, Deavers MT, Kamat AM, Matin SF, Dinney CP, et al. Uroplakin II is a more sensitive immunohistochemical marker than uroplakin III in urothelial carcinoma and its variants. Am J Clin Pathol. 2014;142(6):864-71.
23. Ordóñez NG. Value of thrombomodulin immunostaining in the diagnosis of transitional cell carcinoma: a comparative study with carcinoembryonic antigen. Histopathology. 1997;31(6):517-2.
24. Nielsen OS, Munro AJ, Duncan W, Sturgeon J, Gospodarowicz MK, Jewett MA, et al. Is placental alkaline phosphatase (PLAP) a useful marker for seminoma? Eur J Cancer. 1990;26(10):1049-54.
25. Rabban JT, Zaloudek CJ. A practical approach to immunohistochemical diagnosis of ovarian germ cell tumours and sex cord-stromal tumours. Histopathology. 2013;62(1):71-88.
26. Brahmi U, Srinivasan R, Komal HS, Joshi K, Gupta SK, Rajwanshi A. Comparative analysis of electron microscopy and immunocytochemistry in the cytologic diagnosis of malignant small round cell tumors. Acta Cytol. 2003;47(3):443-9.
27. Pohar-Marinsek Z. Difficulties in diagnosing small round cell tumours of childhood from fine needle aspiration cytology samples. Cytopathology. 2008;19(2):67-79.
28. Rajwanshi A, Srinivas R, Upasana G. Malignant small round cell tumors. J Cytol. 2009;26(1):1-10.

CHAPTER 9

Immunocytochemistry in Head, Neck, Salivary Gland, and Thyroid Lesions

SALIVARY GLAND

The antibody markers of the different components of the salivary glands are:[1]

- *Acinar epithelial cells*
  - CK7: Positive
  - CK18: Positive
  - Epithelial membrane antigen (EMA): Positive
  - DOG1: Positive
- *Myoepithelial cells*
  - p63: Positive
  - Calponin: Positive
  - Alpha smooth muscle actin (SMA): Positive
  - S-100: Positive
- *Duct cells*
  - CK7: Positive
  - CK19: Positive
  - EMA: Positive

Immunocytochemistry (ICC) has a little role in the diagnosis of different salivary gland tumors. The cell morphology of the tumors helps to diagnose most of the salivary tumors. In fine needle aspiration cytology (FNAC) material, the ICC is important only in selected lesions.

Pleomorphic Adenoma

The ICC pattern of this tumor is given in the following text (**Box 1**).[2]

BOX 1 Pleomorphic adenoma.

- It is a biphasic tumor
- Epithelial cells: Positive for PanCK, CK7, CK5/6
- Myoepithelial cells: Calponin and p63 are the most important markers
- Differential diagnosis: Adenoid cystic carcinoma and basal cell adenoma
- Ki67 index is low (mean 1.6%), whereas it is much higher in adenoid cystic carcinoma (>20%)
- p53 positivity is weak and low (mean 1.2%) whereas it is much higher in adenoid cystic carcinoma (>20%)

- The duct cells of the tumor show:
  - PanCK: Positive
  - CK7: Positive
  - CK5/6: Positive
- The myoepithelial cells show:
  - Calponin: Positive
  - Alpha SMA: Positive

Oncocytoma

Oncocytoma shows the following markers:[3]

- CK7: Positive
- CK5/6: Positive
- EMA: Positive
- p63: Positive

Myoepithelioma

The myoepithelial component shows the following[4,5] (**Figs. 1** to **4**):

- Calponin: Positive
- SMA: Positive
- p63: Positive
- S-100: Positive

The epithelial component shows:

- CK7: Positive
- CAM5.2: Positive

So, the tumor shows reactivity of mainly the myoepithelial cell markers (**Box 2**).

Epithelial Myoepithelial Carcinoma

This tumor shows a biphasic pattern of epithelial and myoepithelial cells positivity[6] (**Box 3; Figs. 5** to **10**).

The epithelial component demonstrates:

- CK7: Positive
- CAM5.2: Strongly positive

The myoepithelial component is positive for:

- Calponin: Positive
- SMA: Positive
- p63: Positive

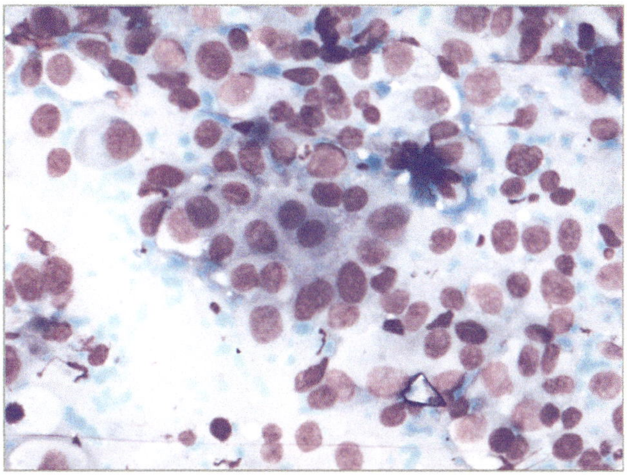

FIG. 1: Myoepithelioma: Cytology smear showing abundant discrete cells with a moderate amount of cytoplasm and eccentric nuclei.

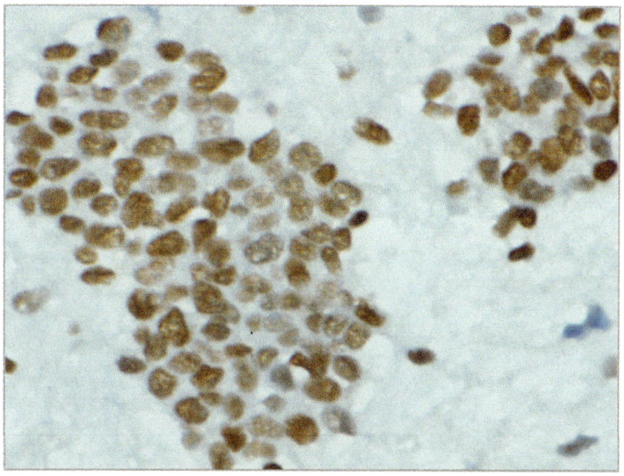

FIG. 2: Myoepithelioma: Cells showing p63 positivity.

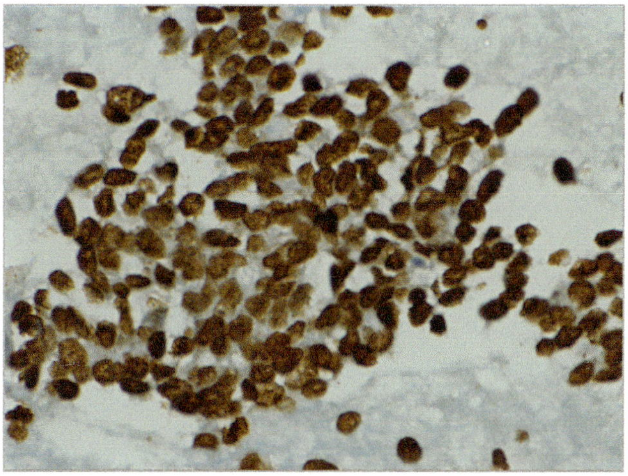

FIG. 3: Myoepithelioma: Cells showing smooth muscle actin (SMA) positivity.

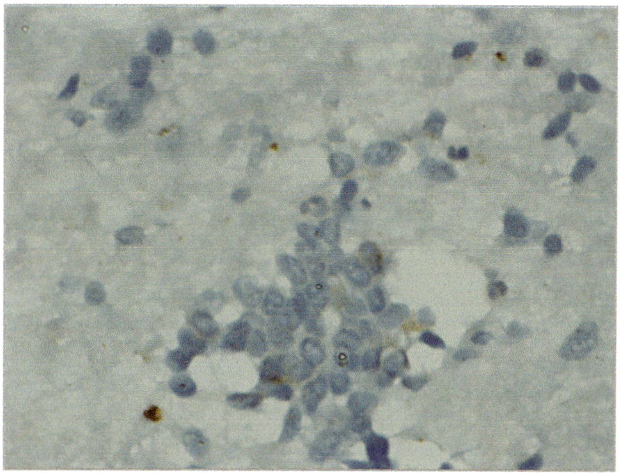

FIG. 4: Myoepithelioma: Cells showing epithelial membrane antigen (EMA) negative cell clusters.

BOX 2 Myoepithelial neoplasm.

- Usually the neoplasm shows mainly myoepithelial cells with striking absence of ductal cells
- Myoepithelial cells: Calponin, SMA, S-100, and p63 positive
- Epithelial cells: CK7 and CAM5.2 positive
- The myoepithelial markers show variable positivity

BOX 3 Epithelial myoepithelial carcinoma.

- Biphasic patter: Ductal cells and myoepithelial cells
  - Ductal cells: PanCK and S-100 positive
  - Myoepithelial cells: Calponin and p63 positive
- Ki67 index is low in both ductal (<1%) in ductal cells

Adenoid Cystic Carcinoma

The ICC pattern of this tumor is given in the following text[1,7] (**Box 4; Figs. 11** to **13**).

The duct cells of the tumor show:

- PanCK: Positive
- CK7: Positive
- CK5/6: Positive

The epithelial luminal cell shows:

- CD117: Positive

The basal epithelial cells show:

- p63: Positive
- SMA: Positive
- Maspin: Positive
- Calponin: Positive
- Alpha SMA: Positive

Acinic Cell Carcinoma

The acinic cell carcinoma shows the following ICC markers[8] (**Box 5**):

- CK7: Positive
- CK19: Negative
- c-Kit: Positive
- Calponin: Negative

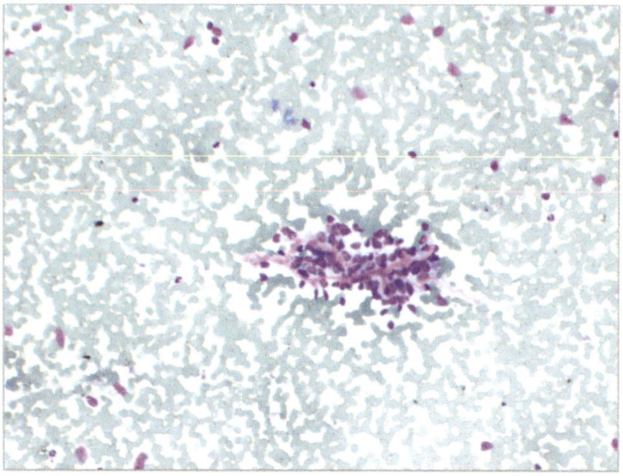

FIG. 5: Epithelial myoepithelial tumor: Fine needle aspiration cytology smear showing loose clusters of tumor cells.

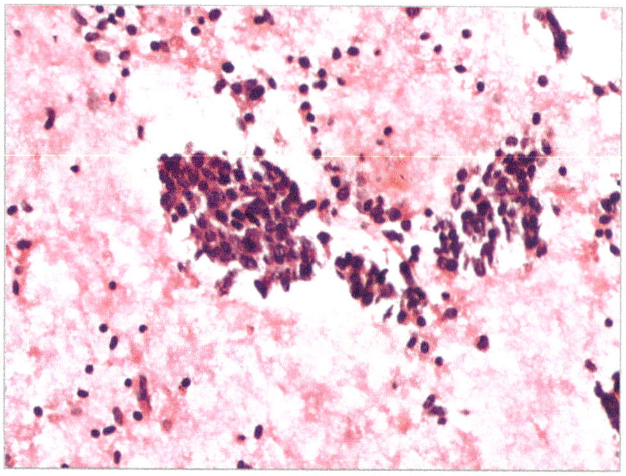

FIG. 6: Epithelial myoepithelial tumor: Fine needle aspiration cytology smear showing loose clusters and discrete round cells.

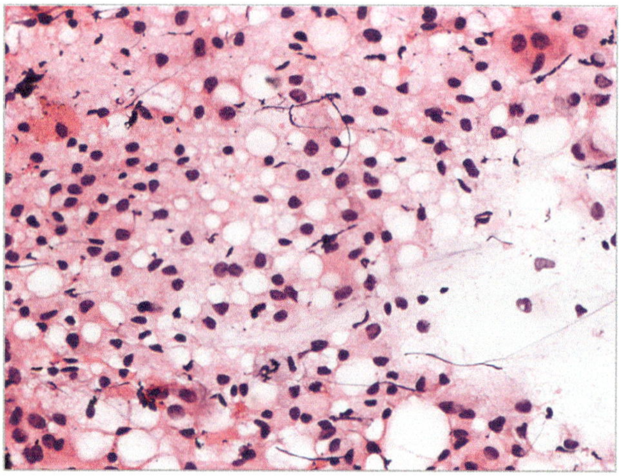

FIG. 7: Epithelial myoepithelial carcinoma: Abundant discrete round cells along with scattered spindle cells in fine needle aspiration cytology smear.

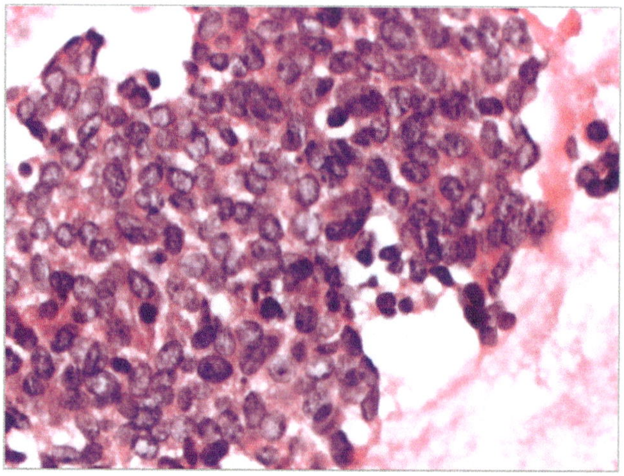

FIG. 8: Epithelial myoepithelial tumor: Cell block section showing a diffuse sheet of cells.

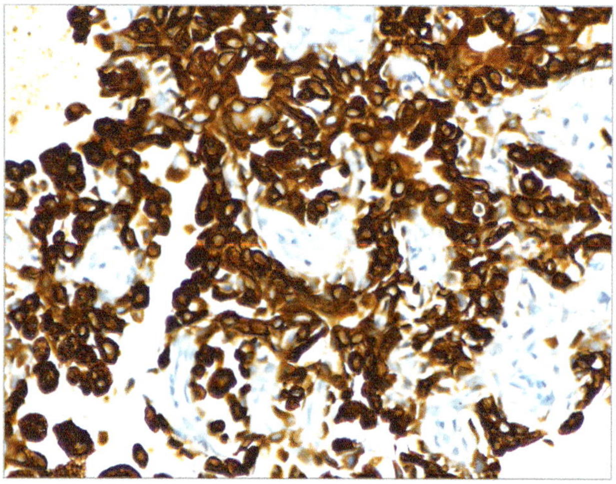

FIG. 9: Epithelial myoepithelial carcinoma: Strong membranous CK7 positivity.

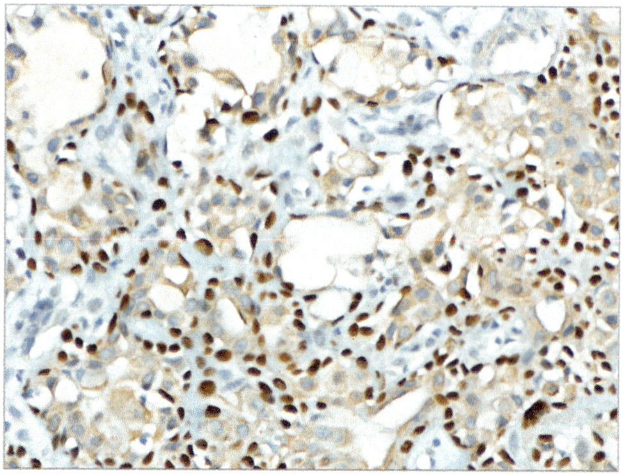

FIG. 10: Epithelial myoepithelial tumor: p63 positive cells.

BOX 4 Adenoid cystic carcinoma.
- The tumor shows characteristic hyaline balls surrounded by tumor cells
- Both epithelial (PanCK, CK5/6, CK7) and myoepithelial markers (calponin and Maspin) are positive
- The basal ductal cells are positive for CD117
- Many tumor cells may express CD43
- Specific chromosomal translocation (6;9)(q22-23:p23-24) is present

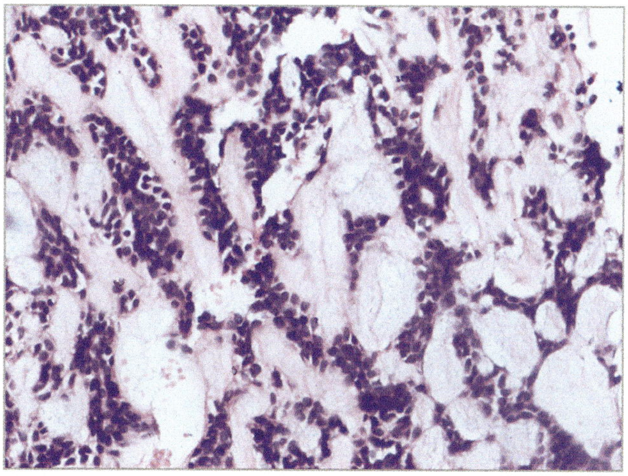

FIG. 11: Adenoid cystic carcinoma: Cell block section.

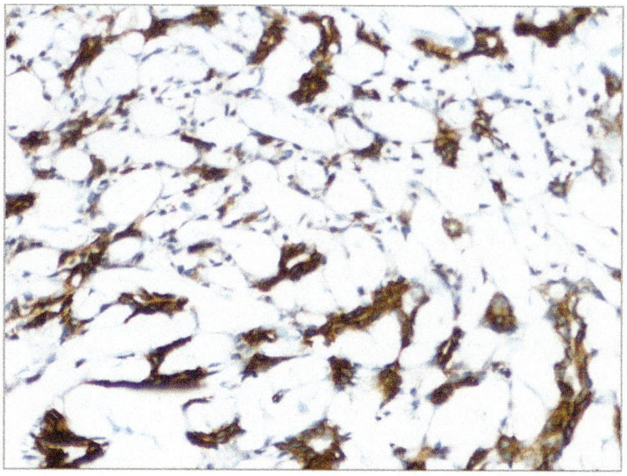

FIG. 13: Adenoid cystic carcinoma: C kit positive cells.

Differential Diagnosis

- *Mammary analogue secretory carcinoma:* This has been discussed later.
- *Benign acinar cells:* ICC does not help.

Mucoepidermoid Carcinoma

Mucoepidermoid carcinoma (MEC) is the most frequent malignant tumor of the salivary gland. The cytological features of this tumor are characteristic, and ICC does not help much in the diagnosis. The tumor shows both epithelial and basal cell marker positivity.[9]

BOX 5 Acinic cell carcinoma.
- Tumor expresses CK7, CEA, amylase, and CD117
- The cells also express DOG-1
- The cells are negative for basal cell marker: Calponin

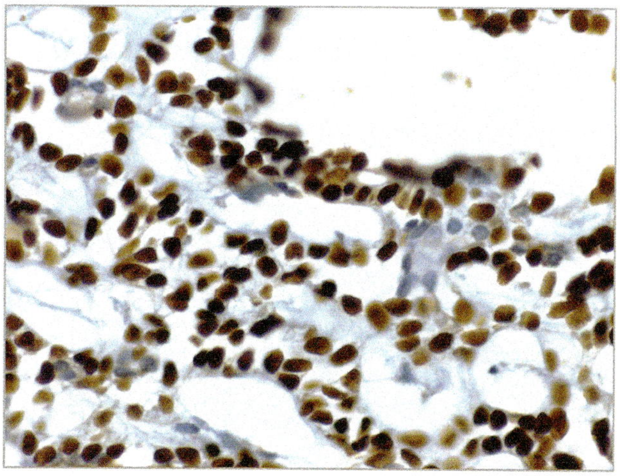

FIG. 12: Adenoid cystic carcinoma: p63 positive cells.

BOX 6 Mucoepidermoid carcinoma.
- The cells are positive
  - Epithelial markers: CK7, CK5/6, and p63
  - Mucin markers: MUC1, MUC4, and MUC5AC
- The cell are negative for myoepithelial markers: Actin and calponin

*The ICC pattern of MEC is as follows (**Box 6**):*
- PanCK: Positive
- CK7: Positive
- CK5/6: Positive
- p63: Positive
- MUC, MUC4, MUC5: Positive

Differential Diagnosis

- *Squamous cell carcinoma:* The squamous cell carcinoma (SQCC) shows malignant squamous epithelial cells and may be mistaken as MEC and vice versa. The cells of SCC are also positive for p63, p40, and CK5/6. Based on cytology and ICC, it is challenging to distinguish SQCC and MEC (**Table 1**).
- *Renal cell carcinoma:* Metastatic renal cell carcinoma (RCC) often shows cells with clear cytoplasm. RCC expresses CD10 and vimentin whereas these two markers are negative in MEC.

TABLE 1: Differentiating markers of high-grade mucoepidermoid carcinoma and squamous cell carcinoma.

Antibody	High-grade mucoepidermoid carcinoma	Squamous cell carcinoma
CD117	Negative	Positive
CK7	Positive	Negative
MUC5AC	Positive	Negative

BOX 7 Polymorphous low-grade adenocarcinoma (PLGA).

- This malignant tumor that mainly affects the minor salivary gland
- No specific immunocytochemistry pattern to confirm the diagnosis of PLGA
- Positive epithelial markers: CK7, CK5/6, p63
- Negative for myoepithelial cell markers

Polymorphous Low-grade Adenocarcinoma

Polymorphous low-grade adenocarcinoma (PLGA) mainly affects the minor salivary glands. This tumor shows round mildly pleomorphic cells along with scanty chondromyxoid stromal tissue. An occasional gland pattern may be noted. Based on ICC, the tumor is difficult to confirm (**Box 7**).

The ICC pattern of PLGA is:[10]

- PanCK: Positive
- CK7: Positive
- CK5/6: Positive
- p63: Positive
- Calponin: Negative

Differential Diagnosis

Pleomorphic adenoma and adenoid cystic carcinoma

Salivary Duct Carcinoma

Salivary duct carcinoma (SDC) is an aggressive malignant tumor. It commonly involves the parotid gland. The tumor cells of SDC show the following ICC pattern[11] (**Box 8**):

- PanCK: Positive
- CK7: Positive
- CK20: Positive
- Androgen receptor (AR): Negative
- Her2-neu: Strongly positive in high-grade SDC
- Estrogen receptor/progesterone receptor (ER/PR): Negative
- p63: Usually negative

Differential Diagnosis

- **Metastatic carcinoma:** Metastatic SQCC expresses p63, p40, and CK5/6.
- **Mucoepidermoid carcinoma:** The epidermoid component of MEC is positive for p63, p40, and CK5/6. Also, the tumor cells of MEC are negative for AR.

Mammary Analogue Secretory Carcinoma

Mammary analogue secretory carcinoma (MASC) is a relatively recent entity with the characteristic *ETV6-NRTK3* gene fusion (**Box 9**).

BOX 8 Salivary duct carcinoma (SDC).

- The cytology features of SDC are similar to those of high-grade ductal carcinoma of breast
- Positive: CK7, CK20, CEA, and EMA
- Majority of the tumors are also characteristically positive for HER-2
- Focally positive: GCDFP
- Ki67 index: High (>20%)

BOX 9 Mammary analogue secretory carcinoma.

- Positive: CK7, CK18, and EMA
- In addition, the cells are characteristically positive for mammaglobin, GCDFP, and GATA3
- Negative: p63 and DOG1
- The characteristic genetic changes: Translocation of t(12;15) (p13;q25) that causes ETV6-*NRTK3* gene fusion

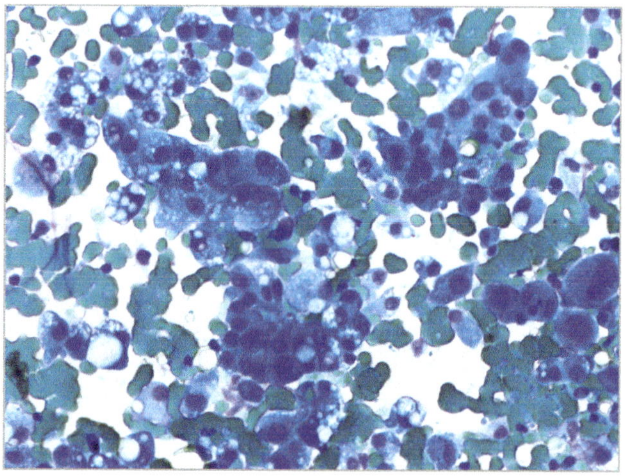

FIG. 14: Mammary secretory analogue carcinoma: Loose clusters of round-to-polygonal cells with a moderate amount of cytoplasm.

The tumor usually occurs in the major salivary gland as a painless swelling. On FNAC, the smear of MASC shows many dissociate and loosely cohesive cells. The individual cells have a moderate amount of vacuolated cytoplasm. The nucleus of the cell is enlarged, having prominent nucleoli.

The ICC pattern of MASC is as follows (**Figs. 14** to **19**):[12,13]

- *Positive*
  - PanCK
  - CK7
  - CK18
  - EMA
  - Mammaglobin
  - GCDFP
  - GATA3
  - MUC1
  - MUC4
- *Negative*
  - AR
  - ER
  - PR

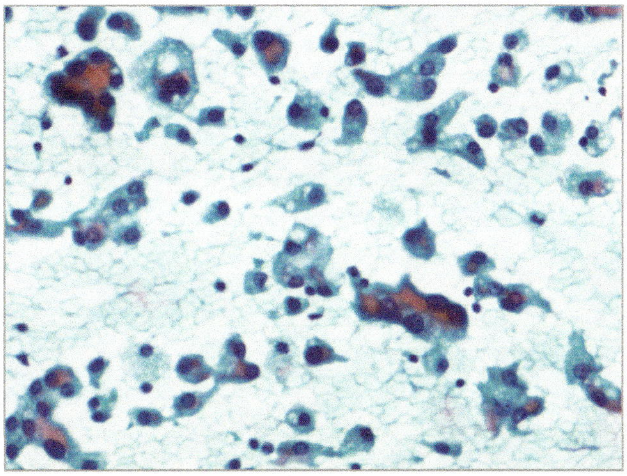

FIG. 15: Mammary secretory analogue carcinoma: Papanicolaou-stained smear showing round cells having vacuolated cytoplasm and central to eccentric nuclei.

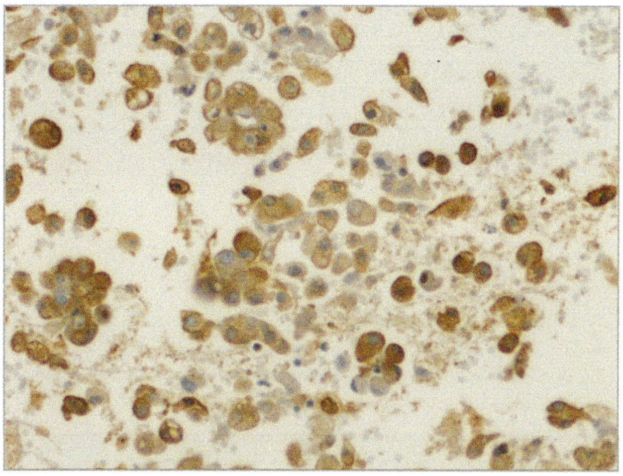

FIG. 16: Mammary secretory analogue carcinoma: The cells showing mammaglobin positivity.

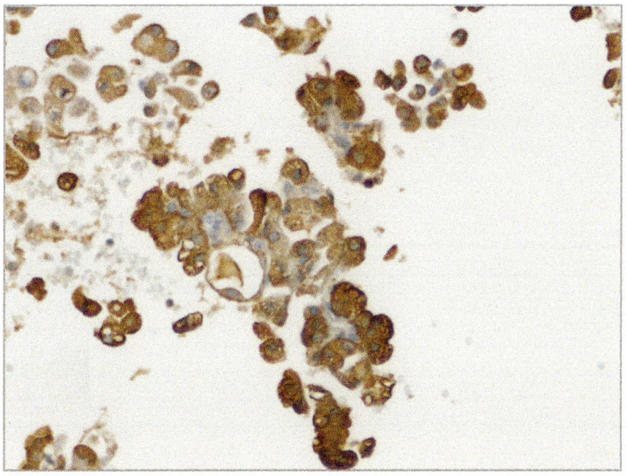

FIG. 17: Mammary secretory analogue carcinoma: The cells showing GCDFP positivity.

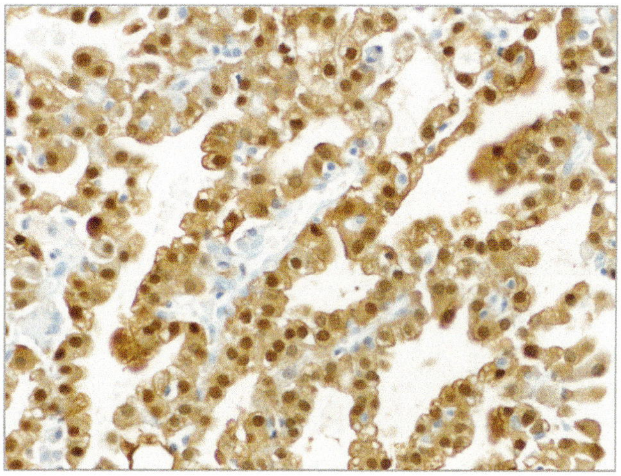

FIG. 18: Mammary secretory analogue carcinoma: The cells showing S-100 positivity.

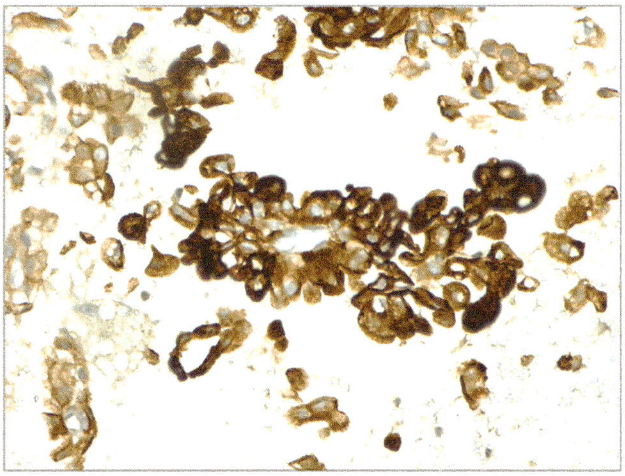

FIG. 19: Mammary secretory analogue carcinoma: The cells showing CK7 positivity.

Besides, the myoepithelial cell markers are also negative:
- Calponin: Negative
- p63: Negative

Differential Diagnosis

Acinic cell carcinoma: The MASC is strongly positive for S-100 and mammaglobin whereas these two markers are negative in acinic cell carcinoma (**Table 2**).

Other Tumors

Non-Hodgkin Lymphoma

Primary non-Hodgkin lymphoma (NHL) of the salivary gland is uncommon, and it represents only 2.4% of all neoplasms of the salivary gland. Most of the NHLs are of B-cell origin consisting of mucosa-associated lymphoid tissue lymphoma, diffuse large B-cell lymphoma, and follicular lymphoma.

TABLE 2: Differentiating immunocytochemistry pattern of mammary analogue secretory carcinoma and acinic cell carcinoma.

Features	Mammary analogue secretory carcinoma	Acinic cell carcinoma
Mammaglobin antibody	+	–
GCDFP	+	–
MUC4	+	–
ETV6-NRTK3 gene fusion	Present	Absent

+ = Positive, – = Negative

TABLE 3: Differentiating lymphoepithelial sialadenitis from MALT lymphoma.

Antibody	Lymphoepithelial sialadenitis	MALT lymphoma
Kappa and Lambda	Both expressed	Only one chain is predominantly expressed
B-cell markers CD19, CD20, and CD3	Variably expressed	Only CD19 and CD 20 positive cells and CD3 positive cells are minimal to nil
EMA	Positive cells	Usually negative

(EMA: epithelial membrane antigen; MALT: mucosa-associated lymphoid tissue)

The ICC pattern of lymphoma depends on the type of NHL. It is essential to differentiate lymphoepithelial sialadenitis (LESA) from MALT lymphoma (**Table 3**).

Plasma Cell Tumor

Solitary plasma cell neoplasm of the salivary gland is uncommon. The ICC of this lesion shows a strongly positive CD38 cell population (**Figs. 20** and **21**).

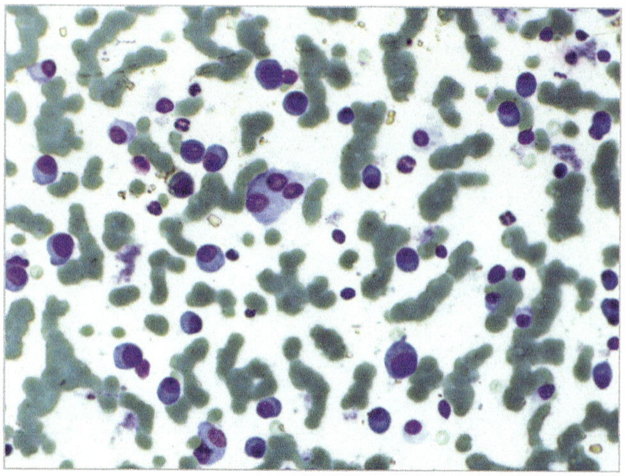

FIG. 20: Plasma cell tumor of the parotid gland showing abundant discrete plasma cells.

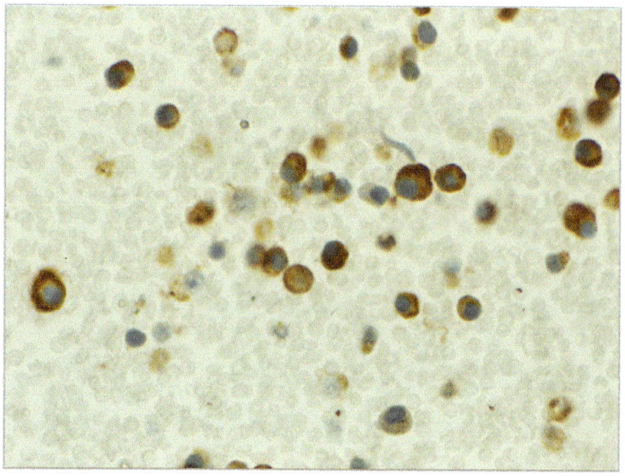

FIG. 21: Plasma cell tumor of the parotid gland showing CD38 positive cells.

THYROID

Normal thyroid cells show the following positive markers:
Thyroid follicular cells:
- Thyroglobulin: Positive
- TTF-1: Positive
- CK7: Positive
- EMA: Positive
- AEI/AE3: Positive

Thyroid-specific antibodies: These are described in detail in Chapter 8.

Hector Battifora Mesothelial Cell 1

Hector Battifora mesothelial cell 1 (HBME-1) is a membrane antigen and is present in the microvilli of the mesothelial cells and respiratory epithelial cells. Adenocarcinomas of lung, breast, and pancreas also express HBME. Benign thyroid follicular cells do not show HBME-1 expression. HBME-1 is positive in nearly 90% of papillary and 84% in follicular thyroid carcinoma (**Table 4**). However, follicular adenoma (FA), colloid goiter, and thyroiditis cases may also show HBME-1 positivity.[14]

TABLE 4: Sensitivity and specificity of HBME-1.

Tumors	Sensitivity	Specificity
Papillary carcinoma	67%	99%
Follicular carcinoma	86%	72.6%

Galectin-3

Galectin-3 is a β-galactoside-binding animal lectin. It is related to the interaction between two cells, apoptosis, tumor development, and metastasis. Normal thyroid follicular cells do not show any expression of galectin-3. The overexpression of galectin-3 is noted in malignant thyroid tumors. The overall sensitivity and specificity of galectin-3 in papillary thyroid carcinoma (PTC) are 70–87% and 83–99%, respectively (**Table 5**). It is a useful marker to differentiate benign from differentiated malignant tumors as the sensitivity and specificity

TABLE 5: Sensitivity and specificity of Galectin-3.

Tumors	Sensitivity	Specificity
Papillary carcinoma	87%	83.6%
Follicular carcinoma	57–100%	88%

TABLE 6: CK19 positivity in papillary and follicular carcinoma.

Tumors	Sensitivity	Specificity
Papillary carcinoma	75%	74%
Follicular carcinoma	44%	63%

TABLE 7: Various essential markers in thyroid tumors.[14-17]

Different tumors	HBME-1	Galectin-3	CK19	Thyroglobulin	TTF-1	Calcitonin
PTC	+	+	+	+	+	–
FC	± (30–80%)	± (50–95%)	± (35%)	+	+	–
FA	–	–	–	+	+	–
MC	+	±	±	–	+	+
PDTC	±	–	±	±	±	–
AC	–	+ (90%)	±	–		–

+ = Positive, – = Negative

(AC: anaplastic carcinoma; FA: follicular adenoma; FC: follicular carcinoma; MC: medullary carcinoma; PDTC: poorly differentiated thyroid carcinoma; PTC: papillary thyroid carcinoma)

of distinguishing benign from differentiated malignant thyroid tumors are 99 and 98%, respectively.[15,16]

CK19

CK19 is an intermediate filament and is usually positive in the glandular epithelial cells. The normal thyroid follicular epithelial cells are generally negative for CK19 immunostaining. Strong CK19 positivity is noted in PTC and follicular carcinoma (FC)[17] (**Table 6**).

A combination of markers consisting of HBME-1, galectin-3, and CK19 is needed:

- To confirm diagnostically challenging cases of PTC.
- To differentiate FC from FA.
- To diagnose malignancy in the indeterminate or gray zone lesion.

The ICC patterns of different thyroid tumors are shown in **Table 7**.

Papillary Thyroid Carcinoma

Papillary thyroid carcinoma usually presents as a painless swelling of the thyroid. It can occur at any age. The cytological diagnosis of PTC is usually straightforward. However, in difficult cases, a panel of ICC marker may help. The key diagnostic ICC markers of PTC are as follows[18,19] (**Box 10**):

- TTF-1: Positive
- HBME-1: Positive
- Galectin-3: Positive
- CK19: Positive

The molecular markers of conventional PTC include:

- *RET*/PTC translocation: 20% cases
- BRAF mutation: 50% cases
- RAS mutation: 15% cases

BOX 10 Papillary thyroid carcinoma (PTC).

- All cases of PTC are positive for thyroglobulin
- Thyroglobulin positivity helps in the diagnosis of metastatic PTC
- Positive markers in PTC: HBME-1, galectin-3, and CK-19 immunostaining
- These markers are unable to distinguish the benign and malignant lesions of thyroid

RET protooncogene is present in the long arm of chromosome 10, and it encodes a cellular tyrosine kinase receptor. Whenever this receptor is activated, a cascade of signal is generated that is ultimately responsible for cell growth, development, and cell proliferation. In *RET*/PTC translocation, there is fusion of 3′ region of *RET* gene with 5′ region of several other genes (*CCD6, NCOA4*) which are named as *RET*/PTC1 or *RET*/PTC3. The *RET*/PTC gene fusion produces a chimeric protein that is able to produce activation of an uncontrolled intracellular tyrosine kinase signaling pathway resulting in the uncontrolled proliferation of the thyroid follicular cells.[20]

Follicular Neoplasms

Follicular adenoma and FC cannot be differentiated based on cytology smear because the histological criteria of capsular invasion and vascular invasion are absent in the cytology smear.

Follicular carcinoma shows the following ICC markers[21] (**Box 11; Figs. 22** and **23**):

- TTF-1: Positive
- HBME-1: Occasional positive
- Galectin-3: Usually negative, occasionally positive

BOX 11 Follicular neoplasms.

Follicular adenoma
- Positive: TTF-1, HBME-1, CK19, and thyroglobulin
- Negative: Galectin-3

Follicular carcinomas
- Positive for HBME-1 and galectin-3 that may help to differentiate them from follicular adenoma

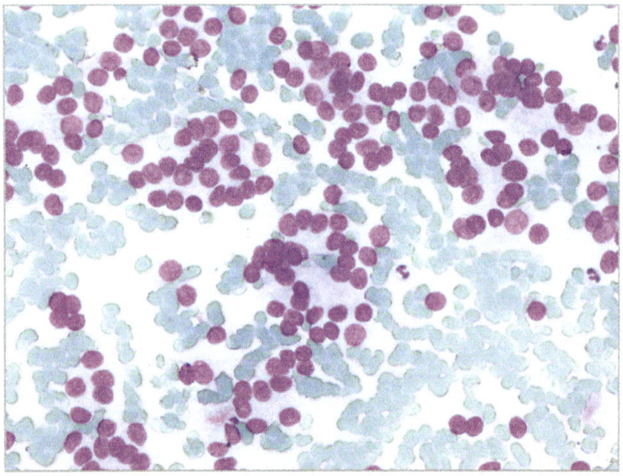

FIG. 22: Follicular carcinoma: Fine needle aspiration cytology smear showing multiple thyroid follicles.

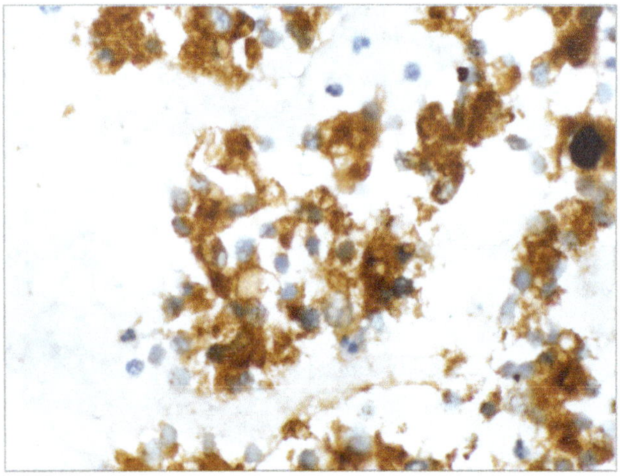

FIG. 23: Follicular carcinoma: The tumor cells are positive for thyroglobulin.

- CK19: Usually negative
- Thyroglobulin: Positive

Based on the ICC pattern, it is difficult to differentiate FA and FC. However, positive HBME-1 and galectin-3 immunostainings are more suggestive of FC.[22]

The molecular markers of FC are:
- *RAS* mutation: 35% of cases
- *RET*/PTC translocation: 0% cases
- *BRAF* mutation: 0% cases

BOX 12 Medullary carcinoma.
- The specific marker of medullary carcinoma is calcitonin
- The tumor is also positive for synaptophysin, chromogranin, and CEA
- TTF-1: may or may not be positive
- The tumor cells are characteristically negative for thyroglobulin

Insular Carcinoma (Poorly Differentiated Thyroid Carcinoma)

The insular carcinoma shows the following ICC markers:[21,22]
- Thyroglobulin: Positive/Negative
- TTF-1: Positive/Negative
- HBME-1: Occasional positive
- Galectin-3: Negative

Anaplastic Carcinoma

The anaplastic carcinoma shows the following ICC markers:
- Thyroglobulin: Negative
- TTF-1: Negative
- HBME-1: Occasional positive
- Galectin-3: Negative, occasional positive

Medullary Carcinoma

Medullary carcinoma (MC) is a tumor of parafollicular C cell. It has characteristic cytomorphology, and for further confirmation, MC needs ICC. The tumor cells are positive for calcitonin. The ICC markers of MC include the following[21-23] (**Box 12; Figs. 24** to **29**):
- Calcitonin: Positive
- Synaptophysin: Positive
- Chromogranin: Positive
- Thyroglobulin: Negative
- TTF-1: Positive
- Galectin-3: Negative
- Carcinoembryonic antigen (CEA): Positive

Calcitonin negativity along with other neuroendocrine markers positivity confirms the diagnosis of MC.

Differential Diagnosis (Table 8)
- Follicular carcinoma
- Paraganglioma
- Metastatic neuroendocrine tumors in the neck.

Parathyroid Tumors

The normal cells of the parathyroid gland are positive for[24] (**Box 13; Figs. 30** and **31**):
- Parathyroid hormone (PTH): Positive
- Chromogranin A: Positive
- GATA-3: Positive

The parathyroid adenoma shows the following ICC pattern:
- PTH: Positive
- Chromogranin A: Positive

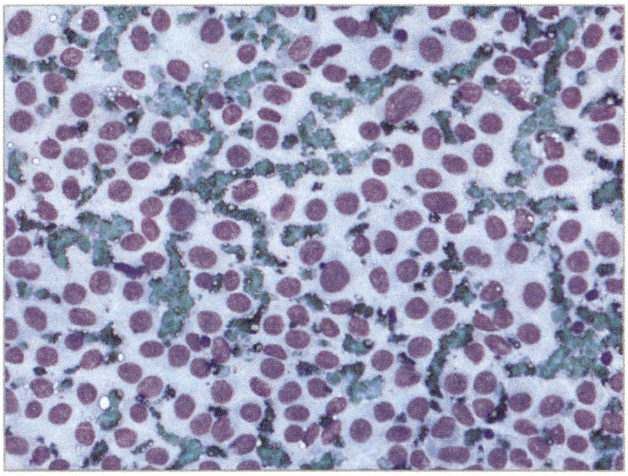

FIG. 24: Medullary carcinoma: Cytology smear showing discrete mildly pleomorphic round cells.

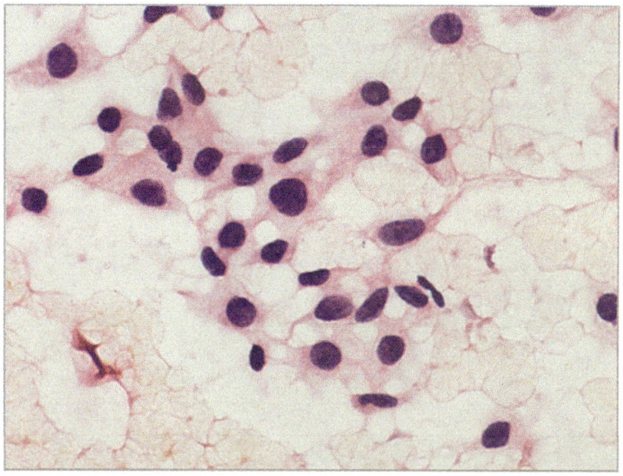

FIG. 25: Medullary carcinoma: Cytology smear showing occasional spindle cells.

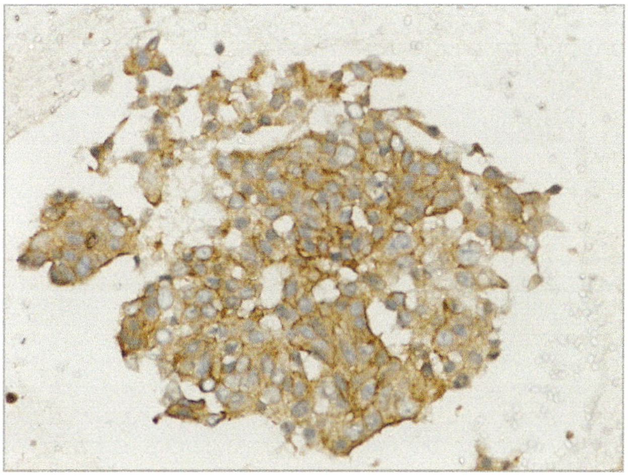

FIG. 26: Medullary carcinoma: Cell block showing calcitonin positive cells.

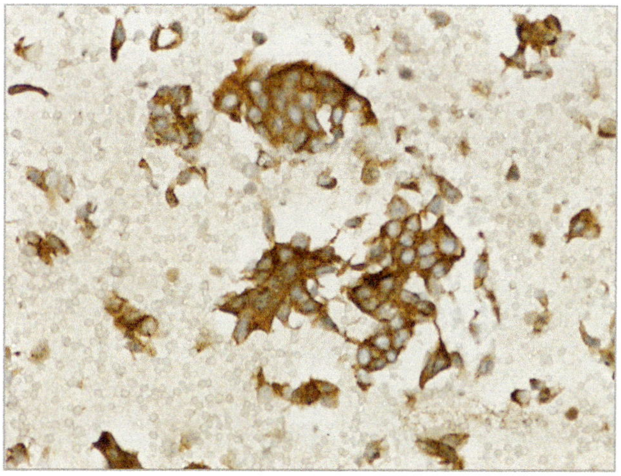

FIG. 27: Medullary carcinoma: Synaptophysin positive cells.

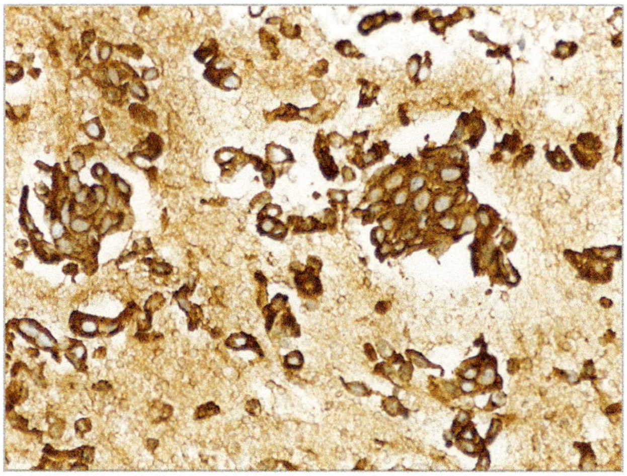

FIG. 28: Medullary carcinoma: Chromogranin positive cells.

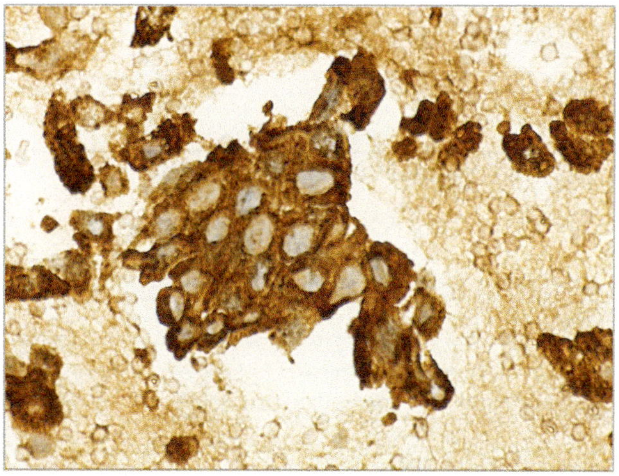

FIG. 29: Medullary carcinoma: Higher magnification of the chromogranin positive cells.

TABLE 8: Differential diagnosis of medullary carcinoma.

Antibodies	MC	FC	Paraganglioma	Metastatic neuroendocrine tumor from lung
TG	–	+	–	–
TTF-1	+	+	–	–
Calcitonin	+	–	–	–
CK	+	+	–	+
Chromogranin	+	–	+	+

+ = Positive, – = Negative

(FC: follicular carcinoma; MC: medullary carcinoma)

BOX 13 Parathyroid tumors.

- The specific marker of parathyroid tumors is parathyroid hormone (PTH)
- Positive: Chromogranin and synaptophysin
- Parathyroid carcinoma shows a high Ki67 index (>5%) that may help in distinguishing it from parathyroid adenoma

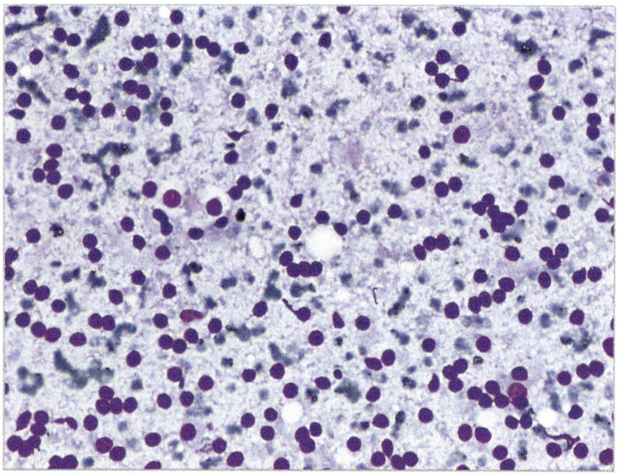

FIG. 30: Parathyroid neoplasm: Cytology smear showing many discrete monomorphic cells with abundant cytoplasm.

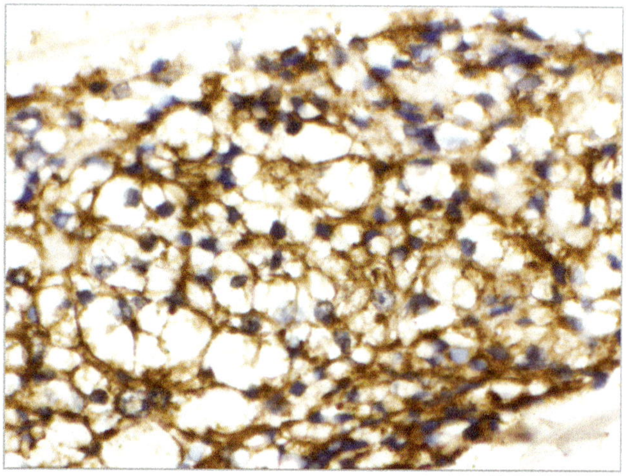

FIG. 31: Parathyroid neoplasm: The cells are positive for parathyroid hormone.

- Galectin-3: Negative
- GATA-3: Positive

Parathyroid carcinomas show the expression of the following markers in ICC:

- PTH: Positive
- Chromogranin A: Positive
- Galectin-3: Positive
- GATA-3: Positive
- Ki67: High index (>5%)

Differentiating markers of parathyroid adenoma versus carcinoma: Galectin-3 positivity and high Ki67 index (>5%) may favor parathyroid carcinoma than adenoma.

Paraganglioma

Paraganglioma of the cervical location mostly develops from the paraganglia of the carotid body. The tumor is composed of predominantly chief cells and sustentacular cells. The paraganglioma shows the following markers on ICC[25] (**Box 14; Figs. 32** and **33**):

- Synaptophysin: Positive
- Chromogranin: Positive
- CD56: Positive
- EMA: Negative
- S-100: Negative

BOX 14 Paraganglioma.

- Positive: Various neuroendocrine markers such as synaptophysin, chromogranin, and CD56
- Negative: Calcitonin and EMA. This is important to differentiate it from medullary carcinoma and metastatic carcinoma respectively

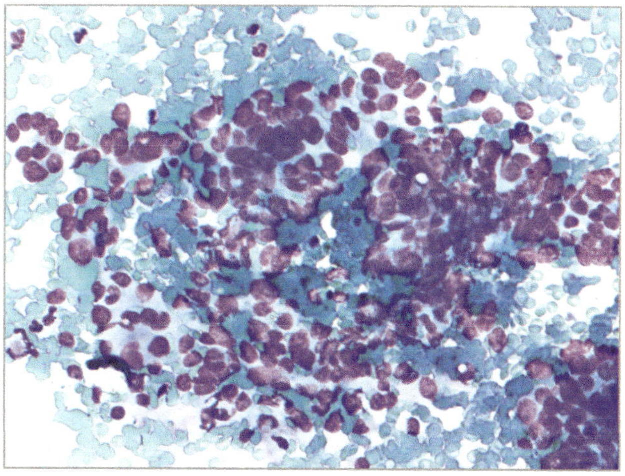

FIG. 32: Paraganglioma: Cytology smear showing loose clusters of round cells.

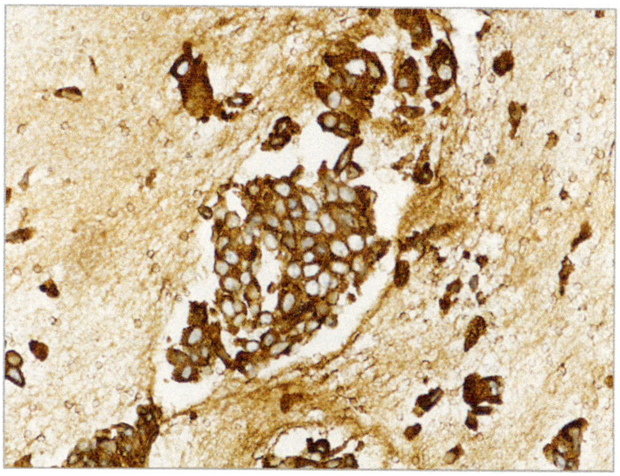

FIG. 33: Paraganglioma: The cells are positive for chromogranin.

TABLE 9: Differentiating the immunocytochemistry pattern of paraganglioma and medullary carcinoma.

Antibodies	Paraganglioma	Medullary carcinoma
Chromogranin	+	+
Synaptophysin	+	Occasionally positive
Calcitonin	–	+
TTF-1	–	+
S-100	+*	–

*Sustentacular cells are positive.

+ = Positive, – = Negative

Differential Diagnosis

Metastatic medullary carcinoma: MCs are positive for calcitonin (**Table 9**).

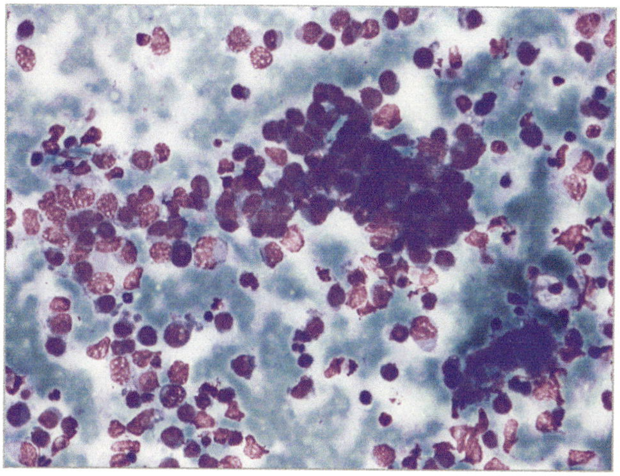

FIG. 34: Olfactory neuroblastoma: Discrete and loose clusters of small round cells.

MALIGNANT ROUND CELL TUMORS IN THE HEAD–NECK REGION

Olfactory Neuroblastoma

Olfactory neuroblastoma (ONB) is one of the small, round blue cell tumors of the head and neck region. It develops from the specialized neuroepithelial cells in the nasal region. The tumor cells are small, round blue cells with an occasional rosette-like arrangement (**Figs. 34** to **37**). Olfactory neuroblastoma shows positive neuroendocrine markers (**Box 15**). The ICC pattern of this tumor is:[26]

- Synaptophysin: Positive
- Chromogranin: Positive
- CD56: Positive
- S-100: Positive

Differential Diagnosis (Table 10)

- Ewing's sarcoma/peripheral neuroectodermal tumor (ES/PNET)
- Rhabdomyosarcoma (RMS)
- Extranodal lymphoma in the nose (NHL)
- Sinonasal undifferentiated carcinoma (UDC)
- Nasopharyngeal carcinoma (NPC)

Meningioma

Rarely extracranial meningioma may be identified in the nasal or orbital cavity. The tumor on cytology smear shows a characteristic whorl appearance, calcification, and psammoma bodies. The cells show abundant cytoplasm with central round nuclei. Intranuclear inclusions are also present. ICC markers of meningioma are as follows[27] (**Box 16**):

- EMA: Positive
- CK7: Positive
- P-CEA: Positive
- Vimentin: Positive
- S-100: Negative

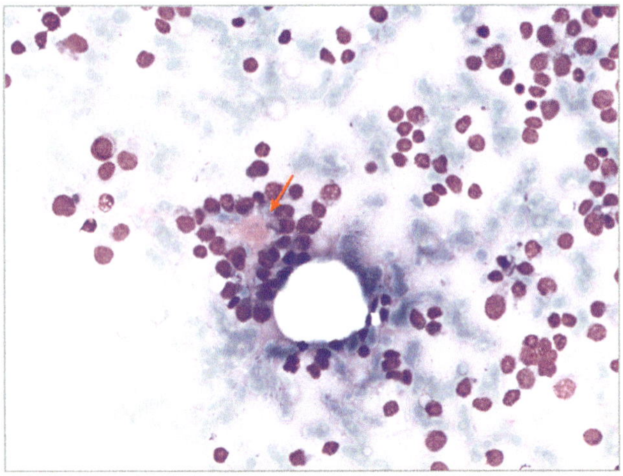

FIG. 35: Olfactory neuroblastoma: Cytology smear showing rosette-like structure (see arrow).

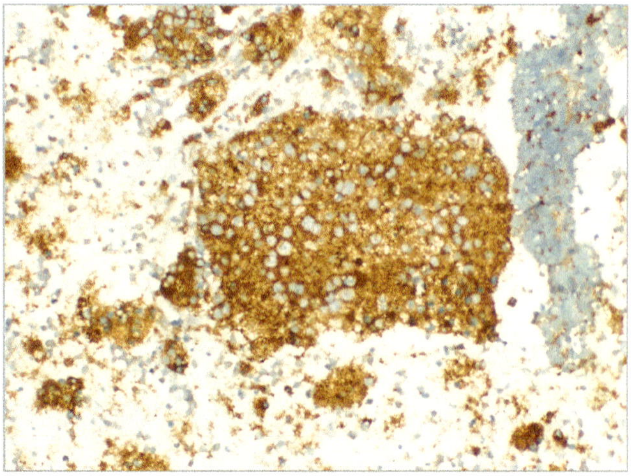

FIG. 36: Olfactory neuroblastoma: The cells showing chromogranin positivity.

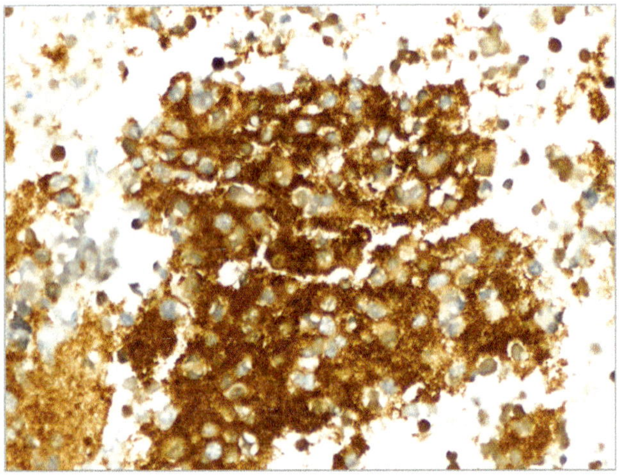

FIG. 37: Olfactory neuroblastoma: The cells showing synaptophysin positivity.

BOX 15 Olfactory neuroblastoma.

- Olfactory neuroblastoma is a malignant round cell tumor
- It is positive for the various neuroendocrine markers: Synaptophysin, chromogranin, neuron-specific enolase, and CD56
- The tumor cells may be positive for calretinin
- Rarely the cells show focal CK positivity

BOX 16 Meningioma.

- Meningioma may be ectopic or may have direct extension with the cranial cavity
- Positive: Vimentin and many epithelial markers such as EMA, CK7, and CAM5

TABLE 10: Differential diagnosis of olfactory neuroblastoma (ONB).

Antibodies	ONB	RMS	ES/PNET	NHL	UDC	NPC
CD45	–	–	–	+	–	–
CD56	+	–	–	–	–	–
CD99	–	–	+	–	–	–
CHG	+	–	–	–	–	–
MyoD1	–	+	–	–	–	–
Myogenin	–	+	–	–	–	–
Desmin	–	+	–	–	–	–
EMA	–	–	–	–	±	+
CK7	–	–	–	–	±	–
S-100	+	–	+	–	–	–

– = Negative, + = Positive

(CHG: chromogranin; ES/PNET: Ewing's sarcoma/peripheral neuroectodermal tumor; NHL: extranodal lymphoma in the nose; NPC: nasopharyngeal carcinoma; RMS: rhabdomyosarcoma; UDC: sinonasal undifferentiated carcinoma)

REFERENCES

1. Nagao T, Sato E, Inoue R, Oshiro H, Takahashi R, Nagai T, et al. Immunohistochemical analysis of salivary gland tumors: application for surgical pathology practice. Acta Histochem Cytochem. 2012;45(5):269-82.
2. Savera AT, Gown AM, Zarbo RJ. Immunolocalization of three novel smooth muscle-specific proteins in salivary gland pleomorphic adenoma: assessment of the morphogenetic role of myoepithelium. Mod Pathol. 1997;10(11):1093-100.
3. Zhou CX, Gao Y. Oncocytoma of the salivary glands: a clinicopathologic and immunohistochemical study. Oral Oncol. 2009;45(12):e232-8.
4. Alós L, Cardesa A, Bombí JA, Mallofré C, Cuchi A, Traserra J. Myoepithelial tumors of salivary glands: a clinicopathologic, immunohistochemical, ultrastructural, and flow-cytometric study. Semin Diagn Pathol. 1996;13(2):138-47.
5. Nagao T, Sugano I, Ishida Y, Tajima Y, Matsuzaki O, Konno A, Kondo Y, Nagao K. Salivary gland malignant myoepithelioma: a clinicopathologic and immunohistochemical study of ten cases. Cancer. 1998;83(7):1292-9
6. Seethala RR, Barnes EL, Hunt JL. Epithelial-myoepithelial carcinoma: a review of the clinicopathologic spectrum and immunophenotypic characteristics in 61 tumors of the salivary glands and upper aerodigestive tract. Am J Surg Pathol. 2007;31(1):44-57. [Erratum in: Am J Surg Pathol. 2008;32(12):1923.]
7. Nagao T, Gaffey TA, Serizawa H, Sugano I, Ishida Y, Yamazaki K, et al. Dedifferentiated adenoid cystic carcinoma: a clinicopathologic study of 6 cases. Mod Pathol. 2003;16(12):1265-72.
8. Limite G, Di Micco R, Esposito E, Sollazzo V, Cervotti M, Pettinato G, et al. Acinic cell carcinoma of the breast: review of the literature. Int J Surg. 2014;12:S35-9.
9. Coca-Pelaz A, Rodrigo JP, Triantafyllou A, Hunt JL, Rinaldo A, Strojan P, et al. Salivary mucoepidermoid carcinoma revisited. Eur Arch Otorhinolaryngol. 2015;272(4):799-819.
10. Darling MR, Schneider JW, Phillips VM. Polymorphous low-grade adenocarcinoma and adenoid cystic carcinoma: a review and comparison of immunohistochemical markers. Oral Oncol. 200238(7):641-5.
11. Jalaly JB, Sanati S, Chernock RD, Dibe DG, El-Mofty SK. Salivary duct carcinoma and invasive ductal carcinoma of the breast: A comparative immunohistochemical study. Head Neck Pathol. 2018;12(4):488-92.
12. Khurram SA, Sultan-Khan J, Atkey N, Speight PM. Cytogenetic and immunohistochemical characterization of mammary analogue secretory carcinoma of salivary glands. Oral Surg Oral Med Oral Pathol Oral Radiol. 2016;122(6):731-42.
13. Ni H, Zhang XP, Wang XT, Xia QY, Lv JH, Wang X, et al. Extended immunologic and genetic lineage of mammary analogue secretory carcinoma of salivary glands. Hum Pathol. 2016;58:97-104.
14. Palo S, Biligi DS. Differential diagnostic significance of HBME-1, CK19 and S100 in various thyroid lesions. Malays J Pathol. 2017;39(1):55-67.
15. Bartolazzi A, Gasbarri A, Papotti M, Bussolati G, Lucante T, Khan A, et al. Application of an immunodiagnostic method for improving preoperative diagnosis of nodular thyroid lesions. Lancet. 2001;357(9269):1644-50.
16. Saggiorato E, Aversa S, Deandreis D, Arecco F, Mussa A, Puligheddu B, et al. Galectin-3: presurgical marker of thyroid follicular epithelial cell-derived carcinomas. J Endocrinol Invest. 2004;27(4):311-7.
17. Dunđerovic D, Lipkovski JM, Boricic I, Soldatovic I, Bozic V, Cvejic D, et al. Defining the value of CD56, CK19, Galectin 3 and HBME-1 in diagnosis of follicular cell derived lesions of thyroid with systematic review of literature. Diagn Pathol. 2015;10:196.
18. Zehani A, Bani A, Chelly I, Rejaibi S, Besbes G, Haouet S, et al. Papillary thyroid tumors: Diagnostic value of CD56 and cytokeratin 19. Tunis Med. 2018;96(2):113-21.
19. Cochand-Priollet B, Dahan H, Laloi-Michelin M, Polivka M, Saada M, Herman P, et al. Immunocytochemistry with cytokeratin 19 and anti-human mesothelial cell antibody (HBME1) increases the diagnostic accuracy of thyroid fine-needle aspirations: preliminary report of 150 liquid-based fine-needle aspirations with histological control. Thyroid. 2011;21(10):1067-73.
20. Romei C, Ciampi R, Elisei R. A comprehensive overview of the role of the RET proto-oncogene in thyroid carcinoma. Nat Rev Endocrinol. 2016;12(4):192-202.
21. Liu H, Lin F. Application of immunohistochemistry in thyroid pathology. Arch Pathol Lab Med. 2015;139(1):67-82.
22. Sack MJ, Astengo-Osuna C, Lin BT, Battifora H, LiVolsi VA. HBME-1 immunostaining in thyroid fine-needle aspirations: a useful marker in the diagnosis of carcinoma. Mod Pathol. 1997;10(7):668-74.
23. Bhanot P, Yang J, Schnadig VJ, Logroño R. Role of FNA cytology and immunochemistry in the diagnosis and management of medullary thyroid carcinoma: report of six cases and review of the literature. Diagn Cytopathol. 2007;35(5):285-92.
24. Erickson LA, Mete O. Immunohistochemistry in diagnostic parathyroid pathology. Endocr Pathol. 2018;29(2):113-29.
25. Cheung VKY, Gill AJ, Chou A. Old, new, and emerging immunohistochemical markers in pheochromocytoma and paraganglioma. Endocr Pathol. 2018;29(2):169-75.
26. Faragalla H, Weinreb I. Olfactory neuroblastoma: a review and update. Adv Anat Pathol. 2009;16(5):322-31.
27. Baxter DS, Orrego A, Rosenfeld JV, Mathiesen T. An audit of immunohistochemical marker patterns in meningioma. J Clin Neurosci. 2014;21(3):421-6.

CHAPTER 10

Immunocytochemistry in Lung Tumors

INTRODUCTION

The recent classification of carcinoma of the lung by the World Health Organization (WHO) has emphasized to classify the group of carcinomas labeled as "non-small cell carcinoma" only.[1] Due to the considerable progress in the therapy of lung carcinomas, it is now essential to identify each of the members of "non small cell carcinoma" of the lung.

Flowchart 1 highlights the diagnostic markers used in the diagnosis of lung carcinoma.

SQUAMOUS CELL CARCINOMA

The cytological features of squamous cell carcinoma (SQCC) have characteristic features. The cells are polyhedral having central hyperchromatic nuclei along with typical orangeophilic cytoplasm in Papanicolaou's stained smear. However, immunocytochemistry (ICC) may help significantly in poorly differentiated SQCC.

The characteristic ICC pattern of SQCC includes the following (**Box 1; Figs. 1** to **7**):

- *Positive*
  - p40
  - p63
  - CK5/6
- *Negative*
  - TTF-1
  - CK7
  - CK20

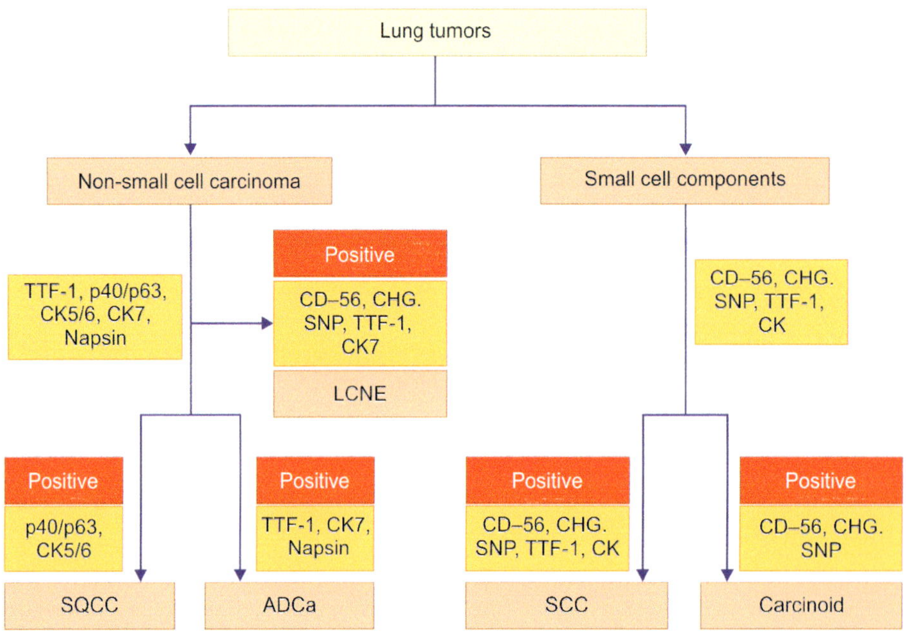

(SQCC: squamous cell carcinoma; ADCa: adenocarcinoma; SCC: small cell carcinoma; LCNE: large cell neuroendocrine carcinoma; CHG: chromogranin; SNP: synaptophysin)

FLOWCHART 1: Immunocytochemistry of lung carcinomas.

BOX 1 Squamous cell carcinoma.

- Cytology smears show polyhedral cells with central hyperchromatic nuclei
- Orangeophilia of the cytoplasm in Papanicolaou's stained smear indicates intracellular keratin
- Characteristic immunocytochemistry:
  - *Positive*: p40/p63 and CK5/6
  - *Negative*: TTF-1, Napsin, CK7, CK20

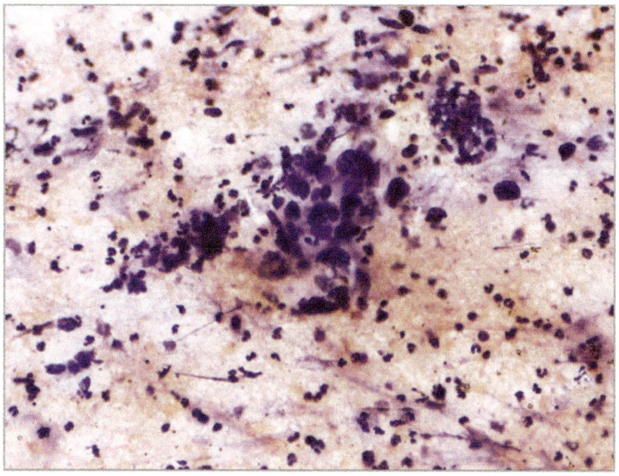

FIG. 1: Squamous cell carcinoma of lung: Cytology smear showing abundant clusters and discrete cells with a moderate amount of cytoplasm and enlarged pleomorphic nuclei.

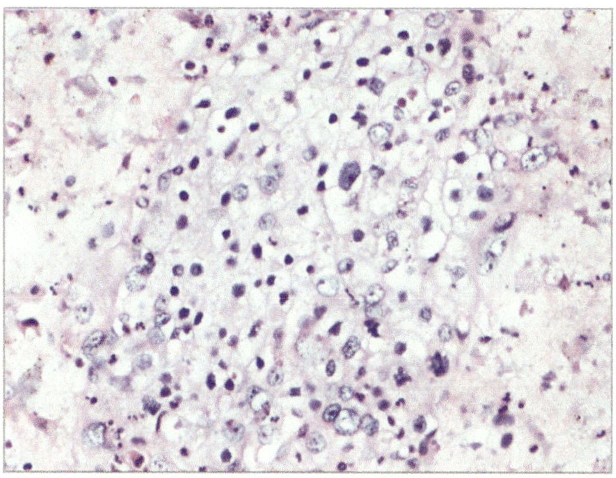

FIG. 2: Squamous cell carcinoma of lung: Cell block of the same showing polyhedral cells having enlarged nuclei.

Out of these markers, p40 is 98% specific and 100% sensitive for SQCC of the lung whereas p63 is only 60% specific and 100% sensitive. Therefore, it is preferable to use p40 for the detection of lung carcinomas.[2]

CK5/6 is the second line of choice for SQCC of the lung. It is 97% sensitive and 90% specific.[3]

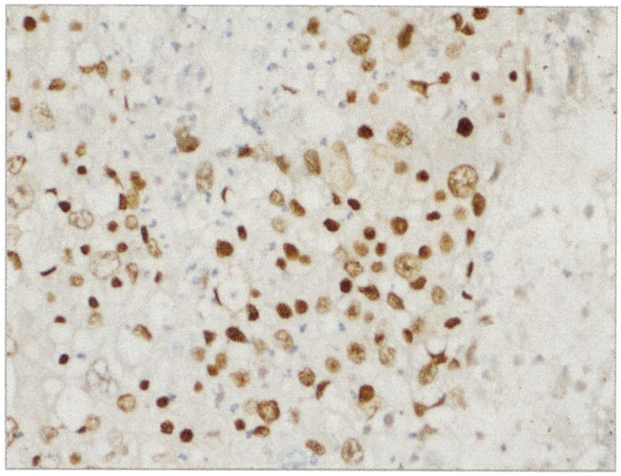

FIG. 3: Squamous cell carcinoma of lung: Nuclear positivity of p40 immunostaining.

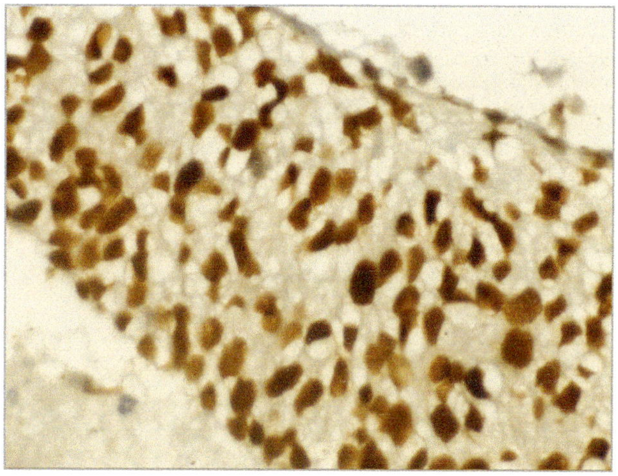

FIG. 4: Squamous cell carcinoma of lung: Nuclear positivity of p40 immunostaining in higher magnification.

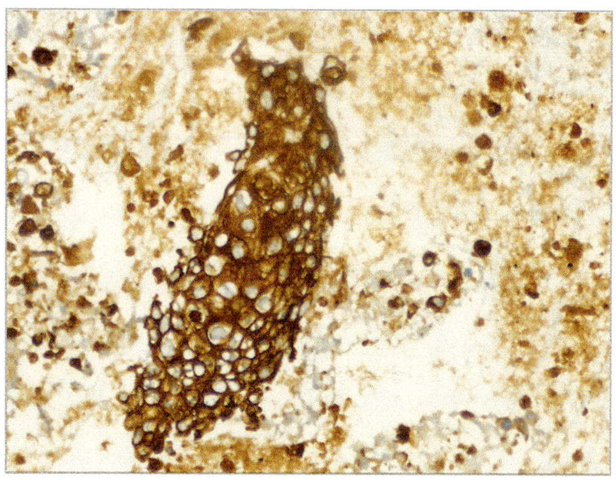

FIG. 5: Squamous cell carcinoma of lung: Membranous positivity of CK5/6.

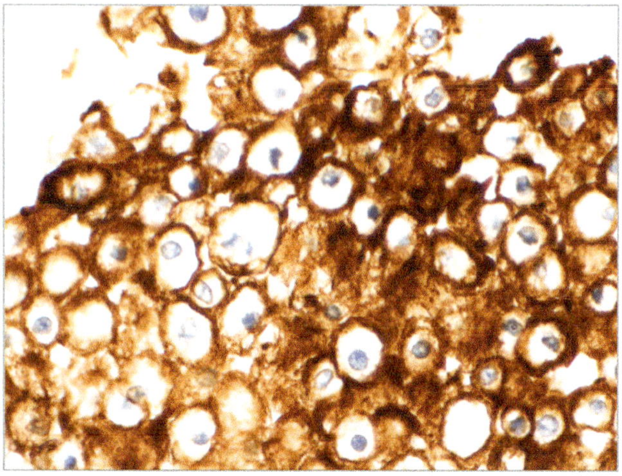

FIG. 6: Squamous cell carcinoma of lung: Membranous positivity of CK5/6 in higher magnification.

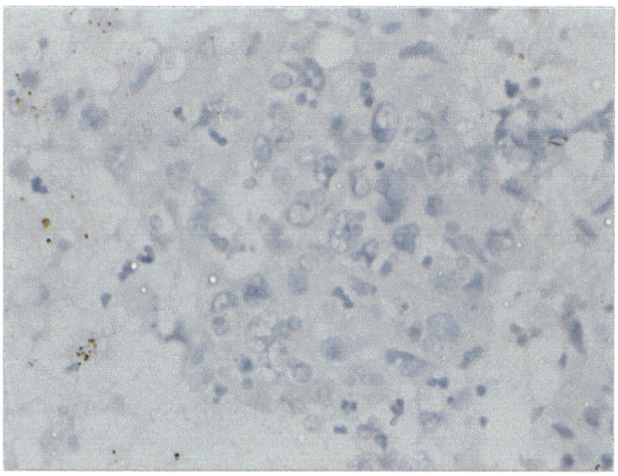

FIG. 7: Squamous cell carcinoma of lung: Negative for TTF-1.

BOX 2 Adenocarcinoma.

- Cytology
  - Glandular arrangement
  - The individual cells: Moderate amount of cytoplasm
  - Cytoplasmic vacuoles
  - Enlarged nuclei and prominent nucleoli
- Immunocytochemistry
  - *Positive*: TTF-1,CK7, Napsin
  - *Negative*: p63/p40, CK5/6, CK20
    - TTF-1 is a good marker
    - Napsin-positive other carcinomas: Renal cell carcinoma, pancreatic adenocarcinoma, gastric carcinoma, and ovarian carcinomas
    - CK7 alone should not be considered to diagnose adenocarcinoma of lung

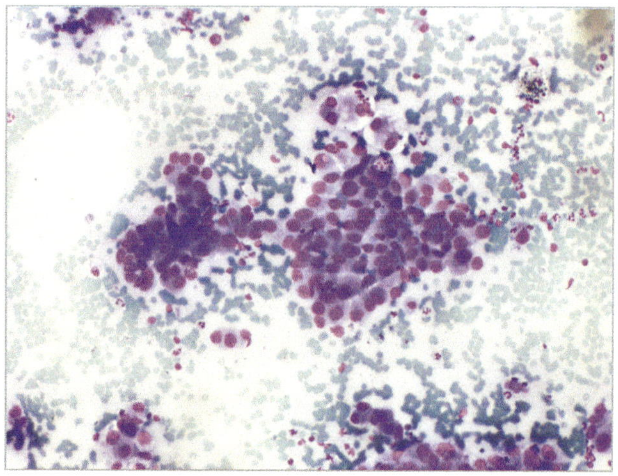

FIG. 8: Adenocarcinoma of lung: Fine needle aspiration cytology smear showing multiple small clusters of cells.

ADENOCARCINOMA

The cytology smears of adenocarcinoma of lung show many gland-like structures. The individual cells have a moderate amount of vacuolated cytoplasm. The nuclei are enlarged with single to multiple prominent nucleoli.

The ICC pattern of adenocarcinoma is as follows (**Box 2; Figs. 8** to **14**):

- *Positive*
  - TTF-1
  - CK7
  - Napsin
- *Negative*
  - p40 or p63
  - CK5/6
  - CK20

TTF-1 is the characteristic marker of adenocarcinoma of the lung. There are two clones of TTF-1: SPT24 and 8G7G3/1. SPT24 clone has high sensitivity and is positive in

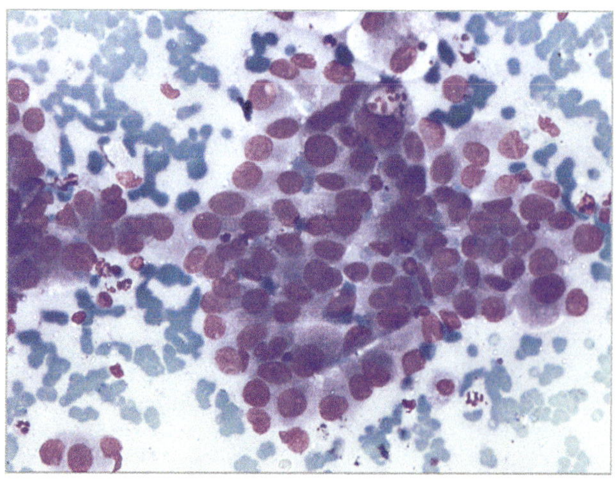

FIG. 9: Adenocarcinoma of lung: Fine needle aspiration cytology smear showing higher magnification of the cells with a moderate amount of cytoplasm and enlarged pleomorphic nuclei.

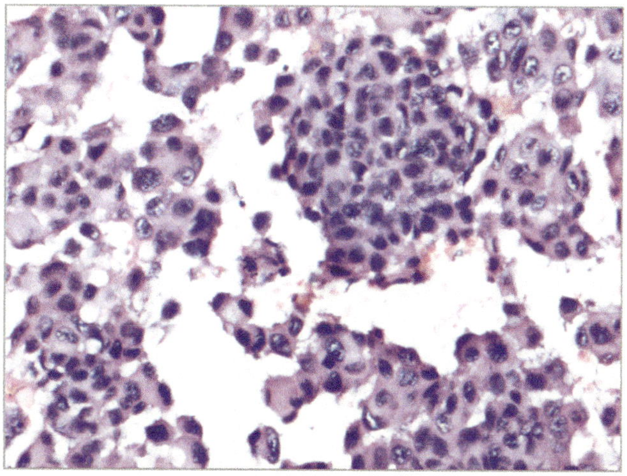

FIG. 10: Adenocarcinoma of lung: Cell block section showing cell with enlarged nuclei having prominent nucleoli.

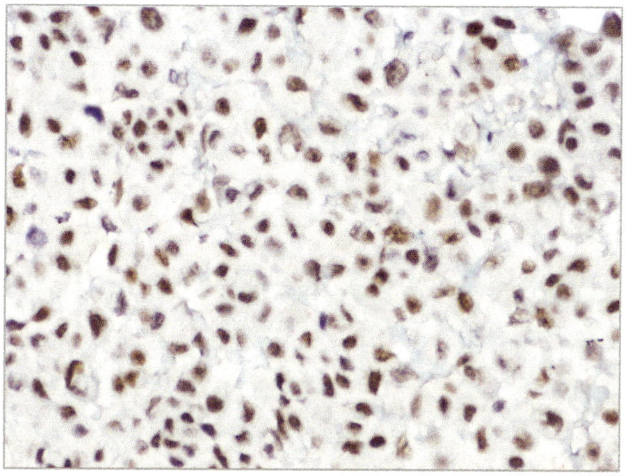

FIG. 11: Adenocarcinoma of lung: Nuclear positivity of TTF-1.

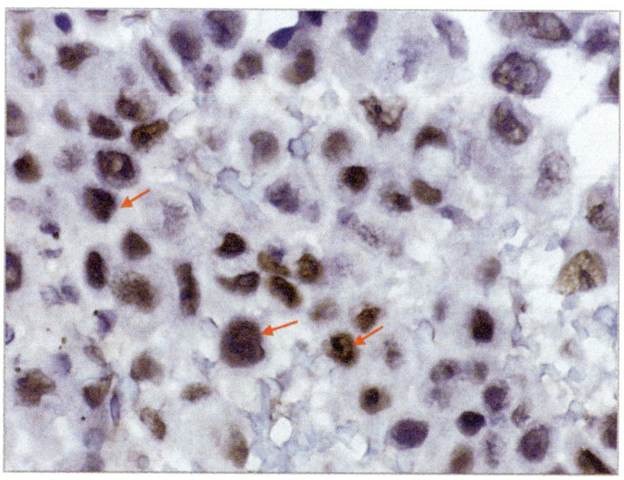

FIG. 12: Adenocarcinoma of lung: Nuclear positivity of TTF-1 in higher magnification.

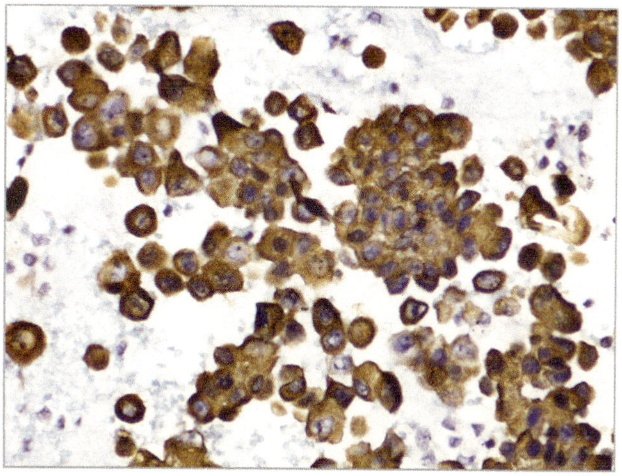

FIG. 13: Adenocarcinoma of lung: Membranous positivity of CK7.

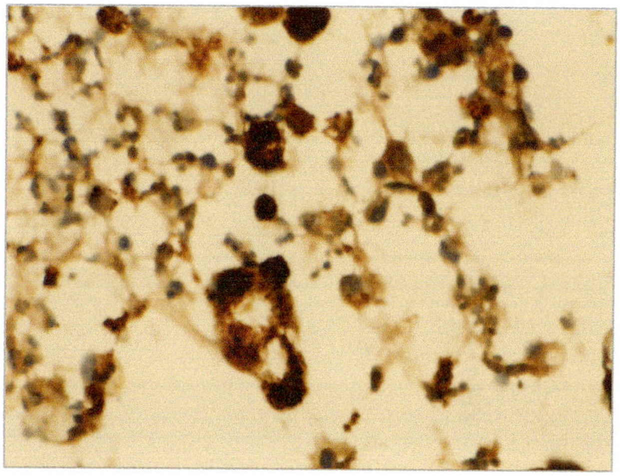

FIG. 14: Adenocarcinoma of lung: Positive for napsin.

the majority of the adenocarcinomas of the lung. However, its specificity is relatively low. In comparison, TTF-1 from 8G7G3/1 clone is highly specific (99%) with relatively less sensitivity (77%).[4,5]

Napsin is also an excellent diagnostic marker of adenocarcinoma of the lung with high sensitivity (87%) and specificity (96%). SQCC of the lung rarely shows napsin positivity (2%). It is important to note that various other carcinomas are also positive for napsin. These include renal cell carcinoma, pancreatic adenocarcinoma, gastric carcinoma, and ovarian carcinomas.[6,7]

One should not depend on CK7 positivity alone to detect adenocarcinoma of the lung as this marker is also positive in the ovary, breast, and cervical carcinomas. CK7 should be used along with other markers of lung adenocarcinoma.

The distinguishing ICC markers of SQCC and adenocarcinoma are highlighted in **Table 1**.

TABLE 1: Distinguishing immunocytochemistry markers of squamous cell carcinoma and adenocarcinoma.

Markers	Squamous cell carcinoma	Adenocarcinoma
TTF-1	–	+
p40	+	–
p63	+	–
CK5/6	+	–
CK7	–	+
Napsin	–	+

– = Negative, + = Positive.

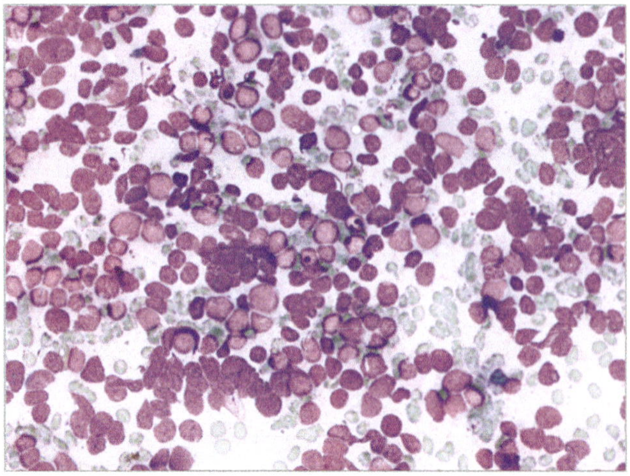

FIG. 15: Small cell carcinoma of lung: Cytology smear showing discrete round-to-oval cells.

Differential Diagnosis

Metastatic adenocarcinoma: It is discussed later in the chapter.

NEUROENDOCRINE TUMORS OF THE LUNG

The WHO classified neuroendocrine tumors of the lung as follows:[1]

- Small cell carcinoma (SCC)
- Large cell neuroendocrine carcinoma
- Carcinoid tumors
- Typical carcinoid
- Atypical carcinoid

Small Cell Carcinoma

Small cell carcinoma (SCC) has the characteristic cytology features in the cytology smear. The individual cells are small, with scanty cytoplasm. The nuclei are round and hyperchromatic with frequent molding. Crushing artifact is also seen.

The ICC pattern of this tumor shows[8] (**Box 3**; **Figs. 15** to **17**):

- *Positive*
  - CD56
  - Synaptophysin
  - Chromogranin: occasionally positive
  - TTF-1
- *Negative*
  - CK7
  - CK5/6

The differentiating ICC markers of SCC versus non-SCC of lung are shown in **Table 2**.

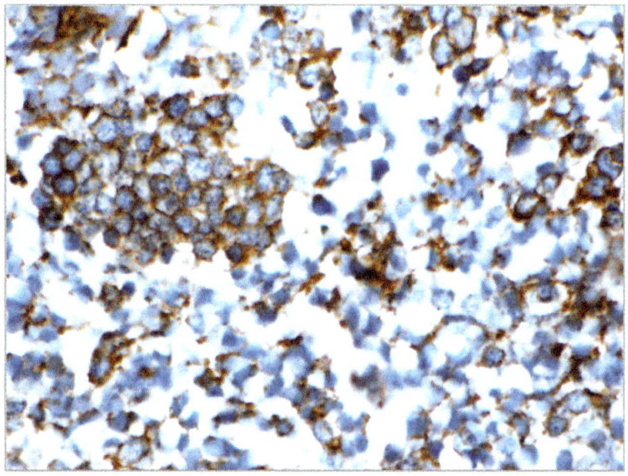

FIG. 16: Small cell carcinoma of lung: Positive for CD56.

BOX 3 Small cell carcinoma.

- Cytology
  - Small, round cells
  - Nuclei round and hyperchromatic
  - Frequent nuclear molding
- Immunocytochemistry
  - *Positive*: CD56, synaptophysin, chromogranin, TTF-1
  - *Negative*: CK7, p63/p40

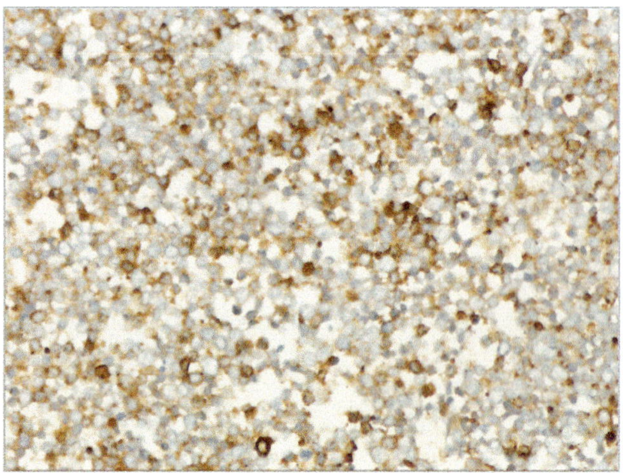

FIG. 17: Small cell carcinoma of lung: Positive for chromogranin.

TABLE 2: Distinguishing immunocytochemistry pattern of small cell versus non-small cell carcinoma of the lung.

Markers	Small cell carcinoma	Non-small cell carcinoma
TTF-1	Negative	Mostly positive except squamous cell carcinoma
CD56	Positive	Negative
p63	Negative	Positive*
Synaptophysin	Positive	Negative

*Positive in squamous cell carcinoma; however, negative in adenocarcinoma.

Large Cell Neuroendocrine Carcinoma

The ICC markers of large cell neuroendocrine carcinoma are as follows:

- *Positive*
    - CD56
    - Synaptophysin
    - Chromogranin: Negative, occasionally positive
    - TTF-1: Negative and sometimes positive
    - CK7: Positive
- *Negative*
    - CK20
    - p63

Ki67: High index

Carcinoid

The ICC markers of carcinoid are as follows:

- *Positive*
    - CD56
    - Synaptophysin
    - Chromogranin
- *Negative*
    - TTF-1
    - CK7
    - CK20

Ki67: Low index

Pulmonary Blastoma

The ICC markers of pulmonary blastoma are given in the following text.

Fetal adenocarcinoma components show:

- Epithelial membrane antigen (EMA): Positive
- Carcinoembryonic antigen (CEA): Positive
- TTF-1: Positive

Stromal cells show:

- Vimentin: Positive
- Desmin: Positive

Cartilage cells show:

- S100: Positive

Adenosquamous Carcinoma

Adenosquamous carcinoma consists of both squamous and adenocarcinoma. Therefore, the ICC shows positivity for both the SQCC and the adenocarcinoma markers.

The ICC pattern is:[9]

- TTF-1: Positive
- Napsin: Positive
- p40: Positive
- p63: Positive

Carcinosarcoma

Carcinosarcoma is an aggressive tumor that contains both carcinoma and sarcoma components **(Figs. 18** to **20)**:

- PanCK: Positive
- EMA: Positive
- Vimentin: Positive
- Desmin: Positive

Metastatic Carcinoma

A large number of adenocarcinomas may metastasize in the lung (**Figs. 21** to **26**). **Table 3** highlights the differentiating features of these tumors.

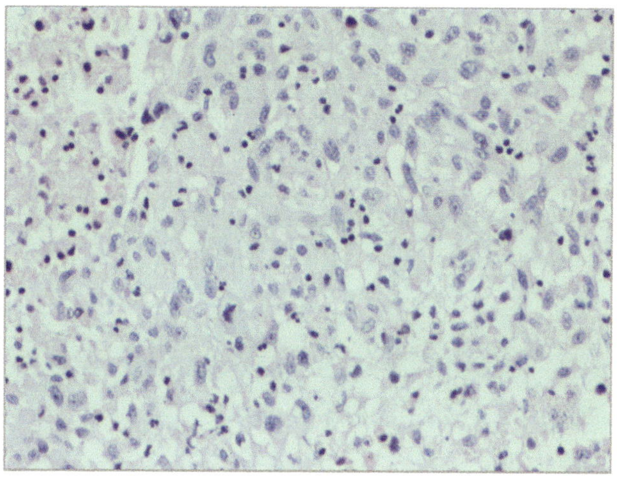

FIG. 18: Carcinosarcoma: Cell block section showing oval-to-elongated cells in sheets.

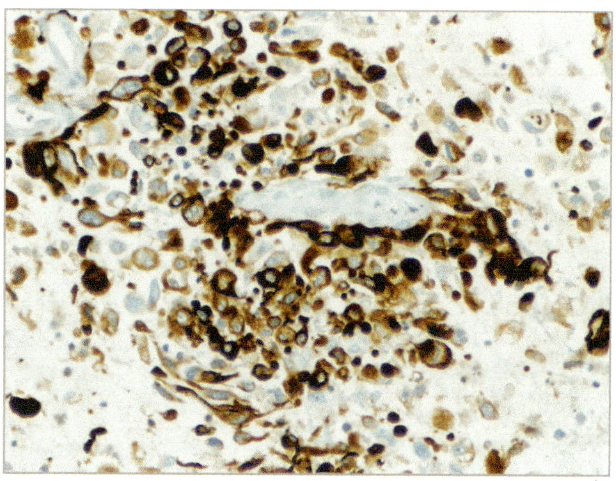

FIG. 19: Carcinosarcoma: Positive for PanCK.

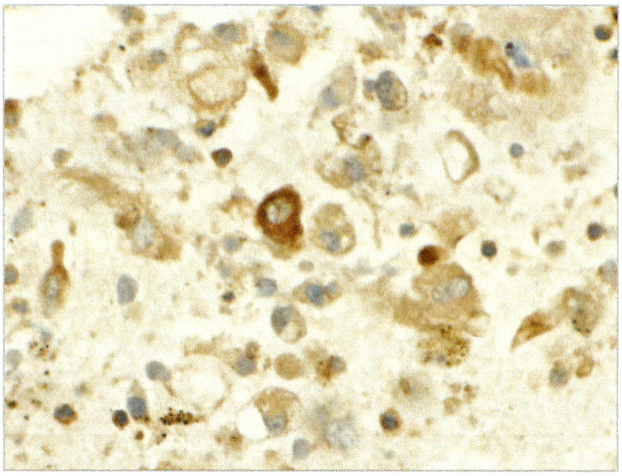

FIG. 20: Carcinosarcoma: Positive for desmin.

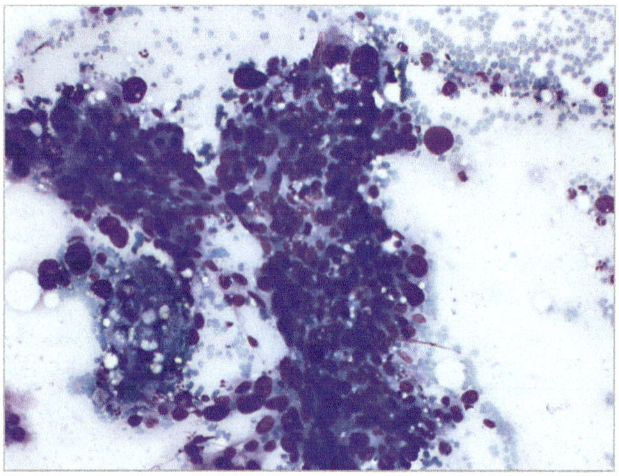

FIG. 21: Metastatic carcinoma from cervix: Cytology smear showing clusters of moderately pleomorphic cells.

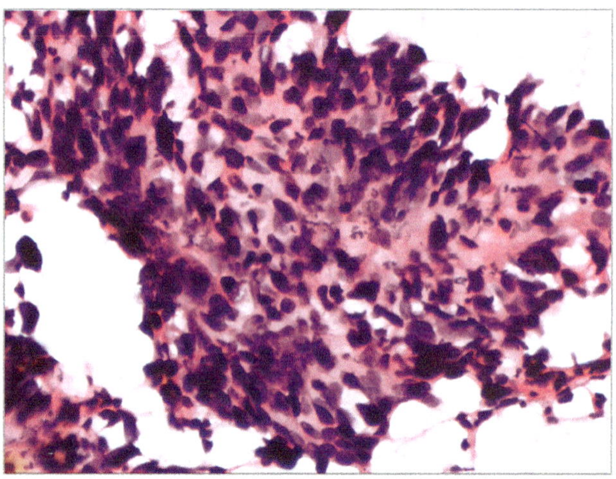

FIG. 22: Metastatic carcinoma from cervix: Hematoxylin and eosin-stained smears showing oval-to-elongated tumor cells.

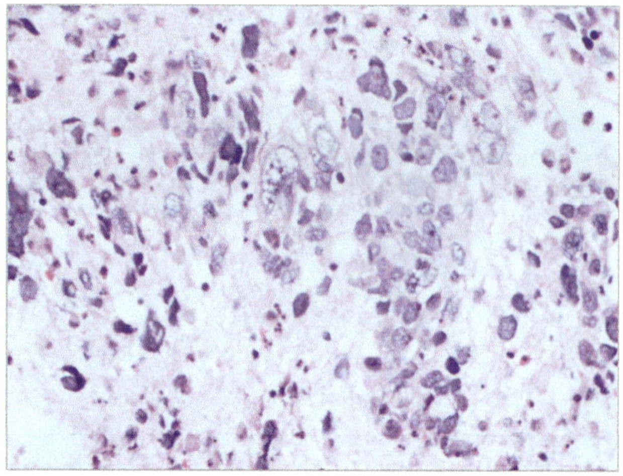

FIG. 23: Metastatic carcinoma from cervix: Cell block section showing moderately pleomorphic cells in sheets.

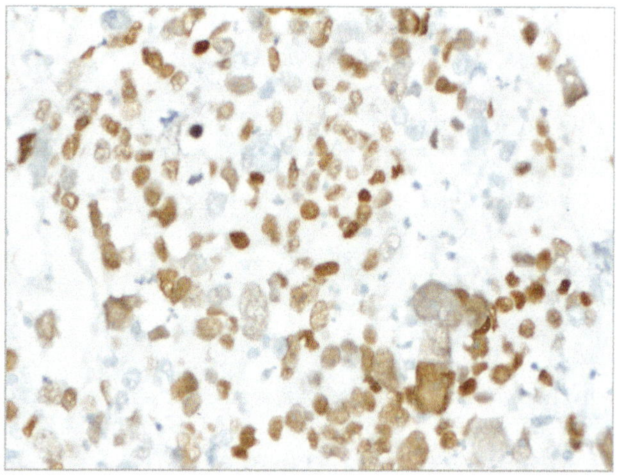

FIG. 24: Metastatic carcinoma from cervix: Positive p63 immunostaining.

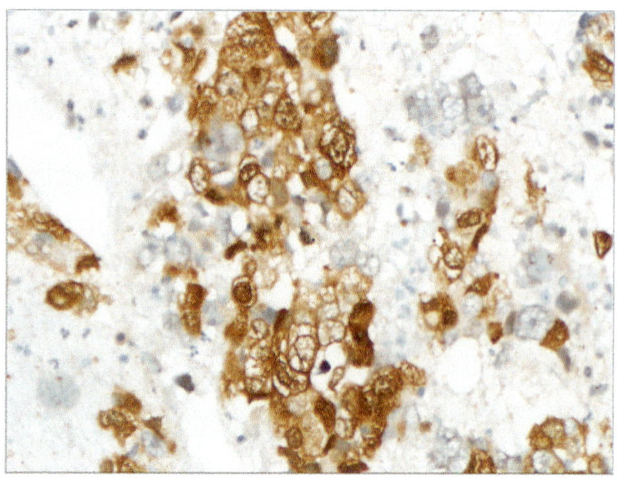

FIG. 25: Metastatic carcinoma from cervix: Positive for p16.

CHAPTER 11

Immunocytochemistry of Lymph Node

INTRODUCTION

A brief overview of the World Health Organization (WHO) classification of lymphoma is shown in **Flowchart 1**.[1,2]

The WHO emphasized three significant findings for classifications of lymphomas:

1. Cytomorphology of the lymphoma cells
2. Immunophenotyping
3. Molecular genetics

Immunocytochemistry (ICC) is often helpful for immunophenotyping of lymphoma cases.

The various markers are used to classify the lymphomas on cell block sections.

Table 1 shows the common markers used in this purpose.

MARKERS THAT ARE POSITIVE IN NONLYMPHORETICULAR NEOPLASMS

Certain markers of lymphoreticular tumor are also positive in nonlymphoreticular neoplasms. These markers are highlighted in **Table 2**.

Table 3 shows the markers to distinguish mature B- and T-cell non-Hodgkin lymphoma (NHL).

Subtyping of B-cell Lymphoma

The immunophenotyping and essential genetic markers of B-NHL have been highlighted in **Table 3**.

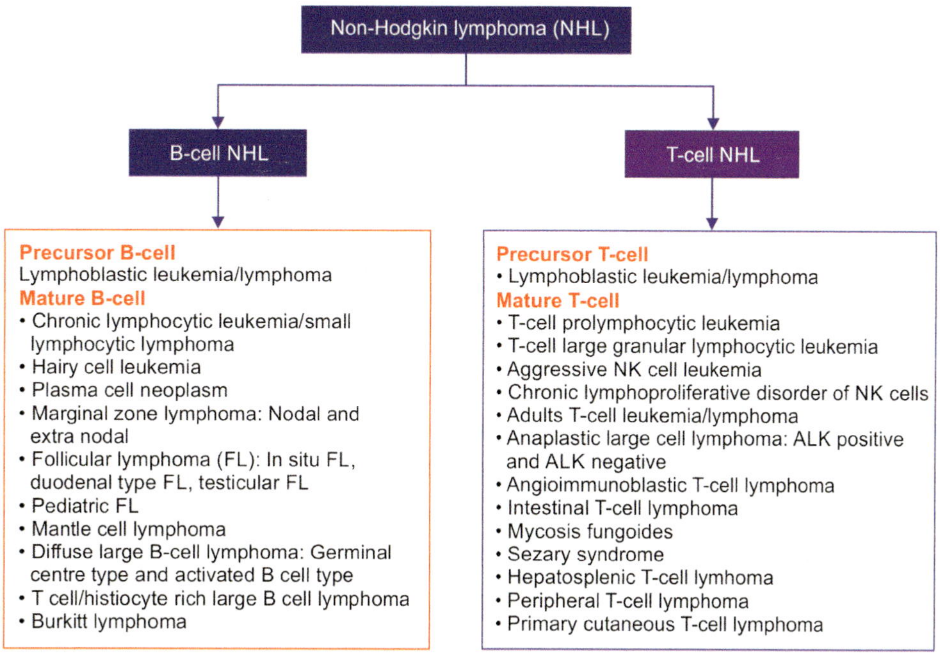

(NK: natural killer; ALK: anaplastic lymphoma kinase)

FLOWCHART 1: Classification of lymphoma by the World Health Organization.

TABLE 1: Antibodies used in lymphoma cases.

Marker	Basic function	Main diagnostic application in lymphomas	Used in other tumors	Localization	Further comments
Basic markers of lymphoma					
CD45	This is a very basic marker of all hematopoietic and lymphoid neoplasms	Lymphoma and leukemia	Granulocytic sarcoma, interdigitating dendritic cell sarcoma, giant cell tumor of tendon sheath	M	It is a specific marker for lymphoid and hematopoietic cells
TdT	It is a DNA polymerase enzyme. It is expressed in immature T and B cells	Lymphoblastic lymphoma: B and T cells	AML, CML and Merkel cell carcinoma	N	It is a DNA polymerase enzyme and specific for precursor T- and B-cell NHL
CD10	It is a cell membrane metalloproteinase related to Zn and is expressed in both lymphomas and other nonlymphoid tumors	Follicular lymphoma, Burkitt's lymphoma, DLBCL, angioimmunoblastic T-cell lymphoma, plasma cell neoplasm	Renal cell carcinoma, hepatocellular carcinoma, mesothelioma, rhabdomyosarcoma, urothelial cell carcinoma, prostatic carcinoma	M	CD10 should be used with other panels of markers as it is expressed in a large variety of carcinomas also
B cell-related markers					
CD19	Pan-B-cell marker. It develops after HLA-DR	All B-cell NHL	AML-M_0	M	It is an excellent B-cell marker
C20	Expressed in late pre-B-cell after CD19. It is a useful marker of B-cell lineage	All B-cell NHL		M	CD20 is a relatively less-specific B-cell marker and at times not expressed in B cells. It is also occasionally positive in peripheral T-lymphoma
PAX5	It is a transcription factor and is expressed in early B cells and mature B cell. It is related to B-cell differentiation. The gene of *PAX5* is in 9p13 chromosome	B-NHL, Reed–Sternberg cell	Merkel cell carcinoma, atypical carcinoids, small cell carcinoma of the lung, Wilms' tumor, neuroblastoma	N	It is often expressed in L&H cells of NLPHL
CD79a	Related to the extracellular domain of Ig receptor complex of B cell. It is positive in the differentiating B cells and in plasma cells	B-NHL, cases of T lymphoblastic lymphoma, AML	AML-M3	M	It is not specific for B cell as it is expressed in T-ALL also
CD23	Certain types of B cell express this marker. It is the Fc receptor of IgE	B-CLL, DLBCL, mediastinal large B-cell lymphoma, hairy cell leukemia		M	Follicular dendritic cells also express this marker. CD23 is often used to discriminate CLL from mantle cell lymphoma
SOX-11	It is a transcription factor and related to the development of CNS	Mantle cell lymphoma	AML-M3 and ovarian carcinoma	N	
Cyclin D1	It is a cell cycle-related protein and controls G1 and S phase of the cell cycle	Mantle cell lymphoma, hairy cell leukemia, occasional cases of DLBCL	Inflammatory pseudotumor, parathyroid adenoma, breast carcinoma, prostatic carcinoma, urothelial cell carcinoma	C	
BCL2	It is related to control of programmed cell death. It is an antiapoptotic protein	Follicular lymphoma	Solitary fibrous tumor, nasopharyngeal carcinoma, synovial sarcoma, rhabdomyosarcoma, neurofibroma, dermatofibrosarcoma	M, C	It is expressed in a large number of epithelial and soft-tissue tumors

Continued

Continued

Marker	Basic function	Main diagnostic application in lymphomas	Used in other tumors	Localization	Further comments
T cell-related markers					
CD2	It is positive in precursor and mature T cell and also NK cells. It is related to the T lymphocyte adhesion and activation	Pan T-cell marker. T-cell NHL and NK cell lymphoma, mastocytosis	Mastocytosis	M	It is a reliable marker of T cell and NK cells
CD3	Pan T-cell marker. It is present in both the cytoplasm and the membrane of the T cell. It is positive in NK cell	T-NHL and NK-cell NHL. Occasional B-NHL related to EBV		M and C	This is the most commonly used T-cell marker
CD5	It is related to the signaling of T lymphocyte. It is positive in both T cell and a subset of B cell	B-cell CLL, mantle cell lymphoma, DLBCL, Burkitt's lymphoma, subset, T-cell NHL often shows loss of this antigen		M	
CD4	CD4 is T helper cells and consisting of a major population of T cells	T-cell lymphoma	AML, histiocytic neoplasm	M	Not a very reliable marker for T cells and should be used along with other T-cell markers
CD8	CD8 is cytotoxic T cell	T-cell lymphoma	CLL, mantle cell NHL		It is also positive in certain B-cell NHL
CD7	A marker of T cell	T-cell lymphoma. Often lost in T-cell lymphoma	AML, CML, pancreatic and bile duct carcinoma	M	It is often expressed in certain carcinomas
CD30 (Ki1)	A transmembrane receptor related to antibody response regulation	Hodgkin's lymphoma	Anaplastic large cell lymphoma	M	
ALK (CD246)	It is a tyrosine kinase receptor. Chimeric ALK protein is synthesized due to translocation of (2:5) causing fusion of nucleophosmin and tyrosine kinase receptor	Anaplastic large cell lymphoma	Non-small cell lung carcinoma, rhabdomyosarcoma, neuroblastoma, Ewing's tumor, PNET, leiomyosarcoma, malignant nerve sheath tumor	C, N and M	
NK cell-related markers					
CD56	NK cell marker. It is a neuronal cell adhesion molecule (NCAM)	NK cell lymphoma	Neuroendocrine tumors, Ewing's sarcoma/PNET angiosarcoma, synovial sarcoma, alveolar rhabdomyosarcoma, schwannoma, melanoma	M	It is not a specific marker of any tumor as it is expressed in a large variety of lymphoid and non-lymphoid tumors

Continued

Continued

Marker	Basic function	Main diagnostic application in lymphomas	Used in other tumors	Localization	Further comments
Plasma cell-related markers					
CD38	A transmembrane glycoprotein. It is positive in pluripotent stem cells, plasma cells, B cells and T cells	Plasma cell tumors	Primary effusion lymphomas	M	
CD138	It is expressed in the different maturational stage of plasma cells and B lymphocytes	Plasma cell tumors and primary effusion lymphoma	Thyroid, breast and lung carcinomas, malignant melanoma, neuroendocrine tumors, thymoma	M	Other than plasma cell tumors, various epithelial tumors are positive for CD138
MUM1	It is a transcriptional activation protein. It is a marker of B cell and plasma cells after they come out from the germinal center	Plasma cell tumors and DLBCL (ABC type)	Hodgkin's lymphoma, malignant melanoma	N, C	It is positive in various tumors, and so the positivity should be interpreted carefully

(AML: acute myeloid leukemia; C: cytoplasmic; CLL: chronic lymphocytic leukemia; CML: chronic myeloid leukemia; CNS: central nervous system; DLBCL: diffuse large B-cell lymphoma; EBV: Epstein–Barr virus; M: membranous; N: nuclear; NLPHL: nodular lymphocyte-predominant Hodgkin lymphoma; NHL: non-Hodgkin lymphoma; PNET: primitive neuroectodermal tumor)

TABLE 2: Lymphoid cell markers that are positive in nonlymphomatous lesions.

Markers	Positive in lymphoreticular neoplasm	Localization
CD10	RCC, endometrial stromal sarcoma	M
CD30	Embryonal carcinoma	M
CD34	GIST, solitary fibrous tumor, hemangioendothelioma, Kaposi sarcoma, angiosarcoma	M
CD56	Neuroendocrine tumor, neuroblastoma, pheochromocytoma	M
CD99	Ewing's sarcoma, neuroblastoma, malignant rhabdoid tumor	M
CD117	GIST, seminoma/dysgerminoma, small cell carcinoma	M and C

(C: cytoplasmic; GIST: gastrointestinal stromal tumor; M: membranous; RCC: renal cell carcinoma)

TABLE 3: Markers to distinguish mature B-cell versus T-cell non-Hodgkin lymphoma.

Markers	Mature B-cell	Mature T-cell
CD20	+	–
CD23	+	–
CD10	±	–
CD79a	+	–
PAX5	+	–
CD5	±	+
CD3	–	+
CD2	–	+
CD7	–	+
CD4	–	+
CD8	–	+

+ = Positive, – = Negative

B-CELL TYPE OF NON-HODGKIN LYMPHOMA

The details of the ICC profile[3] and salient genetic markers of B-NHL are highlighted in **Table 4**.

Precursor Cell Lymphoma

Lymphoblastic Lymphoma[4]

Lymphoblastic lymphoma is composed of either B or T or very rarely NK cells. They are predominantly composed of cells of T lineage (90%).

B-cell Type

Positive ICC:

- *B-cell markers:* Positive for CD19, CD79a, and PAX5. CD20 may be negative in some instances.
- Positive for TdT and CD99.
- HLA-DR and CD34 markers are also positive.

Small Lymphocytic/B-cell Type of Chronic Lymphocytic Lymphoma

Small lymphocytic lymphoma (SLL) is a B-cell NHL. The cells are small, round shaped with scanty cytoplasm. The nuclei

TABLE 4: Immunocytochemistry profile and salient genetic markers of B-cell non-Hodgkin lymphoma.

Diagnosis	Positive	Negative	Molecular genetics
Small lymphocytic/ B-cell type of chronic lymphocytic lymphoma	CD19, CD20, CD5, CD23, CD43, CD79a, PAX5, bcl-2, sIgM	CD10, cyclin D1	• Trisomy 12 • 13q deletion
Mantle cell lymphoma	CD5, CD19, CD 20, CD43, CD79a, PAX5, sIgM, cyclin D1, FMC7	CD10, CD23, bcl6	t(11:14) (*BCL1* gene rearrangement)
Follicular lymphoma	sIg, CD10, CD19, CD20, Bcl2, Bcl6	CD5, CD23, CD43, SOX-11, cyclin D1	t(14:18) and *BCL2* gene rearrangement
Lymphoplasmacytic lymphoma	CD19, CD20, CD22, CD43, CD79a, CD38, CD138	CD5, CD23, CD10	
Nodal marginal zone lymphoma	sIgM, CD19, CD20, CD79a, PAX5	CD5, CD10, CD23, BCL6, SOX-11, cyclin D1	
Extranodal marginal zone	sIgM, CD19, CD20, CD79a, PAX5	CD5, CD10, CD23, SOX-11, cyclin D1, BCL6	
Splenic marginal zone lymphoma	sIgM, sIgD, CD19, CD20, CD21, CD79a, PAX5, BCL2	CD5, CD10, CD23, CD43, CD103, bcl6, cyclin D1	
Hairy cell leukemia	CD19, CD20, CD22, CD103, CD11c, CD25, CD74, cyclin D1, FMC7	CD10, CD3, CD5, CD23, CD43, bcl6	
Plasmacytoma/ multiple myeloma	CD38, CD138, CD79a, CD56, bcl1	CD19, CD10, CD3, CD5, CD23	
Precursor B lymphoblastic	TdT, CD10, CD19, CD20, CD79a, PAX-5	sIg	t(1;9), t(9:22)
Burkitt's lymphoma	sIgM, CD22, CD74, CD79a, CD10, CD19, 20, PAX5, c-myc, CD43	CD5, CD23, TdT, bcl2	t(8:14), t(8:22). Rearrangement of *c-MYC* gene
Diffuse large B cell	CD10±, CD19, CD20, CD79a, CD45, PAX5, CD22, CD74	CD3, CD5, CD200	

BOX 1 **Small lymphocytic/ B-cell type of chronic lymphocytic lymphoma.**

Cytology: Small round cells, coarse chromatin

Immunocytochemistry

- Positive ICC: CD19, CD20, CD5, CD23, CD43, CD79a, PAX5, bcl2, and sIgM
- Negative ICC: CD10 and Cyclin D1

Genetic markers: Deletions of 13q14, 17p, and 11q

are round monomorphic with coarsely clumped chromatin (**Box 1**).

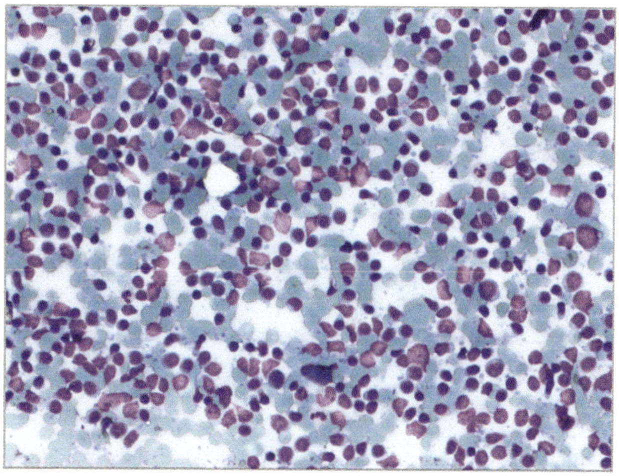

FIG. 1: Small lymphocytic lymphoma: Fine needle aspiration cytology smear showing discrete small monomorphic lymphoid cells.

Immunocytochemistry Findings[3] (Figs. 1 to 9)

- *Positive ICC:* CD19, CD20, CD5, CD23, CD43, CD79a, PAX5, bcl2, sIgM. The cells show a dual expression of CD5 and CD23, which is evident in flow cytometry. This is a helpful feature to distinguish SLL from mantle cell lymphoma (MCL) as the latter shows only CD5 expression.
- *Negative ICC:* CD10 and cyclin D1.
- *Ancillary genetic markers:* SLL/Chronic lymphocytic leukemia (CLL) often shows deletion of 13q14, 17p, and 11q.
- *Prognosis:* Ki67 index of SLL is usually low; however, Ki67 index >25% bears bad prognosis.
- *Pitfalls:* CD20 immunostaining is often weak, and CD43 is often strongly positive in SLL/CLL. It may give a false impression of T-cell NHL. Therefore, a complete panel of ICC is needed for more accurate interpretation.

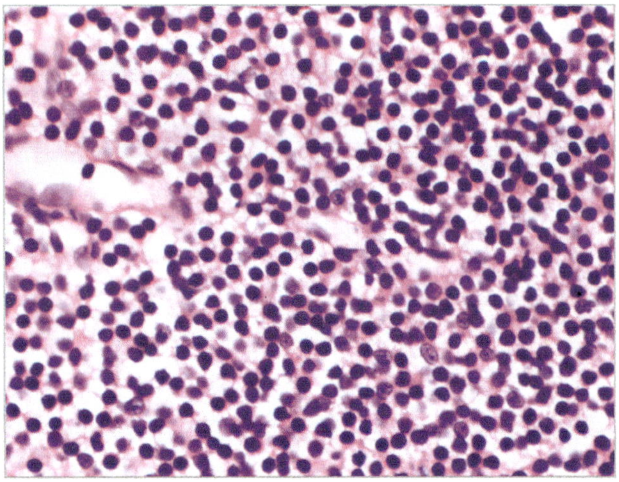

FIG. 2: Small lymphocytic lymphoma: Cell block section showing abundant small round cells with condensed chromatin.

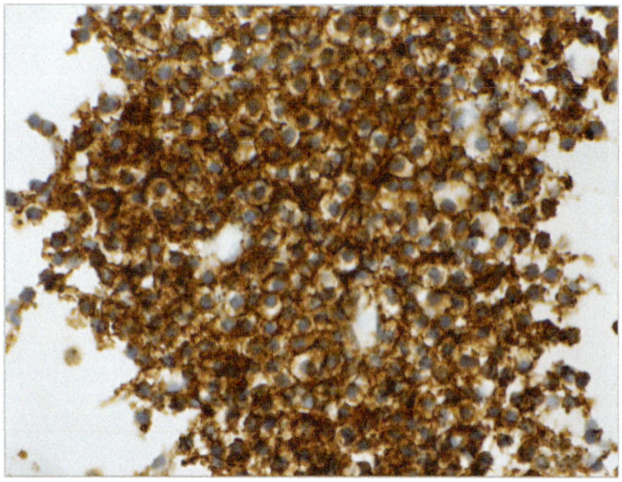

FIG. 3: Small lymphocytic lymphoma: Positive for CD20.

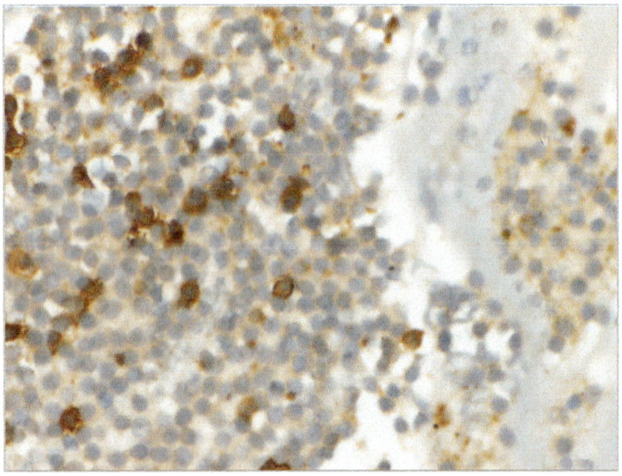

FIG. 4: Small lymphocytic lymphoma: Only occasional cells are positive for CD3.

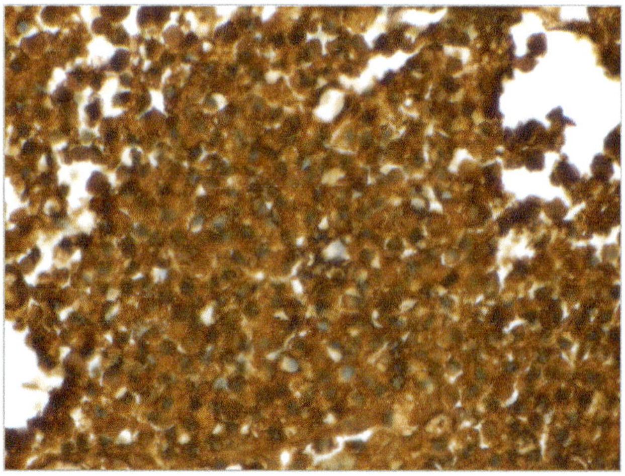

FIG. 5: Small lymphocytic lymphoma: The cells are strongly positive for lambda light chain.

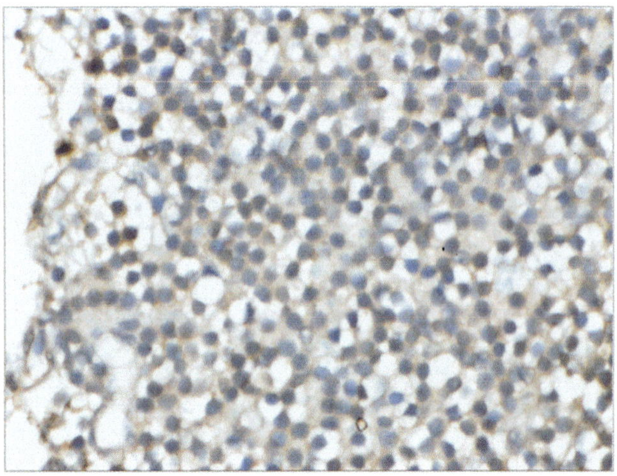

FIG. 6: Small lymphocytic lymphoma: The cells are completely negative for kappa light chain.

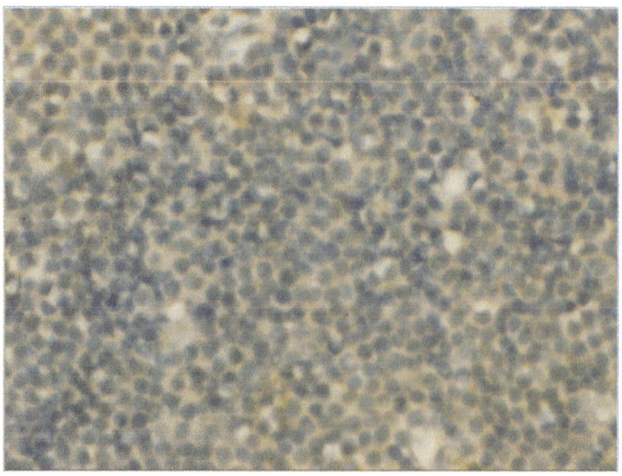

FIG. 7: Small lymphocytic lymphoma: The cells are negative for CD10.

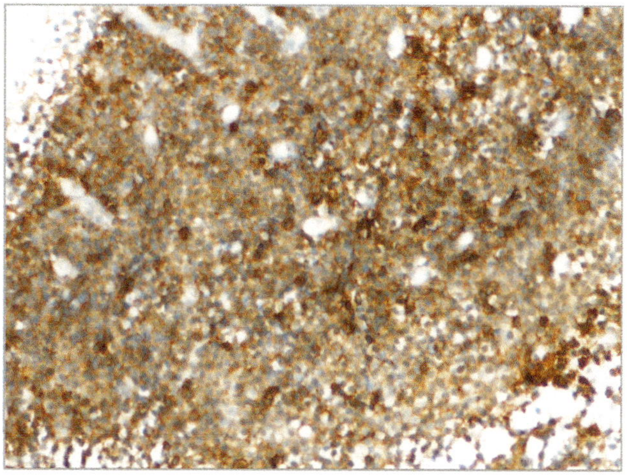

FIG. 8: Small lymphocytic lymphoma: The cells are positive for CD5.

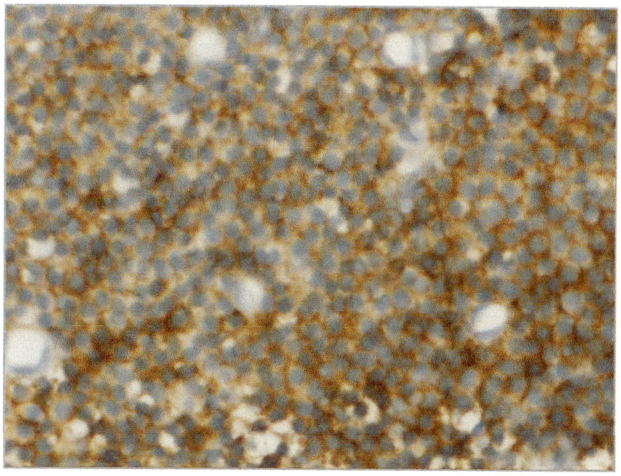

FIG. 9: Small lymphocytic lymphoma: The cells are positive for CD23.

Mantle Cell Lymphoma

Mantle cell lymphoma is a B-NHL with overlapping cytological features with SLL. Immunophenotyping is needed to differentiate this lymphoma from SLL. MCL is more aggressive than SLL, and therefore it is necessary to distinguish SLL from MCL (**Box 2**).

Immunocytochemistry Findings[5] (Figs. 10 to 15)

- *Positive ICC:* CD5, CD19, CD20, CD43, CD79a, PAX5, sIgM, cyclin D1, and FMC7.
  The lymphoid cells are positive for B-cell markers. Almost all cases of MCL show strongly positive cyclin D1, and it is a hallmark of MCL lymphoma.
- *Negative ICC:* CD10, CD23, and BCL6.
- *Ancillary genetic markers:* MCL typically shows t(11:14) (q13:q32) translocation. This results in the translocation of the immunoglobulin heavy chain gene and CCND1.
- *Prognosis:* More than 30% Ki67 index is a poor prognostic marker of MCL.
- *Pitfalls:* Cyclin D1 expression may be seen in plasma cell tumors and some subset of diffuse large B-cell lymphoma (DLBCL). Therefore, cyclin D1 expression should be interpreted carefully.

BOX 2 Mantle cell lymphoma.

Cytology: Small cells, irregular outline of nuclei, coarse chromatin, no nucleoli

Immunocytochemistry

- Positive: CD5, CD19, CD20, CD43, CD79a, PAX5, sIgM, Cyclin D1, and FMC7
- Negative: CD10, CD23, and BCL6

Genetic markers: t(11:14) (q13:q32) translocation

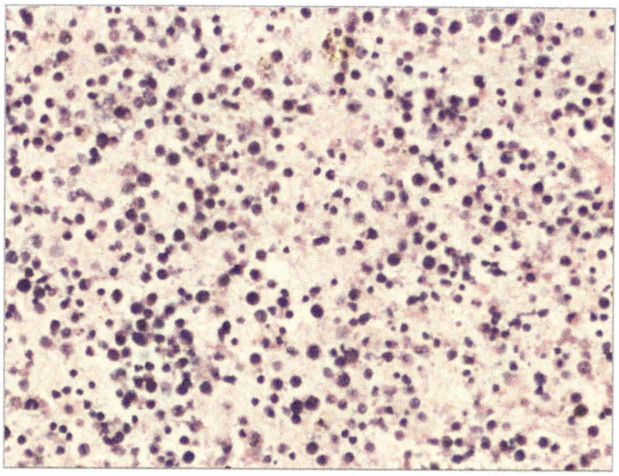

FIG. 10: Mantle cell lymphoma: Cell block section showing abundant immature small round lymphoid cells.

Follicular Lymphoma

Follicular lymphoma (FL) is a B-cell type of NHL and is composed of cells from the follicular center (**Box 3**). On histology, the cells form multiple follicular structures.

Immunocytochemistry Findings[6]

- *Positive ICC:* CD19, CD20, PAX5, BCL2, BCL6, and CD10. Occasionally, the tumor cells may be negative for CD10.
- *Negative ICC:* CD5, CD23, CD43, SOX-11, and cyclin D1.
- *Genetic markers:* FL shows characteristic translocation of t(14:18).
- *Prognosis:* The following factors are related to bad prognosis: (1) Higher age, (2) Low hemoglobin level (<12 g/dL), (3) Ann Arbor staging, (4) Multiple nodal areas (>4), and (5) High lactic acid dehydrogenase (LDH) level.

Lymphoplasmacytic Lymphoma

Lymphoplasmacytic lymphoma is a relatively uncommon neoplasm. This lymphoma consists of plasma cells, lymphocytes, and lymphoplasmacytic cells (**Box 4**). There is no characteristic ICC pattern of this lymphoma.

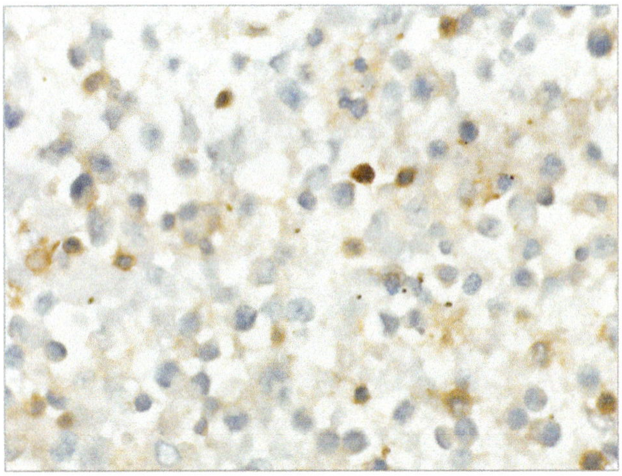
FIG. 11: Mantle cell lymphoma: The cells are negative for CD3.

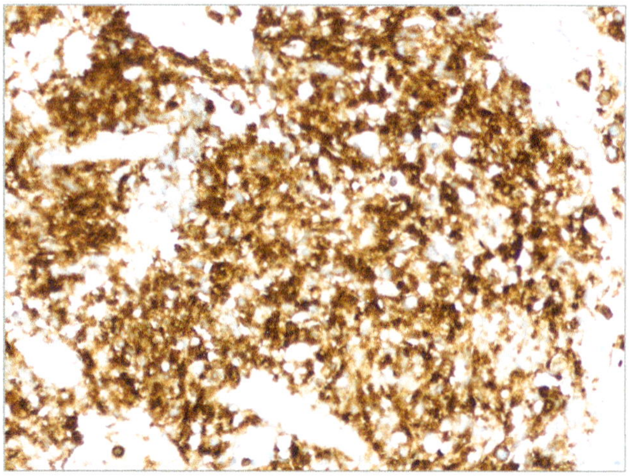
FIG. 12: Mantle cell lymphoma: Predominant CD20 positive lymphoid cells.

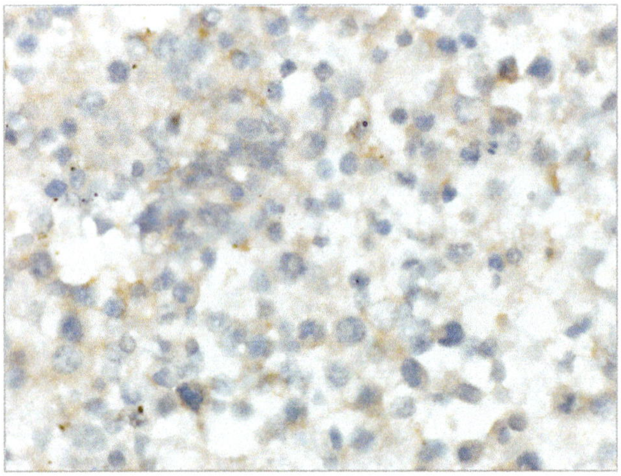
FIG. 13: Mantle cell lymphoma: The cells are negative for CD10.

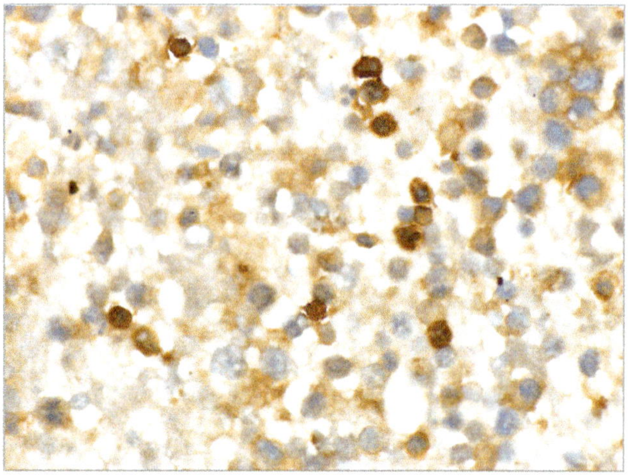
FIG. 14: Mantle cell lymphoma: Many cells are positive for CD5.

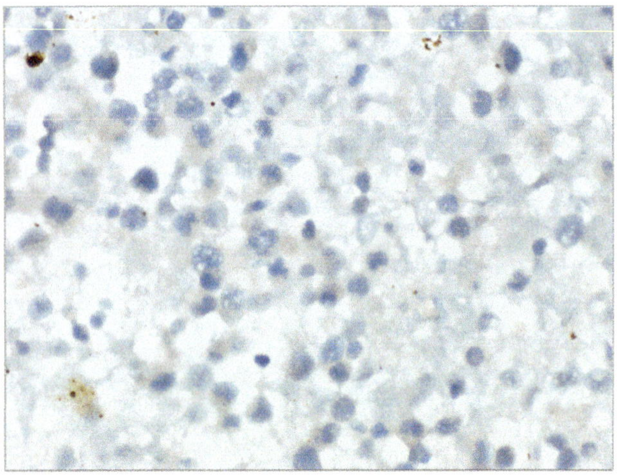
FIG. 15: Mantle cell lymphoma: The cells are negative for CD23.

BOX 3 Follicular lymphoma.

Cytology: Follicular center cells (small and large cells). Small cleaved cells admixed with large noncleaved cells

Immunocytochemistry

- *Positive ICC:* CD19, CD20, PAX5, Bcl2, Bcl6, and CD10
- *Negative ICC:* CD5, CD23, CD43, SOX-11, and cyclin D1

Genetic markers: Translocation of t(14:18)

BOX 4 Lymphoplasmacytic lymphoma.

Cytology: Smears show plasma cells, lymphocytes and lymphoplasmacytic cells.

Immunocytochemistry

- *Positive:* CD19, CD20, CD79a, CD38, and CD138
- *Negative:* CD5, CD23, and CD10

Cytogenetics: Translocation of t(9;14)I in 50% cases

Immunocytochemistry Findings

- *Positive ICC:* CD I9, CD20, and CD79a. Besides, this lymphoma also shows CD38 and CD138.
- *Negative ICC:* The lymphoplasmacytic lymphoma usually shows negative CD5, CD23, and CD10. However, occasionally the cells are positive for C5 and CD10.
- *Prognosis:* ICC findings are not related to prognosis of the patients.

Nodal Marginal Zone Lymphoma

Nodal marginal zone lymphoma (NMZL) is an uncommon B-cell NHL that consists of small B cells.

Immunocytochemistry Findings[7]

- *Positive ICC:* CDI9, CD20, CD79a, and CD43
- *Negative ICC:* CD5, CD10, CD23, BCL6, SOX-11, and cyclin D1
- *Genetic markers:* NMZL does not show any characteristic genetic changes. The tumor may show triosomy 3, 712, and 18.

Extranodal Marginal Zone Lymphoma

Extranodal marginal zone lymphoma (EMZL) is noted in the extranodal sites such as lung, small intestine, stomach, salivary gland, and thyroid. This lymphoma does not bear any characteristic ICC markers (**Box 5**).

- *Positive ICC:* CD19, CD20, CD79a, and PAX-5. They also show sIgM positivity.
- *Negative ICC:* CD5, CD10, CD23, BCL6, SOX-11, and cyclin D1.
- *Genetic markers:* EZML shows Ig gene rearrangement and somatic hypermutations.

Splenic Marginal Zone Lymphoma

Splenic marginal zone lymphoma (SMZL) is a rare NHL that involves the white pulp of spleen.

- *Positive ICC:* CD19, CD20, CD21, CD79a, and PAX-5. In addition, the cells are also positive for sIgM, sIgD, and BCL2.
- *Negative ICC:* CD5, CD10, CD23, CD43, CD103, Bcl6, and cyclin D1.
- *Genetic markers:* Deletion 7q is a common finding.

Hairy Cell Leukemia

Hairy cell leukemia (HCL) is a very rare type of B-NHL that mainly involves spleen and bone marrow and very rarely the lymph node.

Immunocytochemistry Findings[8]

- *Positive ICC:* HCL is mainly recognized by flow cytometry. The tumor cells are positive for B-cell markers such as CD19 and CD20. The cells are typically positive for CD22, CD103, CD11c, CD25, CD74, cyclin D1, and FMC7.
- *Immunohistochemistry:* The cells of HCL show the expression of tartrate -resistant-acid-phosphatase. Most of the cases are also positive for Annexin B.

BOX 5 Extranodal marginal zone lymphoma.

Cytology: Small cells, irregular nuclear margin, fine chromatin, no nucleoli

Immunocytochemistry

- *Positive:* CD19, CD20, CD79a, PAX5, and sIgM
- *Negative:* CD5, CD10, CD23, BCL6, SOX-11, and cyclin D1

Genetic markers: Ig gene rearrangement and somatic hypermutations

BOX 6 Plasmacytoma/multiple myeloma.

Cytology: Abundant plasma cells, central to eccentric nuclei, cartwheel chromatin, many bi- and multinucleated cells

Immunocytochemistry

- *Positive:* CD38, CD138, CD79a, CD56, and bcl1
- *Negative:* CD20, CD19 AND pax5. The cells may also be negative for CD45

BOX 7 Burkitt's lymphoma.

Cytology: Medium-sized cells, moderate amount of deep basophilic cytoplasm. Cytoplasmic vacuolation, coarse chromatin, single to multiple prominent nucleoli. Many mitotic figures, starry sky pattern due to scattered macrophages

Immunocytochemistry

- *Positive:* CD19, CD20, CD79a, PAX5, and CD10
- *Negative:* CD5, CD23, TdT, and bcl-2

The Ki67 index: Nearly 100%

Genetic markers

- Specific translocation of t(8:14)(q22;q32)
- Translocations such as t(8:22)(q24;q11) and t(2:8)(p12;q24)

- *Negative ICC:* HCL cells are negative for CD10, CD3, CD5, CD23, CD43, and bcl6.

Plasmacytoma/Multiple Myeloma

Plasmacytoma is a neoplastic lesion of plasma cells that forms localized tumor-like collections of plasma cells (**Box 6**). Solitary plasmacytoma usually progresses to multiple myeloma in half of the cases.

Immunocytochemistry Findings[9]

- *Positive ICC:* The tumor cells express the markers of plasma cells such as CD38 and CD138. These cells are also positive for CD79a, CD56, and bcl1.
- *Negative ICC:* In the majority of the cases, the plasma cells do not show any expression of B-cell markers such as CD20, CD19, and pax5. The cells may even do not express CD45.

Burkitt's Lymphoma

Burkitt's lymphoma is an aggressive B-NHL with characteristic cytomorphology. ICC of this lymphoma may not be very helpful for its diagnosis (**Box 7**).

Immunocytochemistry Findings[10]

- *Positive ICC:* The tumor cells express pan B-cell markers: CD19, CD20, CD79a, and PAX5. The cells are strongly

positive for CD10. The Ki67 index is usually very high (near about 100%).
- *Negative ICC:* The cells are negative for CD5, CD23, TdT, and bcl2.
- *Genetic markers:* Burkitt's lymphoma shows specific translocation of t(8:14)(q22;q32) that causes MYC and heavy chain gene rearrangement. Occasionally, other translocations such as t(8:22)(q24;q11) and t(2:8)(p12;q24) also occur.

Diffuse Large B-cell Lymphoma

Diffuse large B-cell lymphoma is an aggressive B-NHL which is composed of large B cells (**Box 8**). Based on the gene expression profile, this lymphoma is classified into germinal center B-cell like (GCB) and activated B-cell like (ABC). A third subgroup was also identified that does not show any specific gene expression profile.

Hans algorithmic approach[11] to classify DLBCL is shown in **Flowchart 2**.

Immunocytochemistry Findings (Figs. 16 to 20)

- *Positive ICC:*
  - *B-cell indicators:* CD19, CD20, CD79a, and PAX5
  - The tumor cells show occasional CD10 positivity.
  - A panel of CD10, BCL6, and MUM1 helps to classify DLBCL further.

BOX 8 Diffuse large B-cell lymphoma.

Cytology: Large cells with enlarged nuclei, irregular nuclear contour, prominent nucleoli, many mitotic figures

Immunocytochemistry
- *Positive ICC:* CD19, CD20, CD22, CD74, CD79a, and PAX5, occasional CD10 positivity
- *Negative:* CD3, CD5, CD200
- A panel of CD10, BCL6, and MUM1 mandatory

Genetics: t(14:18) in one-third of cases

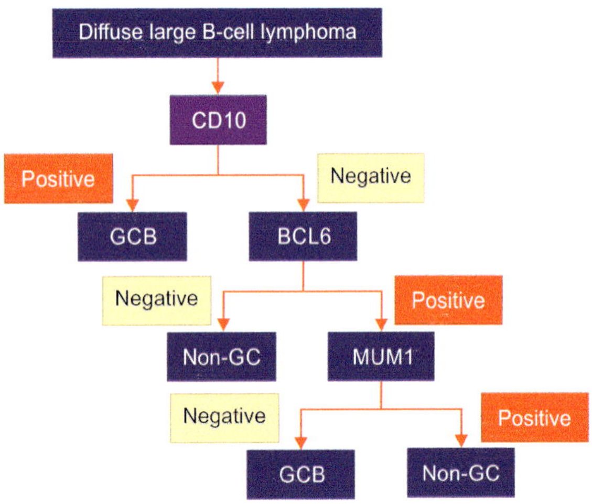

FLOWCHART 2: Schematic diagram showing Hans algorithmic approach to classify diffuse large B-cell lymphoma.

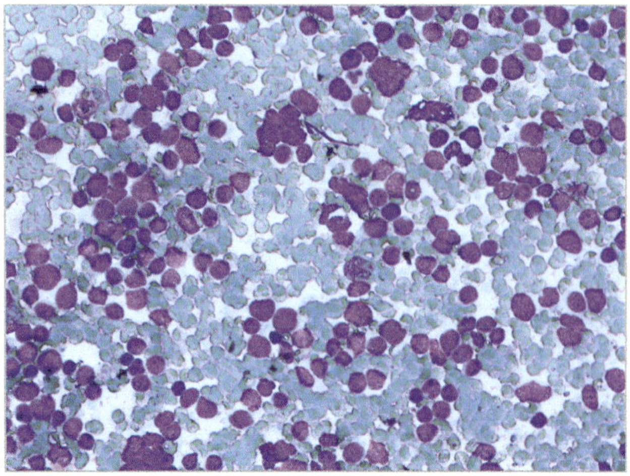

FIG. 16: Diffuse large B-cell lymphoma: Discrete large cells with scanty cytoplasm.

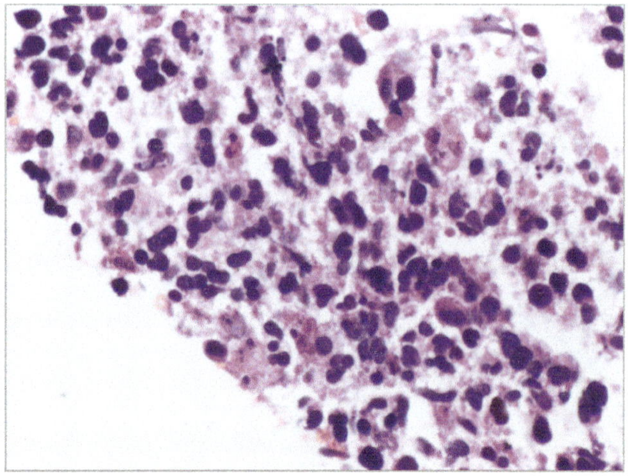

FIG. 17: Diffuse large B-cell lymphoma: Cell block section showing discrete large cells.

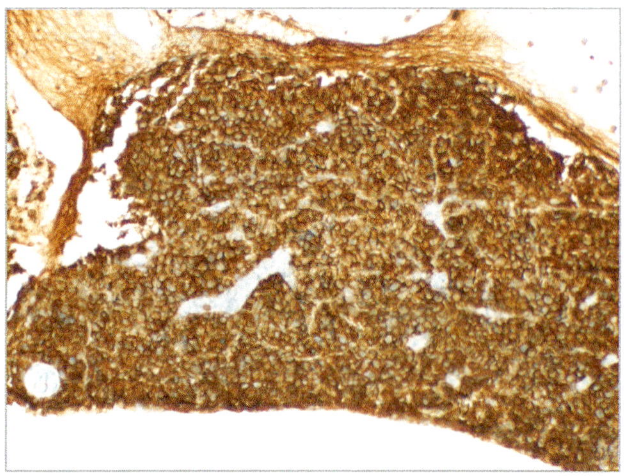

FIG. 18: Diffuse large B-cell lymphoma: The cells are positive for CD20.

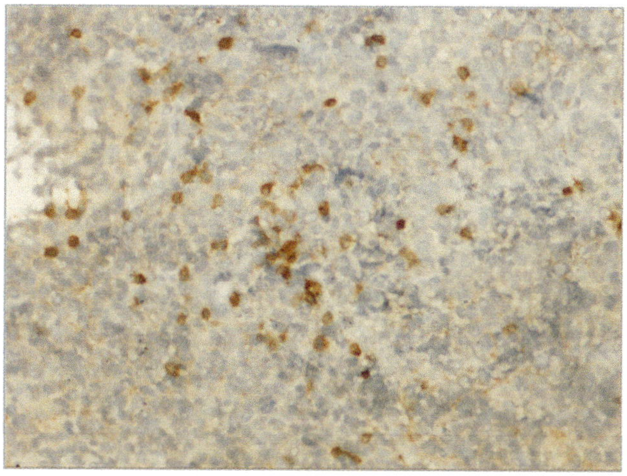

FIG. 19: Diffuse large B-cell lymphoma: The predominant cells are negative for CD3. Only occasional scattered cells are positive for CD3.

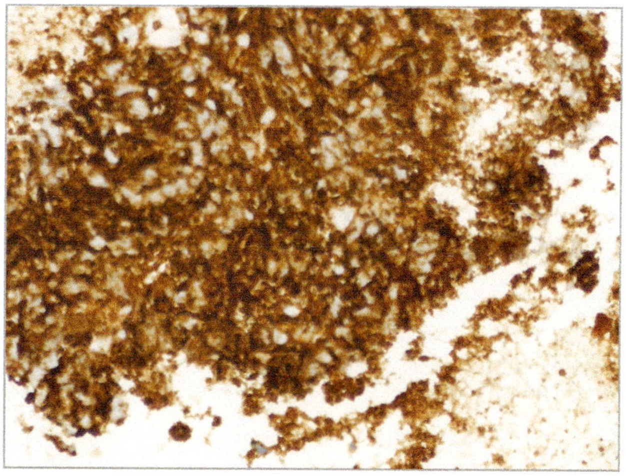

FIG. 20: Diffuse large B-cell lymphoma: The cells are positive for CD10.

- *Negative ICC:* DLBCL cells are negative for CD3, CD5, and CD200.
- *CD5 positive DLBCL:* The cases of CD5 positive DLBCL are noted. These cases have a poorer prognosis.
- *CD30 positive DLBCL:* Certain types of DLBCL show CD30 positivity. These cases are treated with anti-CD30 antibody.

Primary Effusion Lymphoma

Primary effusion lymphoma (PEL) is a rare B-NHL that is commonly seen in men.

Immunocytochemistry Findings[12] (Box 9)

- *Positive ICC:*
  - CD45, CD38, CD138
  - ALK1
- *Negative ICC:*
  - Usually shows variable expression or loss of expression of pan B-cell markers: CD19, CD20, PAX5

BOX 9 Primary effusion lymphoma.

Rare B-cell non-Hodgkin lymphoma

- **Cytology:** Large nuclei with irregular margin, prominent nucleoli, plasmacytoid appearance
- **Immunocytochemistry**
  - *Positive:* CD45, CD38, CD138, ALK1, MUM1, EMA
  - *Negative:* CD19, CD20, PAX5

TABLE 5: Immunocytochemistry of T-cell non-Hodgkin lymphoma.

Diagnosis	Positive	Negative	Molecular genetics
T lymphoblastic	TdT, CD99, CD3, CD7, CD4±, CD8		
Peripheral T-cell lymphoma	CD3, CD7 and CD4/CD8		
Anaplastic large cell lymphoma	CD30 (Ki1), EMA, ALK, CD43	CD20, PAX5, CD45±, CD3±	t(2:5), causing fusion of ALK and NPM
Angioimmunoblastic T-cell lymphoma	CD2, CD3, CD4, CD5, CD7, CD10, PD1, bcl6	CD56	
Mycosis fungoides	CD2, CD3, CD5, CD4, CD45RO	CD8, CD25	

T-CELL NON-HODGKIN LYMPHOMA

The details of ICC[3] of T-cell NHL are highlighted in **Table 5**.

Lymphoblastic Lymphoma

T-cell Type

Immunocytochemistry Findings

- *Positive ICC:* CD3, CD2 CD7, TdT, and CD99. The cells variably express CD4, CD8, CD43, and CD10.

Peripheral T-cell Lymphoma

Peripheral T-cell lymphoma (PTCL) constitutes the significant bulk of T-cell lymphoma.

Immunocytochemistry Findings

- *Positive ICC:*
  - T-cell pointers: CD2, CD3
  - CD5 and CD7 markers are often lost.
  - Variable expression of CD30 and HLA-DR
- *Negative ICC:*
  - CD10 and BCL6 are not expressed.

Anaplastic Large Cell Lymphoma

Anaplastic large cell lymphoma (ALCL) is a type of T-NHL with characteristic ICC findings (**Box 10**).

Immunocytochemistry Findings[13,14]

- *Positive ICC:*
  - CD30 in almost all cases.

BOX 10 Anaplastic large cell lymphoma.

Cytology: Large cell, multiple lobated nuclei, large cell with horseshoe-shaped nuclei ("Hallmark" cells) and doughnut cells

Immunocytochemistry

- Positive: CD30, ALK1, CD3, CD2, CD5 and CD7, CD43, EMA. The cells may or may not be positive for CD45

Genetic markers: Translocation of t(2;5)(q23;q35)

BOX 11 Angioimmunoblastic T-cell lymphoma.

Cytology: Medium to large cells, irregular nuclear margin, coarse chromatin, prominent nucleoli

Immunocytochemistry

- Positive: T-cell markers: CD2, CD3, and CD5; follicular dendritic cell markers: CD23; follicular helper cells: CD10, PD1, BCL6
- Negative: Typically negative for CD56

    - ALK: Positive
    - T lineage markers are positive or may be absent. CD3 is a variable positive. The tumor cells may express CD2, CD5, and CD7.
    - CD43 is positive in majority of the cases.
    - CD45 may or may not be positive.
    - In >80% of cases, epithelial membrane antigen (EMA) is positive.
- *Genetic markers:* ALCL shows typical translocation of t(2;5)(q23;q35). It causes *ALK* and *NPM* gene fusion. It produces the nuclear expression of ALK.

Angioimmunoblastic T-cell Lymphoma

Angioimmunoblastic T-cell lymphoma (AITL) is a T follicular helper cell NHL. It is characterized by skin rash, hypergammaglobulinemia, and generalized lymphadenopathy. The ICC pattern of AITL is characteristic (**Box 11**).

Immunocytochemistry Findings[15]

- *Positive ICC:*
    - *T-cell markers:* CD2, CD3, and CD5,
    - *Follicular dendritic cell markers:* CD23
    - *Follicular helper cells:* CD10, PD1, and BCL6
- *Negative ICC:* Typically negative for CD56

Mycosis Fungoides

GHKL

Immunocytochemistry Findings

- *Positive ICC:*
    - *Positive T-cell markers:* CD2, CD3, CD5, CD4, and CD45RO
- *Negative ICC:*
    - Loss of CD7

HODGKIN LYMPHOMA

Hodgkin lymphoma (HL) is a B-cell neoplastic lesion characterized by the presence of Reed–Sternberg (RS) cell (**Box 12**).

BOX 12 Hodgkin lymphoma.

Classical Hodgkin lymphoma

- Immunocytochemistry
    - Positive: Reed–Sternberg cells are CD30 and CD15 positive, weak positive PAX5. It also shows MUM1, HLA DR, and CD25 expression

Nodular lymphocyte predominant Hodgkin lymphoma (NLPHL)

- Immunocytochemistry
    - Lymphocyte and histiocytic cells
        - Positive: CD45, CD20, CD79a, occasionally CD30 (1020%), OCT2 and BOB1 and EMA
        - Negative: Always negative for CD15

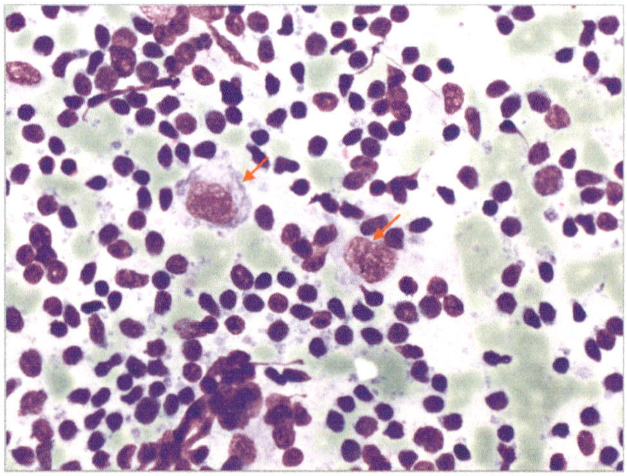

FIG. 21: Hodgkin lymphoma: Fine needle aspiration cytology smear showing polymorphic population of cells along with classical Reed–Sternberg cells (arrows).

Classical Hodgkin Lymphoma (CHL)

Immunocytochemistry Findings[16,17] *(Figs. 21 to 27)*

- *Positive ICC:*

  Reed–Sternberg (RS) cells show:
    - CD30 positivity in all cases and CD15 positivity in 85% cases.
    - Weak positive PAX5
    - Reed–Sternberg cells also express MUM1, HLA DR, and CD25.

Nodular Lymphocyte-predominant Hodgkin Lymphoma (NLPHL)

Immunocytochemistry Findings[18]

Lymphocyte and histiocytic (L&H) cells

- *Positive ICC:*
    - CD45 positive
    - *Positive B-cell markers:* CD20, CD79a
    - Occasionally CD30 positivity (10–20%)
    - OCT2 and BOB1 positive
    - EMA positive
- *Negative ICC:*
    - Always negative for CD15

The ICC findings of NLPHL and classical HL are compared in **Table 6**.

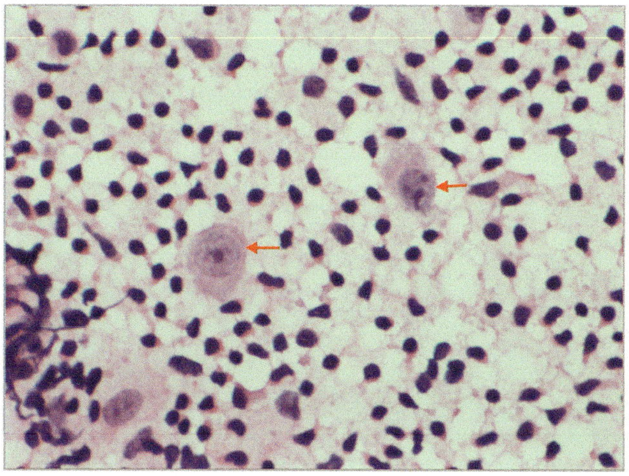

FIG. 22: Hodgkin lymphoma: Fine needle aspiration cytology smear showing large mononuclear Hodgkin cells (arrows).

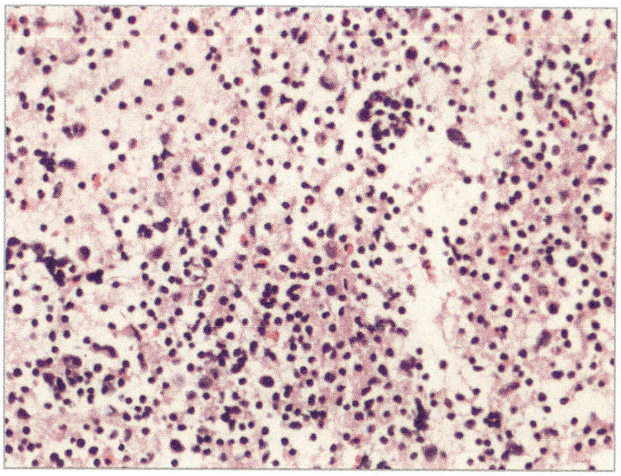

FIG. 23: Hodgkin lymphoma: Cell block section showing discrete cell population of mature lymphoid cells and large atypical cells (hematoxylin and eosin stain).

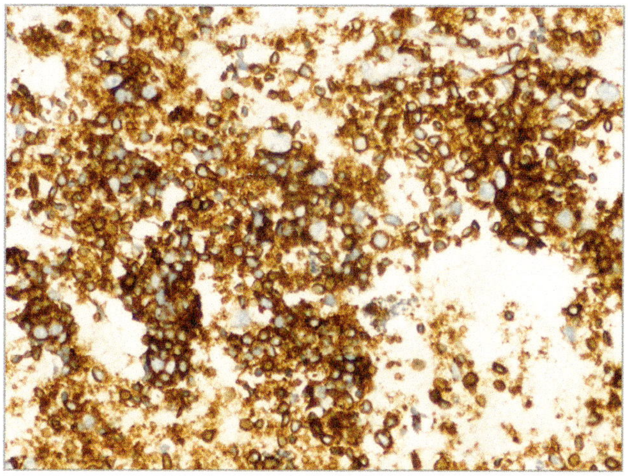

FIG. 24: Hodgkin lymphoma: CD45 positive cells along with large atypical cells that are negative for CD45.

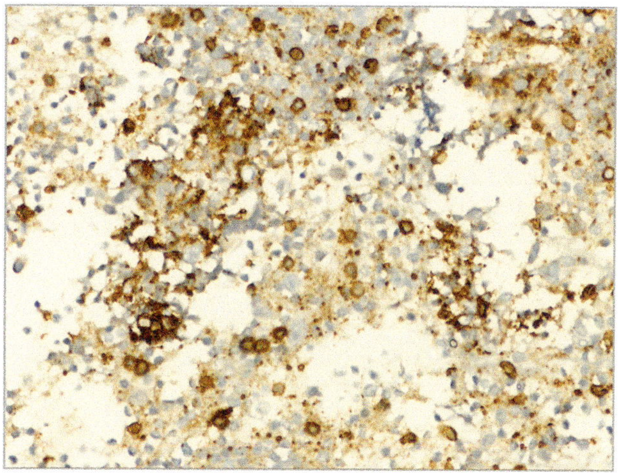

FIG. 25: Hodgkin lymphoma: The large cells are CD20 positive.

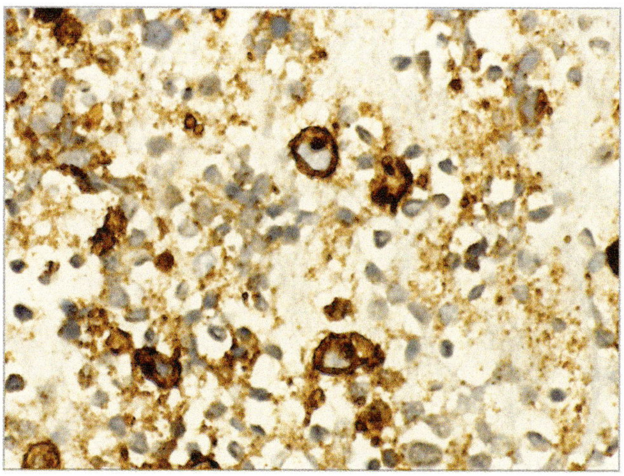

FIG. 26: Hodgkin lymphoma: The large atypical cells are positive for CD15.

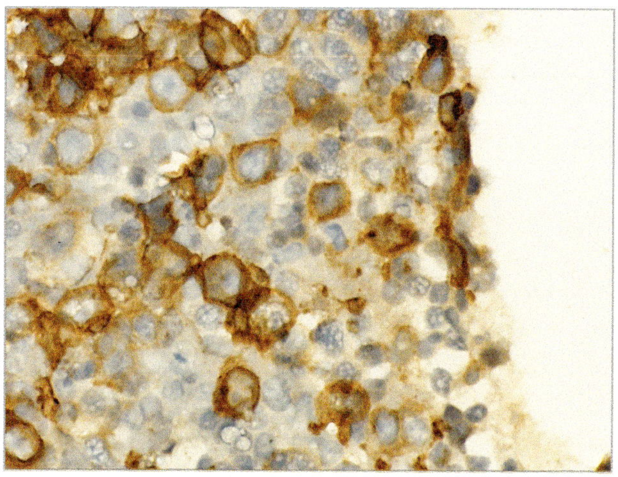

FIG. 27: Hodgkin lymphoma: The large atypical cells are positive for CD30.

TABLE 6: Difference of immunocytochemistry in NLPHL and classical HL.

Markers	NLPHL	CHL
C15	Negative	Positive
CD30	Occasionally positive	Positive
CD20	Positive	Occasionally positive
CD79a	Positive	Negative
EMA	Positive	Negative
BCL6	Always positive	Mostly negative

(CHL: classical Hodgkin lymphoma; NLPHL: nodular lymphocyte predominant Hodgkin lymphoma)

REFERENCES

1. Cazzola M. Introduction to a review series: the 2016 revision of the WHO classification of tumors of hematopoietic and lymphoid tissues. Blood. 2016;127(20):2361-4.
2. Hsi ED. 2016 WHO Classification update-What's new in lymphoid neoplasms. Int J Lab Hematol. 2017;39:14-22.
3. Higgins RA, Blankenship JE, Kinney MC. Application of immunohistochemistry in the diagnosis of non-Hodgkin and Hodgkin lymphoma. Arch Pathol Lab Med. 2008;132(3):441-61.
4. Al Gwaiz LA, Bassioni W. Immunophenotyping of acute lymphoblastic leukemia using immunohistochemistry in bone marrow biopsy specimens. Histol Histopathol. 2008;23(10):1223-8.
5. Katzenberger T, Kienle D, Stilgenbauer S, Höller S, Schilling C, Mäder U et al. Delineation of distinct tumour profiles in mantle cell lymphoma by detailed cytogenetic, interphase genetic and morphological analysis. Br J Haematol. 2008;142(4):538-50.
6. El-Sayed AM, El-Borai MH, Bahnassy AA, El-Gerzawi SM. Flow cytometric immunophenotyping (FCI) of lymphoma: correlation with histopathology and immunohistochemistry. Diagn Pathol. 2008;3:43.
7. van den Brand M, van Krieken JH. Recognizing nodal marginal zone lymphoma: recent advances and pitfalls. A systematic review. Haematologica. 2013;98(7):1003-13.
8. Kreitman RJ, Arons E. Update on hairy cell leukemia. Clin Adv Hematol Oncol. 2018;16(3):205-15.
9. Sukpanichnant S, Cousar JB, Leelasiri A, Graber SE, Greer JP, Collins RD. Diagnostic criteria and histologic grading in multiple myeloma: histologic and immunohistologic analysis of 176 cases with clinical correlation. Hum Pathol. 1994;25(3):308-18.
10. Agarwal R, Lade S, Liew D, Rogers TM, Byrne D, Feleppa F et al. Role of immunohistochemistry in the era of genetic testing in MYC-positive aggressive B-cell lymphomas: a study of 209 cases. J Clin Pathol. 2016;69(3):266-70.
11. Hans CP, Weisenburger DD, Greiner TC, Gascoyne RD, Delabie J, Ott G et al. Confirmation of the molecular classification of diffuse large B-cell lymphoma by immunohistochemistry using a tissue microarray. Blood. 2004;103:275-82.
12. Patel S, Xiao P. Primary effusion lymphoma. Arch Pathol Lab Med. 2013;137(8):1152-4.
13. Tsuyama N, Sakamoto K, Sakata S, Dobashi A, Takeuchi K. Anaplastic large cell lymphoma: pathology, genetics, and clinical aspects. J Clin Exp Hematop. 2017;57(3):120-42.
14. Ng WK, Ip P, Choy C, Collins RJ. Cytologic and immunocytochemical findings of anaplastic large cell lymphoma: analysis of ten fine-needle aspiration specimens over a 9-year period. Cancer. 2003;99(1):33-43.
15. Rai MP, Bedi PS, Marinas EB, Khan NNS. Angioimmunoblastic T-cell lymphoma: a rare subtype of peripheral T-cell lymphoma. Clin Case Rep. 2018;6(4):750-2.
16. Jogai S, Al-Jassar A, Dey P, Adesina AO, Amanguno HG, Francis I M. Fine needle aspiration cytology of Hodgkin's lymphoma: A cytohistologic correlation study from a cancer center in Kuwait. Acta Cytol. 2006;50(6):656-62.
17. Zeppa P, Cozzolino I. Hodgkin lymphoma. Monogr Clin Cytol. 2018;23:52-9.
18. Smith LB. Nodular lymphocyte predominant Hodgkin lymphoma: diagnostic pearls and pitfalls. Arch Pathol Lab Med. 2010;134(10):1434-9.

CHAPTER 12

Immunocytochemistry of Breast Tumors

INTRODUCTION

The commonly used immunocytochemistry (ICC) markers of breast tumors include GATA3, estrogen receptor (ER), progesterone receptor (PR), mammaglobin, gross cystic disease fluid protein 15 (GCDFP-15), and human epidermal growth factor receptor 2 (HER-2).

GATA3

GATA3 is described in detail in Chapter 16.

MAMMAGLOBIN

Mammaglobin is a low-molecular-weight secretory protein and is expressed in the breast epithelial cells. It is relatively sensitive (80%) and highly specific for the identification of metastatic breast carcinoma. This marker is also positive in other nonbreast tumors that include mammary secretory analogue carcinoma of the salivary gland and endometrioid carcinoma (**Table 1**).[1,2]

GROSS CYSTIC DISEASE FLUID PROTEIN 15

Gross cystic disease fluid protein 15 is an androgen and prolactin-inducing protein. This protein was initially noted in the cystic fluid of breast. GCDFP-15 is a highly specific marker (99%) for the breast carcinoma (**Table 2**).[3]

ESTROGEN RECEPTOR

Estrogen receptor is a type of steroid receptor and is expressed in breast, prostate, and endometrial epithelial cells. There are two different types of ER: ER-alpha and ER-beta. They are produced by two different genes that are located in two different chromosomes. ER is expressed in breast and endometrial carcinomas (**Table 3**).[4]

TABLE 1: Mammaglobin expression.

Marker	Positive in breast lesions	Other tumor positivity	Localization	Comments
Mammaglobin	Carcinoma of breast	• Mammary analogue secretory carcinoma of the salivary gland • Sweat gland carcinoma • Endometrial adenocarcinoma • Endocervical adenocarcinoma	Cytoplasmic	• Highly sensitive with low specificity • A panel with other markers should be used

TABLE 2: Gross cystic disease fluid protein 15 (GCDFP-15) expression.

Marker	Positive in breast lesions	Other tumor positivity	Localization	Comments
GCDFP-15	Carcinoma of breast	• Salivary gland carcinoma • Skin adnexal tumor • Lung adenocarcinoma • Renal cell carcinoma • Ovarian carcinoma • Endometrial adenocarcinoma	Cytoplasmic	• Highly specific marker • It can be used for the identification of the metastatic breast carcinoma in a difficult situation

TABLE 3: Estrogen receptor (ER) expression.

Marker	Positive in breast lesions	Other tumor positivity	Localization	Comments
Estrogen receptor	Carcinoma of breast	• Endometrial adenocarcinoma • Serous and mucinous carcinoma of ovary • Adenocarcinoma of stomach • Leiomyosarcoma • Hepatocellular carcinoma	Nuclear	• Any pattern other than nuclear should be ignored

TABLE 4: McCarty scoring system.

Strength of positivity	Percentage positivity	Total
Strong 3	A%	3A
Moderate 2	B%	2B
Weak 1	C%	C
Total score (out of 300)		3A + 2B + C

TABLE 5: Progesterone receptor (PR) expression.

Marker	Positive in breast lesions	Other tumor positivity	Localization	Comments
Progesterone receptor	Carcinoma of breast	• Endometrial adenocarcinoma • Adnexal tumor of the skin • Stromal tumor of kidney and prostate	Nuclear	• This is a relatively specific marker than estrogen receptor • It is an important prognostic marker for the imitation of hormone therapy

There are many types of scoring system of ER expression. The McCarty scoring system[5] is described in the following text (**Table 4**).

Interpretation

- *Negative:* <50
- *Weak positive:* 51–100
- *Moderate positive:* 101–200
- *Strong positive:* 201–300

PROGESTERONE RECEPTOR

Progesterone receptor is also a member of the steroid family receptor and is expressed in breast and endometrial cells. It is mainly used as a favorable prognostic marker of breast carcinoma (**Table 5**).[6]

HUMAN EPIDERMAL GROWTH FACTOR RECEPTOR 2

Human epidermal growth factor receptor 2 (HER-2) is also called c-erb-B2. It is one of the members of the epidermal growth factor receptors. *HER-2* gene is overexpressed in breast and gastric carcinoma. HER-2 protein expression is mainly related to the prognosis of breast carcinoma. HER-2 positive patients need adjuvant chemotherapy, otherwise the tumor behaves aggressively. Targeted therapy against HER-2 by anti-HER-2 monoclonal therapy is helpful in metastatic carcinoma of breast (**Table 6**).[7]

Human epidermal growth factor receptor 2 scoring is described in **Table 7**.

IMMUNOCYTOCHEMISTRY IN BREAST LESIONS

Infiltrating Duct Carcinoma of the Breast

The ICC pattern of the infiltrating duct carcinoma of the breast is described as follows (**Figs. 1** to **5**):

- *Positive*
  - CK7
  - GATA3
  - E cadherin
  - Beta-catenin
  - Mammaglobin

GATA-3 positivity can be used as a marker of metastatic carcinoma of breast along with other markers (**Figs. 6** and **7**).[8,9]

On the basis of the ICC pattern, the genomic classification of carcinoma of breast can be done (**Table 8**).[10]

TABLE 6: Human epidermal growth factor receptor 2 (HER-2) expression.

Marker	Positive in breast lesions	Other tumor positivity	Localization	Comments
HER-2	Carcinoma of breast	• Endometrial adenocarcinoma • Adenocarcinoma of stomach • Adenocarcinoma of lung	Membranous	• Only the membranous positivity should be considered • Proper positive control is needed

TABLE 7: Human epidermal growth factor receptor 2 (HER-2) scoring.

Score	HER-2 interpretation	Pattern of staining
0	Negative	No membranous staining
1+	Negative	Incomplete membranous staining
2+	Weak positive	Incomplete circumferential membranous staining or moderate staining in <10% tumor cells
3+	Positive	Homogenous intense circumferential stain in >10% cells

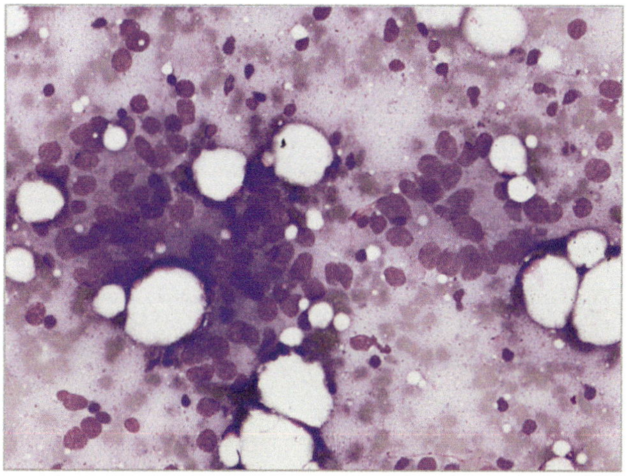

FIG. 1: Ductal carcinoma: Fine needle aspiration cytology smear showing loose clusters and discrete malignant cells.

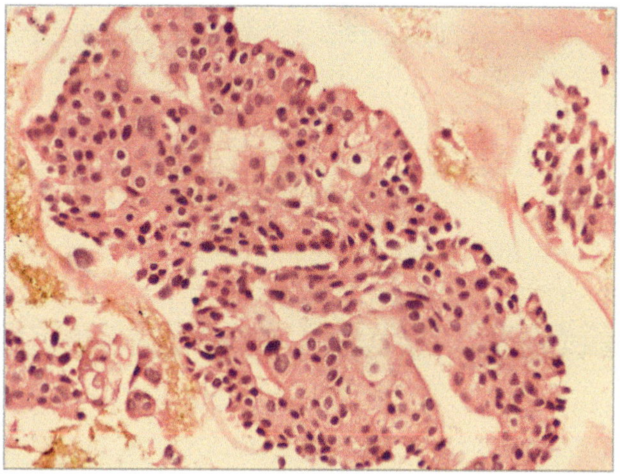

FIG. 2: Ductal carcinoma: Cell block section showing diffuse sheet of malignant cells.

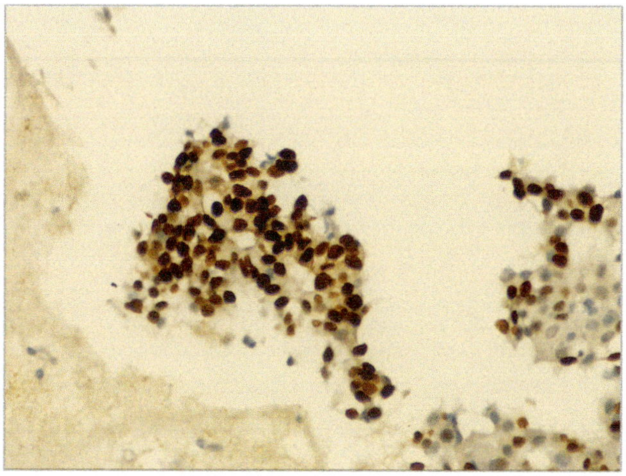

FIG. 3: Ductal carcinoma: The tumor cells showing a strongly positive estrogen receptor.

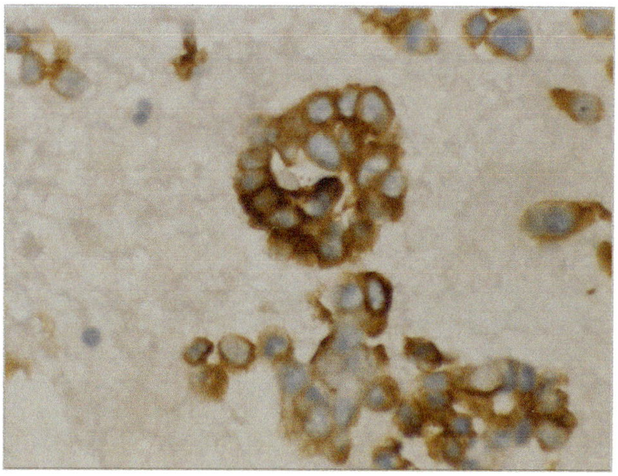

FIG. 4: Ductal carcinoma: The tumor cells showing membranous positive CK7.

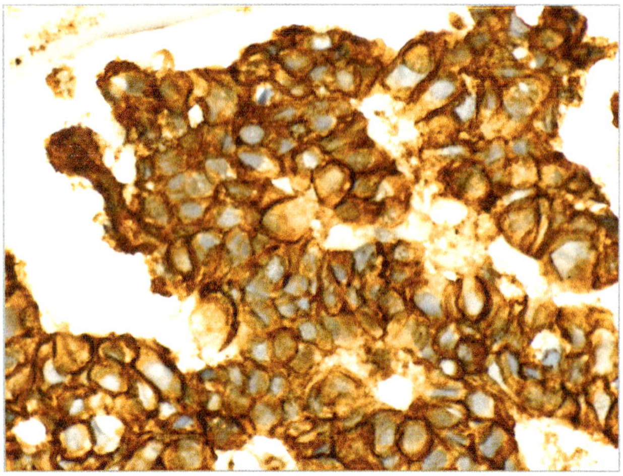

FIG. 5: Ductal carcinoma: The tumor cells showing membranous positive HER-2.

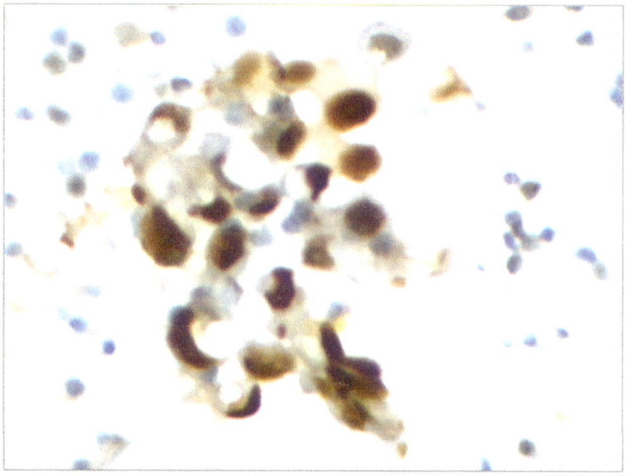

FIG. 6: Ductal carcinoma metastasis in effusion cytology: GATA3 positive cells.

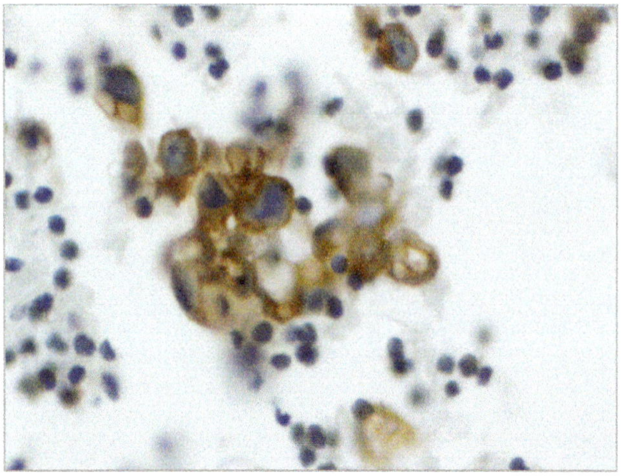

FIG. 7: Ductal carcinoma metastasis in effusion cytology: HER-2 positive cells.

Infiltrating Lobular Carcinoma of the Breast

- *Positive*
    - CK7
    - GATA3
    - Beta-catenin
    - Mammaglobin

Apocrine Carcinoma

- *Positive*
    - GCDFP-15

TABLE 8: Molecular phenotype of carcinoma of the breast.

Type	ER	PR	HER-2	CK5/6	Ki67 index
Luminal A	+	+	–	–	<14%
Luminal B	+	– (or <20%)	–	–	>14%
HER-2 enriched	–	–	+	–	High
Basal-like	–	–	–	+	High

+ = Positive, – Negative

(ER: estrogen receptor; HER-2: human epidermal growth factor receptor 2; PR: progesterone receptor)

- Androgen receptor
- CK8,18,19

Mucinous Carcinoma

- *Positive*
    - CK7
    - Carcinoembryonic antigen (CEA)
    - Neuron-specific enolase (NSE)

Phyllodes Tumor

Stromal Cells

- *Positive*
    - Vimentin

Epithelial Cells

- CK5/6
- EMA
- Pan CK

REFERENCES

1. Li C, Zhang T. Human mammaglobin: A specific marker for breast cancer prognosis. J BUON. 2016;21(1):35-41.
2. Soares CD, de Lima Morais TM, Carlos R, Martins MD, de Almeida OP, Mariano FV, et al. Immunohistochemical expression of mammaglobin in salivary duct carcinomas de novo and salivary duct carcinoma ex pleomorphic adenoma. Hum Pathol. 2019;92:59-66.
3. Darb-Esfahani S, von Minckwitz G, Denkert C, Ataseven B, Högel B, Mehta K, et al. Gross cystic disease fluid protein 15 (GCDFP-15) expression in breast cancer subtypes. BMC Cancer. 2014;14:546.
4. Heubner A, Beck T, Grill HJ, Pollow K. Comparison of immunocytochemical estrogen receptor assay, estrogen receptor enzyme immunoassay, and radioligand-labeled estrogen receptor assay in human breast cancer and uterine tissue. Cancer Res. 1986;46:4291s-5s.
5. McCarty KS Jr, Miller LS, Cox EB, Konrath J, McCarty KS Sr. Estrogen receptor analyses. Correlation of biochemical and immunohistochemical methods using monoclonal antireceptor antibodies. Arch Pathol Lab Med. 1985;109(8):716-21.
6. Ricketts D, Turnbull L, Ryall G, Bakhshi R, Rawson NS, Gazet JC, et al. Estrogen and progesterone receptors in the normal female breast. Cancer Res. 1991;51(7):1817-22.
7. Wu VS, Kanaya N, Lo C, Mortimer J, Chen S. From bench to bedside: What do we know about hormone receptor-positive and human epidermal growth factor receptor 2-positive breast cancer? J Steroid Biochem Mol Biol. 2015;153:45-53.
8. Davis DG, Siddiqui MT, Oprea-Ilies G, Stevens K, Osunkoya AO, Cohen C, et al. GATA-3 and FOXA1 expression is useful to differentiate breast carcinoma from other carcinomas. Hum Pathol. 2016;47(1):26-31.
9. Hou Y, Shen R, Chaudhary S, Tonkovich D, Li Z. Utility of different immunostains for diagnosis of metastatic breast carcinomas in both surgical and cytological specimens. Ann Diagn Pathol. 2017;30:21-7.
10. Heng YJ, Lester SC, Tse GM, Factor RE, Allison KH, Collins LC, et al. The molecular basis of breast cancer pathological phenotypes. J Pathol. 2017;241(3):375-91.

Immunocytochemistry of Gastrointestinal Tract Tumors

INTRODUCTION

The conventional markers that are used in the various epithelial tumors of the gastrointestinal tract (GIT) tumors are given in the following text.

CYTOKERATIN

- CK7
- CK19
- CK20

The cytokeratin is an intermediate filament. Various subtypes of cytokeratin are very important for the primary assessment of the origin of the epithelial tumors.[1,2] Usually, a combination of CK7, CK20, and CK9 is used. **Table 1** highlights the distribution pattern of various CKs to the GIT.

CDX2

Caudal-related homeobox 2 (CDX2) is a transcription factor associated with maturation and proliferation of the epithelial cells of the GIT. It is a reliable marker of the colonic and rectal adenocarcinoma (**Table 2**).[3,4]

SATB2

Special AT-rich sequence binding protein 2 (SATB2) is a transcription protein associated with the nuclear matrix attachment region. It is positive in colonic carcinoma, rectal carcinoma, and appendicular carcinoma (**Table 3**).[5]

CARCINOEMBRYONIC ANTIGEN

Carcinoembryonic antigen (CEA) is expressed in the adenocarcinoma of colon, stomach, and pancreas. Both CEA and CK20 positive tumors are seen in colonic adenocarcinoma (**Table 4**).[6,7]

COMMONLY USED ANTIBODIES FOR NEUROENDOCRINE TUMORS OF THE GIT

Synaptophysin

Synaptophysin (SNP) is the component of the neurosecretory granules. It is a sensitive but not specific marker of neuroendocrine tumors (**Table 5**).[8]

TABLE 1: Distribution pattern of various cytokeratins in relation to the gastrointestinal tract.

Markers	Positivity in gastrointestinal tract	Other tumors showing positivity	Localization
CK7	Pancreatobiliary tumors	Lung, breast, ovary, urothelial	Cytoplasmic
CK19	Gastric adenocarcinoma		Cytoplasmic and membranous
CK20	Carcinomas of the small and large intestines	Merkel cell carcinoma	Cytoplasmic

TABLE 2: CDX2 positive tumors.

Marker	Positivity in gastrointestinal tract	Other tumors showing positivity	Localization
CDX2	Colonic and rectal adenocarcinoma	• Mucinous tumors of the ovary • Neuroendocrine tumors in GIT • Pancreatic and biliary carcinoma	Nuclear

(CDX2: caudal-related homeobox 2; GIT: gastrointestinal tract)

TABLE 3: Special AT-rich sequence binding protein 2 (SATB2) positive tumors.

Marker	Positivity in the gastrointestinal tract	Other tumors showing positivity	Localization
SATB2	• Colonic carcinoma • Rectal carcinoma • Appendicular carcinoma	• Merkel cell carcinoma • Osteosarcoma	Nuclear

TABLE 4: Carcinoembryonic antigen (CEA) positive tumors.

Marker	Positivity in gastrointestinal tract	Other tumors showing positivity	Localization
CEA	• Colonic carcinoma • Stomach adenocarcinoma	• Breast • Salivary gland • Lung	Cytoplasmic

TABLE 5: Synaptophysin positive tumors.

Marker	Positivity in gastrointestinal tract	Other tumors showing positivity	Localization
Synaptophysin	Neuroendocrine tumors	• Large cell undifferentiated neuroendocrine carcinoma of the lung • Adrenocortical adenoma and carcinoma • Neuroblastoma • Pheochromocytoma • Desmoplastic small round cell tumor	Cytoplasmic

TABLE 6: Chromogranin.

Marker	Positivity in gastrointestinal tract	Other tumors showing positivity	Localization
Chromogranin	Neuroendocrine tumors	• Desmoplastic small round cell tumor • Hepatoblastoma fetal type • Parathyroid neoplasm • Medullary carcinoma of the thyroid • Small cell carcinoma	Cytoplasmic

TABLE 7: CD117/c-kit.

Marker	Positivity in gastrointestinal tract	Other tumors showing positivity	Localization
CD117	Gastrointestinal stromal tumors	• Peripheral neuroectodermal tumor • Seminoma • Melanoma • Renal cell carcinoma • Granulocytic sarcomas	Cytoplasmic

CHROMOGRANIN

Chromogranin (CHG) is present in the neurosecretory granules. It is a relatively specific marker of the neuroendocrine tumors (**Table 6**).[8,9]

MARKERS FOR GASTROINTESTINAL STROMAL TUMORS

The markers for gastrointestinal stromal tumors (GISTs) are described below.

CD117 (C-KIT)

CD117 is a tyrosine kinase receptor and is related to the growth and differentiation of the hematopoietic cells, germ cells, and cells of Cajal.[10] Majority of GIST show CD117 positivity (**Table 7**).

DOG-1

DOG-1 is a marker of GIST. It is expressed in 80% of cases of GIST (**Table 8**).[10]

TABLE 8: DOG-1 expression.

Marker	Positivity in gastrointestinal tract	Other tumors showing positivity	Localization
DOG-1	Gastrointestinal stromal tumors	• Synovial cell sarcoma • Leiomyosarcoma of uterus • Oncocytoma of kidney • Hepatocellular carcinoma • Squamous cell carcinoma of esophagus	Cytoplasmic and membranous

IMMUNOCYTOCHEMISTRY PATTERN OF DIFFERENT TUMORS OF THE GIT

Adenocarcinoma of the Esophagus

The immunocytochemistry (ICC) pattern of esophageal adenocarcinoma includes the following:

- *Positive*
  - CK7
  - CK19
  - AE1/AE3
  - MOC-31
  - Ber-EP4

Squamous Cell Carcinoma of the Esophagus

- *Positive*
  - AE1/AE3
  - p63
  - p40
  - CK5/6

Adenocarcinoma of the Stomach

- *Positive*
  - AE1/AE3
  - CK7
  - CK19
  - CK20
  - CEA
  - MOC-31
  - Ber-EP4

Adenocarcinoma of the Small Intestine

- *Positive*
  - CK7
  - CK19
  - CDX2

The differential diagnosis of small intestinal adenocarcinoma and bile duct carcinoma is highlighted in **Table 9**.

TABLE 9: Differential diagnosis of small intestinal adenocarcinoma and bile duct carcinoma.

Markers	Small intestine	Bile duct carcinoma
HepPer1	Positive (occasionally negative)	Negative
CK17	Negative	Positive
CDX2	Usually positive	Occasionally positive

Adenocarcinoma of the Colon and Rectum (Figs. 1 to 4)

- *Positive*
  - CK20
  - CK19
  - CDX2
  - SATB2
  - CEA
  - MUC2

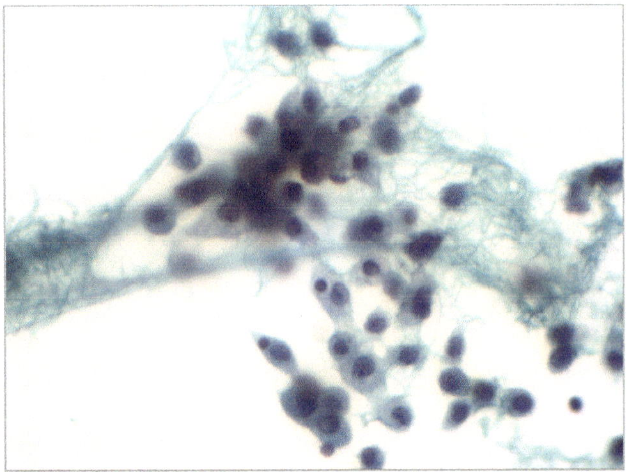

FIG. 1: Adenocarcinoma of colon: Fine needle aspiration cytology smear of the colonic mass showing clusters of tumor cells.

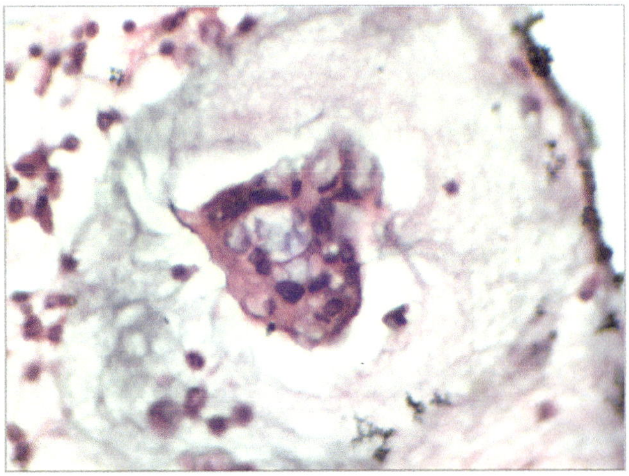

FIG. 2: Adenocarcinoma of colon: Cell block section showing clusters of tumor cells in a mucinous background.

The differentiating ICC pattern of adenocarcinoma of the upper GIT and large intestine is highlighted in **Table 10**.

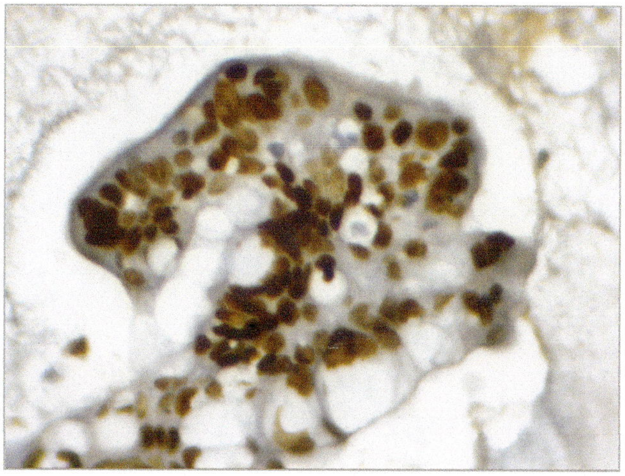

FIG. 3: Adenocarcinoma of colon: Tumor cells are positive for CDX2.

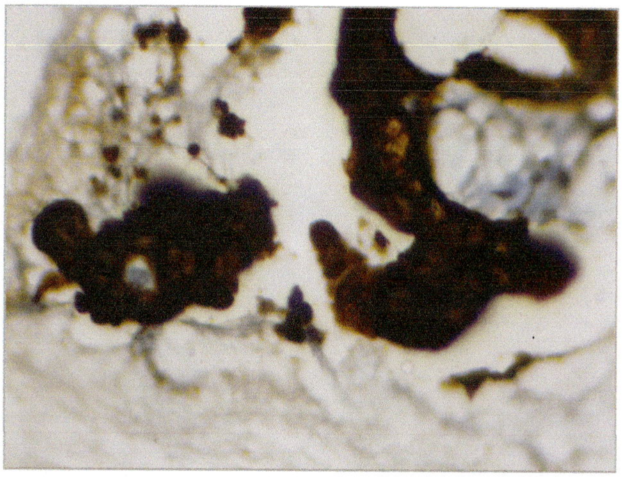

FIG. 4: Adenocarcinoma of colon: Strong membranous positivity of CK20.

TABLE 10: Differentiating the immunocytochemistry pattern of adenocarcinoma of the upper gastrointestinal tract (GIT) and large intestine.

Markers	Adenocarcinoma of upper GIT	Large intestinal adenocarcinoma	Comments
CDX2	Occasionally negative	Mostly positive	Overlapping
CK7	Positive	Negative	Occasionally negative in upper GIT
SATB2	Negative	Positive	The most useful marker

Neuroendocrine Tumors of the GIT Region (Box 1; Figs. 5 to 9)[11,12]

- *Positive*
  - Synaptophysin
  - Chromogranin
  - Neuron-specific enolase (NSE)
  - CD56
  - Ki67 index
  - G1 tumor: <2%
  - G2 tumor: 3–20%
  - G3 tumor: >20%

Gastrointestinal Stromal Tumors (Box 2; Figs. 10 to 15)[10,13]

- *Positive*
  - CD117
  - CD34
  - DOG-1
- *Negative*
  - S100
  - Desmin
  - HMB45

The differential diagnosis of other mesenchymal tumors with GIST is given in **Table 11**.

BOX 1 **Neuroendocrine tumors (NETs) of the gastrointestinal tract.**

- NET of the gastrointestinal tract develop from the neuroendocrine cells of these regions
- NETs are potentially malignant tumor
- The diagnostic markers are synaptophysin and chromogranin
- These tumors are graded according to Ki67 index as grade 1 (<2%), grade 2 (3–20%), and grade 3 (>20%)

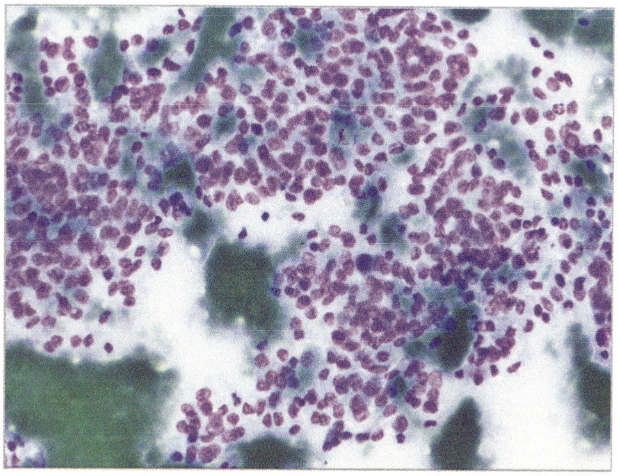

FIG. 5: Neuroendocrine tumor: Fine needle aspiration cytology smear of an appendicular tumor showing monomorphic tumor cells.

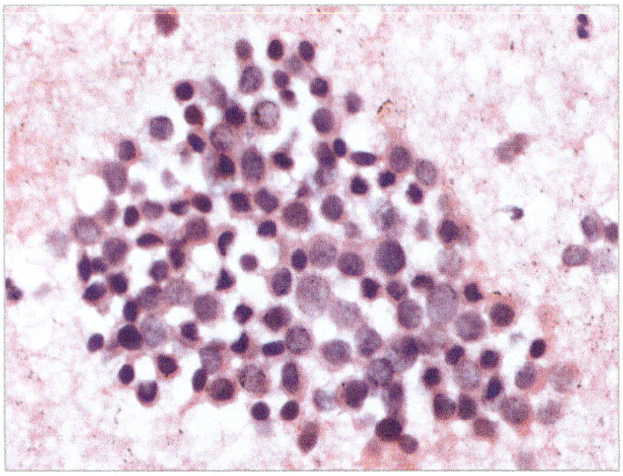

FIG. 6: Neuroendocrine tumor: Cell block section showing clusters of round monomorphic tumor cells.

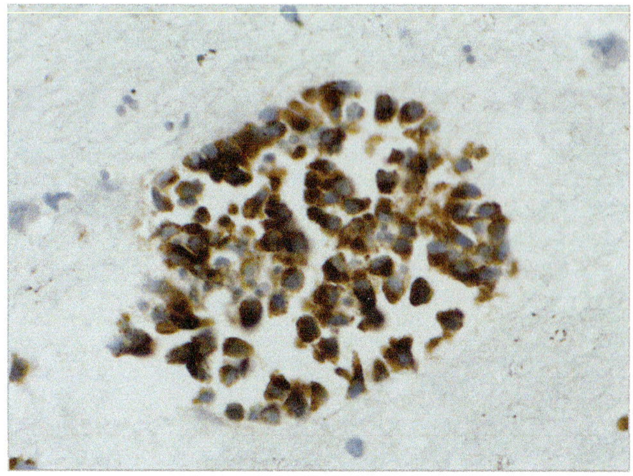

FIG. 7: Neuroendocrine tumor: The cells showing cytoplasmic positivity of chromogranin.

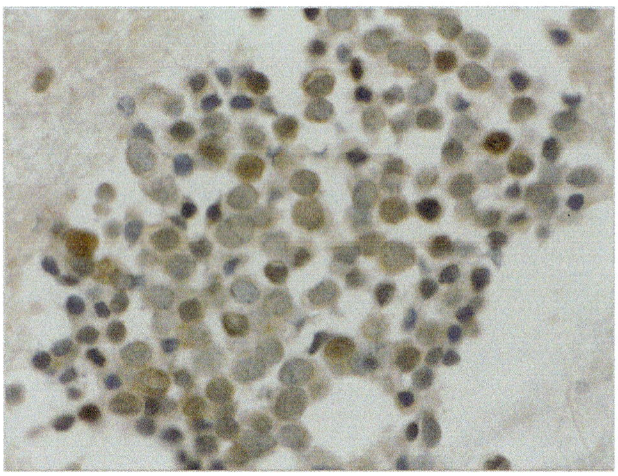

FIG. 8: Neuroendocrine tumor: Occasional cells showing neuron-specific enolase (NSE) positivity.

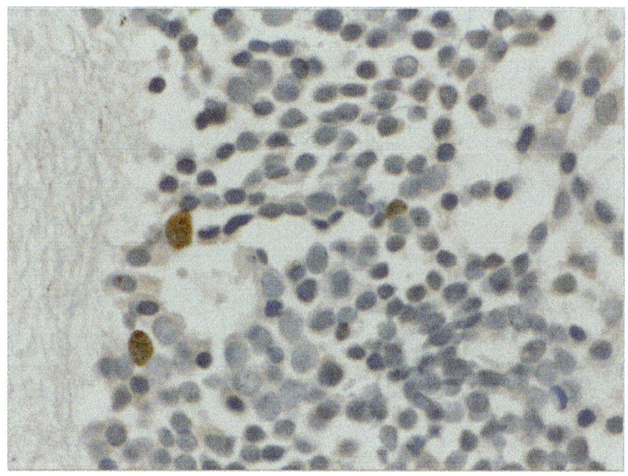

FIG. 9: Neuroendocrine tumor: The tumor cells showing <2% Ki67 positive cells.

BOX 2 Gastrointestinal stromal tumors (GISTs).

- GISTs arise from the cells of Cajal of the intestinal tract
- Majority of GISTs are positive for CD117, DOG-1, and CD34
- DOG-1 is specific for GIST; however, only 50% of CD117 positive cases show DOG-1 positivity

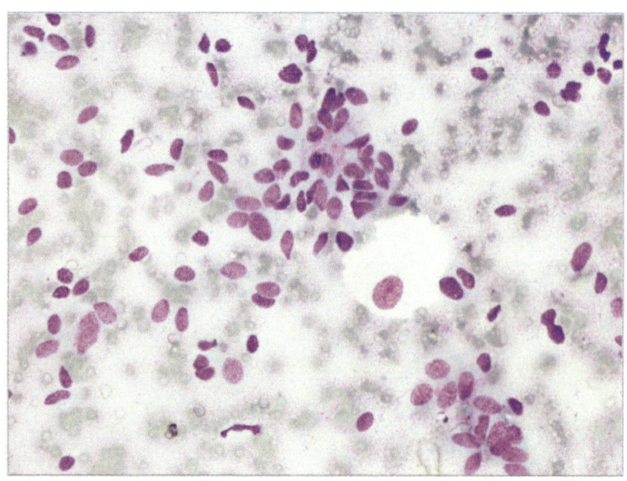

FIG. 10: Gastrointestinal stromal tumor: Fine needle aspiration cytology smear showing discrete oval to spindle cells.

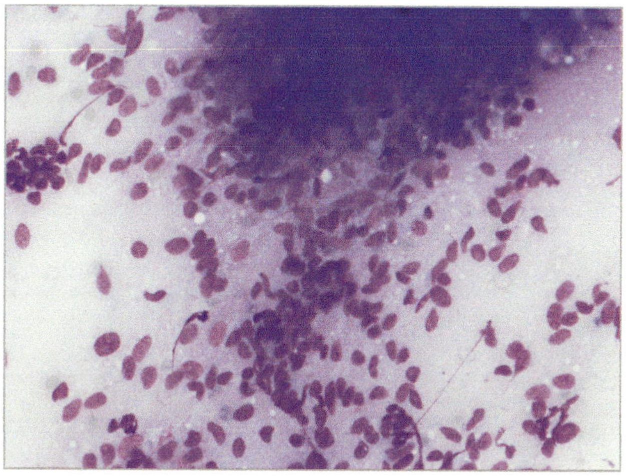

FIG. 11: Gastrointestinal stromal tumor: Fine needle aspiration cytology smear showing cohesive clusters of tumor cells.

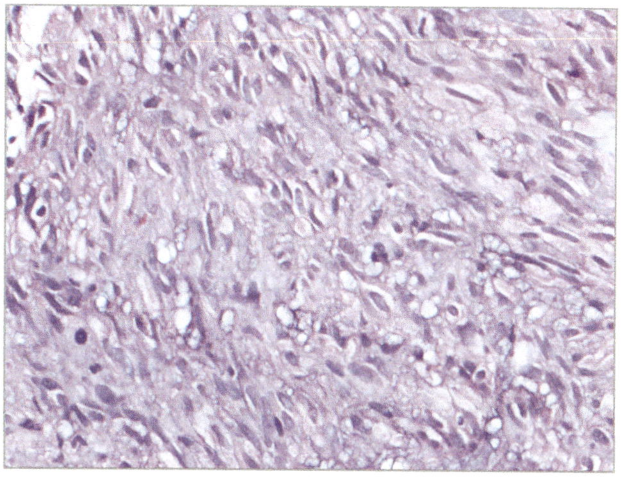

FIG. 12: Gastrointestinal stromal tumor: Cell block sections showing spindle cells.

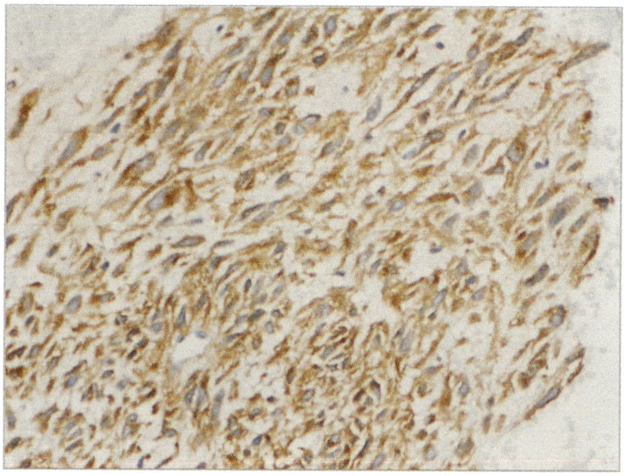

FIG. 13: Gastrointestinal stromal tumor: The tumor cells showing cytoplasmic positivity of CD117.

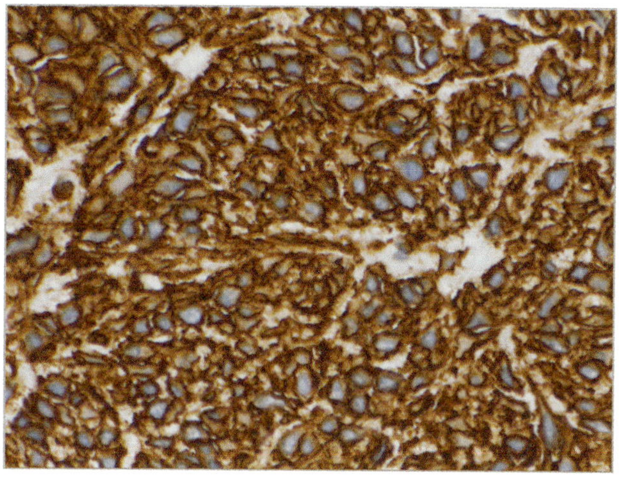

FIG. 14: Gastrointestinal stromal tumor: Cytoplasmic passivity of CD34.

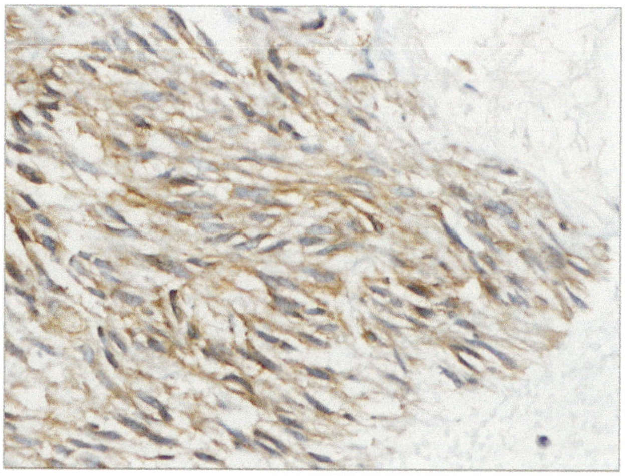

FIG. 15: Gastrointestinal stromal tumor: Cytoplasmic positivity of DOG-1.

TABLE 11: Differential diagnosis of other mesenchymal tumors with GIST.

Markers	GIST	Schwannoma	Leiomyoma	Mesenteric fibromatosis
Desmin	Negative	Negative	Positive	Negative
S100	Negative	Positive	Negative	Negative
CD117	Positive	Negative	Negative	Negative
CD34	Positive	Negative	Negative	Negative

(GIST: gastrointestinal stromal tumors)

REFERENCES

1. Chu P, Wu E, Weiss LM. Cytokeratin 7 and cytokeratin 20 expression in epithelial neoplasms: a survey of 435 cases. Mod Pathol. 2000;13(9): 962-72.
2. Dey P, Togra J, Mitra S. Intermediate filament: structure, function, and applications in cytology. Diagn Cytopathol. 2014;42(7):628-35.
3. Stelow EB, Yaziji H. Immunohistochemistry, carcinomas of unknown primary, and incidence rates. Semin Diagn Pathol. 2018;35(2):143-52.
4. Saad RS, Silverman JF, Khalifa MA, Rowsell C. CDX2, cytokeratins 7 and 20 immunoreactivity in rectal adenocarcinoma. Appl Immunohistochem Mol Morphol. 2009;17(3):196-201.
5. Zhao LH, Chen C, Mao CY, Xiao H, Fu P, Xiao HL, et al. Value of SATB2, ISL1, and TTF1 to differentiate rectal from other gastrointestinal and lung well-differentiated neuroendocrine tumors. Pathol Res Pract. 2019;215(7):152448.
6. Zhang B, Sun Z, Song M, Ma S, Tian Y, Kong Q. Ultrasound/CT combined with serum CEA/CA19.9 in the diagnosis and prognosis of rectal cancer. J BUON. 2018;23(3):592-7.
7. Hammarström S. The carcinoembryonic antigen (CEA) family: structures, suggested functions and expression in normal and malignant tissues. Semin Cancer Biol. 1999;9(2):67-81.
8. Lloyd RV. Practical markers used in the diagnosis of neuroendocrine tumors. Endocr Pathol. 2003;14(4):293-301.
9. Oberg K, Stridsberg M. Chromogranins as diagnostic and prognostic markers in neuroendocrine tumours. Adv Exp Med Biol. 2000;482:329-37.
10. Miettinen M, Lasota J. KIT (CD117): a review on expression in normal and neoplastic tissues, and mutations and their clinicopathologic correlation. Appl Immunohistochem Mol Morphol. 2005;13(3):205-20.
10. Parab TM, DeRogatis MJ, Boaz AM, Grasso SA, Issack PS, Duarte DA, et al. Gastrointestinal stromal tumors: a comprehensive review. J Gastrointest Oncol. 2019;10(1):144-54.
11. Wilander E. Diagnostic pathology of gastrointestinal and pancreatic neuroendocrine tumours. Acta Oncol. 1989;28(3):363-9.
12. Chejfec G, Falkmer S, Askensten U, Grimelius L, Gould VE. Neuroendocrine tumors of the gastrointestinal tract. Pathol Res Pract. 1988;183(2):143-54.
13. Miettinen M, Majidi M, Lasota J. Pathology and diagnostic criteria of gastrointestinal stromal tumors (GISTs): a review. Eur J Cancer. 2002;38:S39-51.

CHAPTER 14

Immunocytochemistry of Liver and Bile Duct Tumors

INTRODUCTION

The commonly used markers in hepatocellular tumors are: Hepatocyte paraffin-1 (HepPar-1), arginase-1, alpha-fetoprotein (AFP), and glypican-3.

HEPATOCYTE PARAFFIN 1

Hepatocyte paraffin-1 is a urea cycle enzyme. It is present on the membrane of mitochondria. It is a reliable marker with high sensitivity and specificity for the identification of hepatocellular carcinoma (HCC)[1,2] (**Table 1**).

GLYPICAN-3

Glypican-3 is a heparan sulfate proteoglycan. It is present both in the cell membrane of the fetal hepatocytes and extracellularly. Adult liver cells do not express glypican-3, and therefore, it is helpful to distinguish malignant from the benign hepatocytes[3,4] (**Table 2**).

ARGINASE-1

Arginase-1 is a urea cycle enzyme. It helps in the conversion of arginine to ornithine. It is a specific (90–100%) marker of hepatocytes and also HCC (**Table 3**). The sensitivity and specificity of arginase-1 are better than HepPar-1.[2]

ALPHA-FETOPROTEIN

Alpha-fetoprotein is an oncofetal glycoprotein. It is present in fetal hepatocytes, epithelial cells of the gastrointestinal tract (GIT), and yolk sac. About 70% of HCC express AFP.[5,6] The other tumors with hepatoid differentiation and yolk sac tumors also show AFP positivity (**Table 4**).

TABLE 1: Hepatocyte paraffin-1 (HepPar-1) expression.

Markers	Positivity in the liver tumors	Expression in extrahepatic tumors	Localization
HepPar-1	Hepatocellular carcinoma	• Signet carcinoma of the stomach • Ovarian carcinoma with hepatoid differentiation • Intestinal adenocarcinoma • Yolk sac tumor	Cytoplasmic

TABLE 2: Glypican-3 expression.

Markers	Positivity in the liver tumors	Expression in extrahepatic tumors	Localization
Glypican-3	Hepatocellular carcinoma	• Yolk sac tumor of ovary • Neuroblastoma • Acinar cell carcinoma of pancreas • Nephroblastoma • Rhabdomyosarcoma • Liposarcoma	Cytoplasmic and membranous

TABLE 3: Arginase-1 expression.

Markers	Positivity in the liver tumors	Expression in extrahepatic tumors	Localization
Arginase-1	Hepatocellular carcinoma	• Myeloid cells • Macrophages	Cytoplasmic and membranous

TABLE 4: Alpha-fetoprotein (AFP) expression.

Markers	Positivity in the liver tumors	Expression in extrahepatic tumors	Localization
Alpha-fetoprotein	Hepatocellular carcinoma	• Yolk sac tumor • Hepatoid differentiation in ovary, pancreas, stomach, lung	Cytoplasmic

IMMUNOCYTOCHEMISTRY EXPRESSION PATTERN IN HEPATOCELLULAR TUMORS

Hepatocellular Adenoma

The immunocytochemistry (ICC) pattern of hepatocellular adenoma is as follows:

- *Positive*
  - Arginase-1
  - HepPar-1
  - CD34
- *Negative*
  - Glypican-3
  - Heat shock protein 70 (HSP70)

Hepatocellular Carcinoma

The ICC pattern of HCC[6] is (**Box 1**; **Figs. 1** to **6**):

- *Positive*
  - Arginase-1
  - HepPar-1
  - Glypican-3
  - AFP
  - CD34
  - CD10 and polyclonal-carcinoembryonic antigen (CEA): Canalicular staining pattern
- *Negative*
  - Epithelial membrane antigen (EMA)
  - Epithelial cell adhesion molecule (EpCAM)

Glypican-3 and HepPar-1 are the most commonly used markers of HCC.

To differentiate benign from the malignant hepatocytes is often very difficult; however, glypican-3 and HSP70 positivity may be helpful in this aspect (**Table 5**).

The liver is a rich source of the metastatic tumor, and ICC may help to confirm the metastatic from the HCC (**Figs. 1** to **11**). A combination of different antibodies may be useful in this respect[6,7] (**Table 6**).

BOX 1 **Hepatocellular carcinoma (HCC).**

- Hepatocellular carcinoma is positive for: Arginase-1, hepatocyte paraffin-1 (HepPar-1), glypican-3, AFP, CD34, CD10, and polyclonal-CEA
- HepPar-1 stains both benign and neoplastic hepatocytes
- Glypican-3 stains only neoplastic hepatocytes and more sensitive than HepPar-1
- Arginase-1 is highly sensitive and specific marker

(AFP: alpha-fetoprotein; CEA: carcinoembryonic antigen)

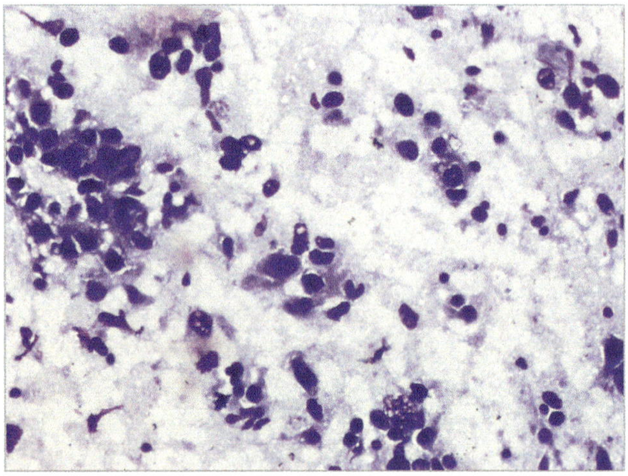

FIG. 1: Hepatocellular carcinoma (HCC): Fine needle aspiration cytology smear showing discrete oval to polyhedral tumor cells.

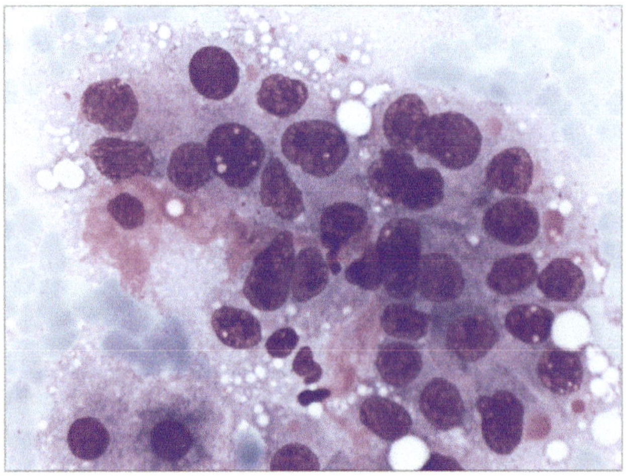

FIG. 2: Hepatocellular carcinoma (HCC): Fine needle aspiration cytology smear showing large polygonal cells with enlarged pleomorphic nuclei.

Hepatoblastoma[8]

The ICC pattern of hepatoblastoma is:

- *Positive*
  - HepPar-1
  - Glypican-3
  - AE1/AE3

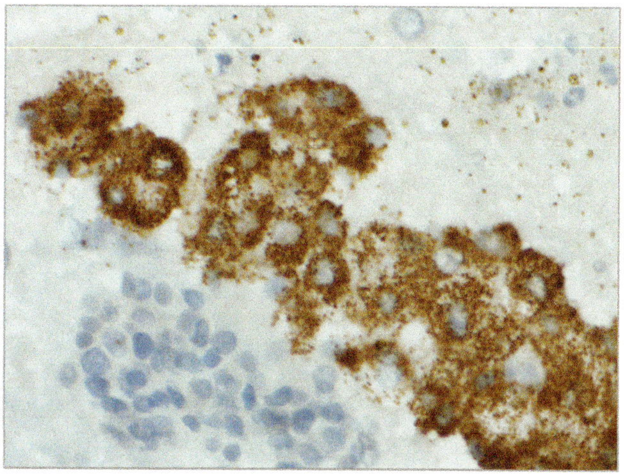

FIG. 3: Hepatocellular carcinoma (HCC): Hepatocyte paraffin-1 (HepPar-1) positive.

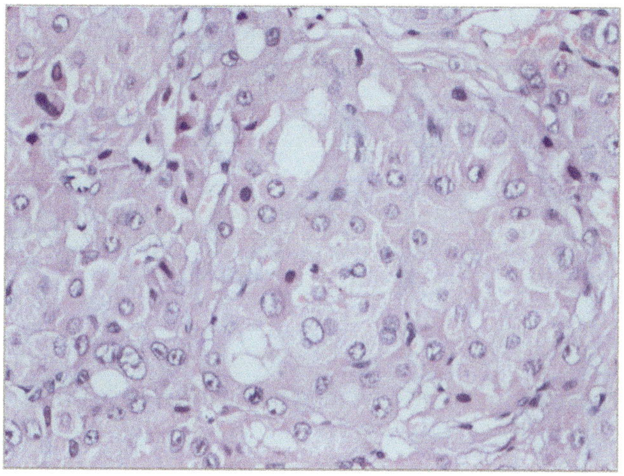

FIG. 4: Hepatocellular carcinoma (HCC): Cell block sections showing tumor cells.

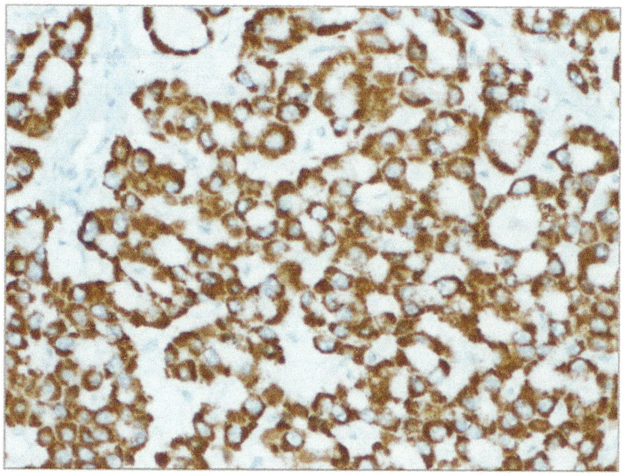

FIG. 5: Hepatocellular carcinoma (HCC): Hepatocyte paraffin-1 (HepPar-1) positive tumor cells.

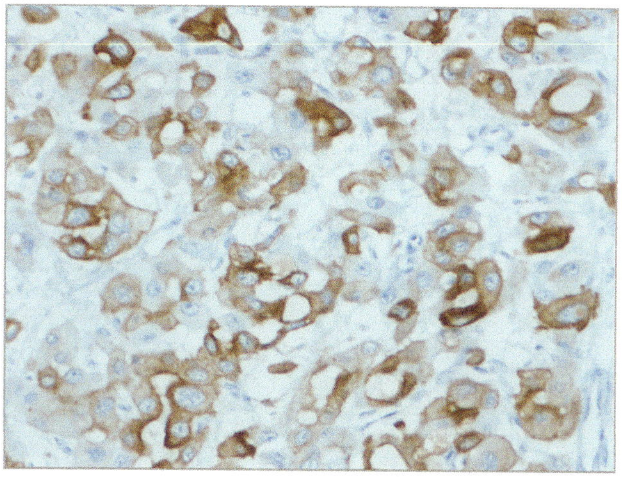

FIG. 6: Hepatocellular carcinoma (HCC): CK7 positive tumor cells.

TABLE 5: Benign versus malignant hepatocytes.[1-6]

Markers	Benign	Malignant
Glypican-3	Negative	Positive
HSP70	Negative	Positive
CD34	Sinusoidal pattern focally present	Sinusoidal pattern diffusely present
Glutamine synthetase	Occasionally positive	Positive

(HSP70: heat shock protein 70)

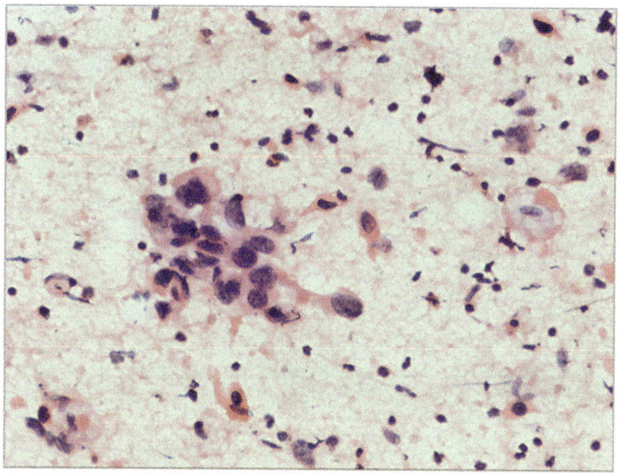

FIG. 7: Metastatic squamous cell carcinoma in liver: Fine needle aspiration cytology smear showing discrete and loose clusters of polyhedral tumor cells.

- Arginase-1
- Beta-catenin
- Vimentin

Cholangiocarcinoma

The ICC pattern of cholangiocarcinoma is[9,10] (**Box 2**; **Figs. 12 to 14**):

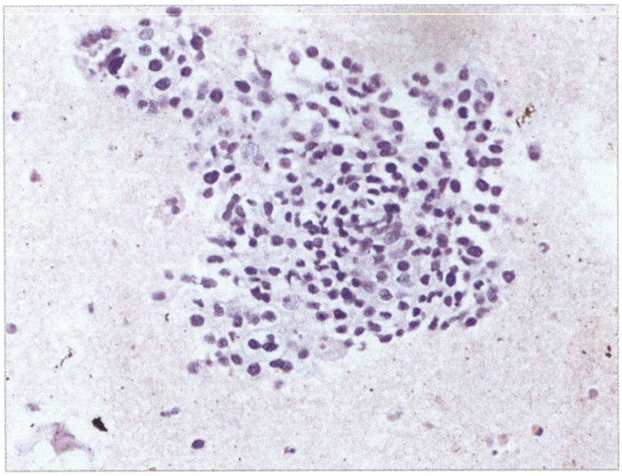

FIG. 8: Metastatic squamous cell carcinoma in liver: Cell block section showing clusters of malignant cells.

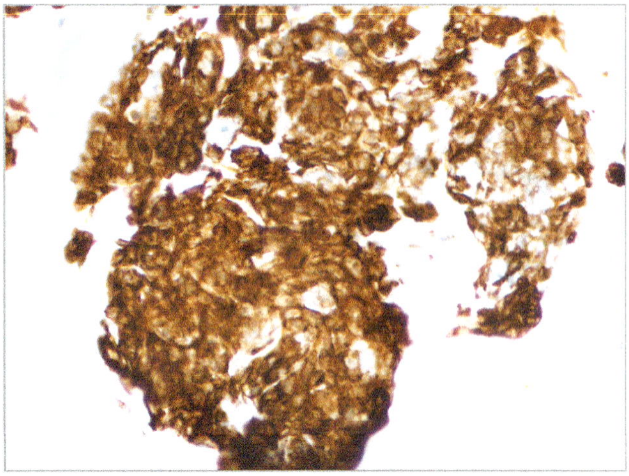

FIG. 10: Metastatic squamous cell carcinoma in liver: The malignant cells showing membranous positivity of CK5/6.

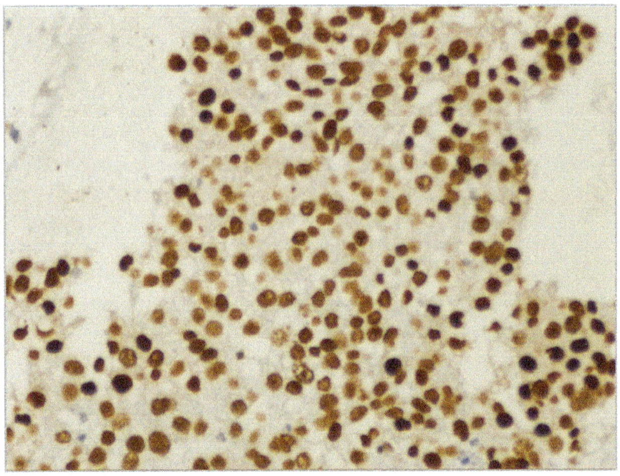

FIG. 9: Metastatic squamous cell carcinoma in liver: Nuclear positive p63.

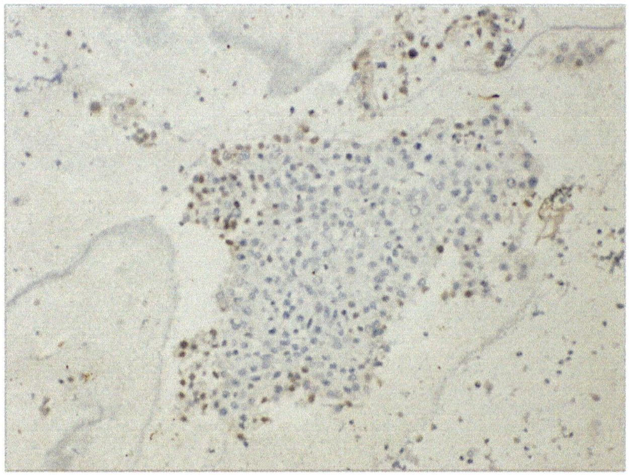

FIG. 11: Metastatic squamous cell carcinoma in liver: The cells are negative for CK7.

TABLE 6: Differentiating features of hepatocellular carcinoma (HCC) and metastatic adenocarcinomas from the different organs.

	HCC	Lung	GIT	RCC	Breast	Endocervix
Glypican-3	Positive	Negative	Negative	Negative	Negative	Negative
Arginase-1	Positive	Negative	Negative	Negative	Negative	Negative
PAX8	Negative	Negative	Negative	Positive	Negative	Positive
CD10	Negative	Negative	Negative	Positive	Negative	Negative
TTF1	Negative	Positive	Negative	Negative	Negative	Negative
Napsin	Negative	Positive	Negative	Negative	Negative	Negative
CDX2	Negative	Negative	Positive	Negative	Negative	Negative
GATA3	Negative	Negative	Negative	Negative	Positive	Negative
Mammaglobin	Negative	Negative	Negative	Negative	Positive	Negative
p16	Negative	Negative	Negative	Negative	Negative	Positive

(GIT: gastrointestinal tract; HCC: hepatocellular carcinoma; RCC: renal cell carcinoma)

BOX 2 Cholangiocarcinoma.

- Positive: CK7 and CK19
- Positive for MOC-31 and Ber-EP4
- This tumor is always negative for HepPar-1
- Immunocytochemistry may not be helpful to differentiate cholangiocarcinoma and metastatic adenocarcinoma

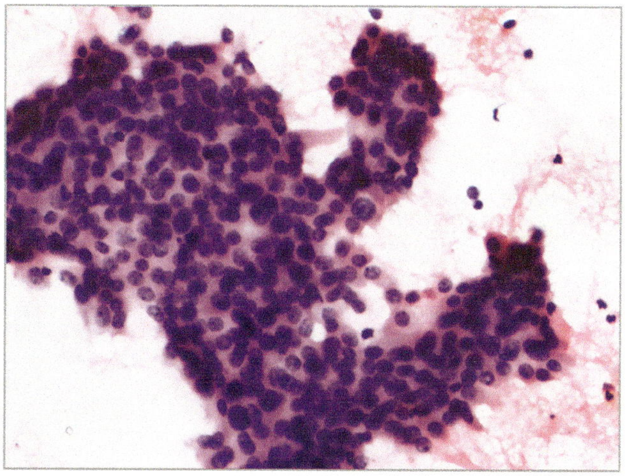

FIG. 12: Cholangiocarcinoma: Fine needle aspiration cytology smear showing large cluster of tumor cells.

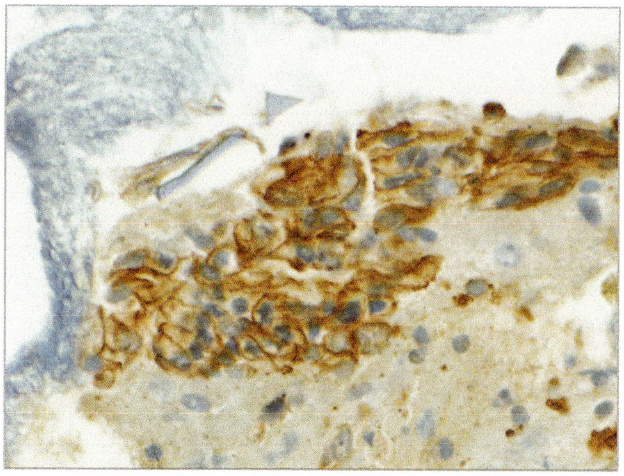

FIG. 13: Cholangiocarcinoma: The cells are positive for CK7.

- *Positive*
  - CK7
  - CK19
  - CAM5.2
  - MOC-31
  - CEA
  - PDX-1
  - Claudin

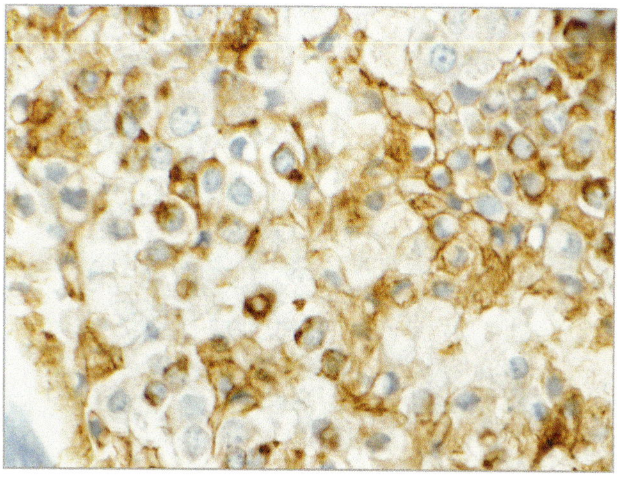

FIG. 14: Cholangiocarcinoma: Membranous positivity of CK19.

TABLE 7: Distinguishing immunocytochemistry (ICC) of cholangiocarcinoma versus hepatocellular carcinoma (HCC).

Markers	Cholangiocarcinoma	Hepatocellular carcinoma
CK7	Positive	Negative
CK19	Positive	Negative
HepPar-1	Negative	Positive
Arginase-1	Negative	Positive
Glypican-3	Negative	Positive
Alpha-fetoprotein	Negative	Positive

Hepatocyte paraffin-1, CK7, and CK19 combination may be helpful to differentiate cholangiocarcinoma from HCC (**Table 7**).

Gall Bladder Adenocarcinoma

- *Positive*
  - CK7
  - CK19
  - CEA
  - EMA
  - Maspin

REFERENCES

1. Ibrahim TR, Abdel-Raouf SM. Immunohistochemical study of glypican-3 and HepPar-1 in differentiating hepatocellular carcinoma from metastatic carcinomas in FNA of the Liver. Pathol Oncol Res. 2015;21(2):379-87.
2. Radwan NA, Ahmed NS. The diagnostic value of arginase-1 immunostaining in differentiating hepatocellular carcinoma from metastatic carcinoma and cholangiocarcinoma as compared to HepPar-1. Diagn Pathol. 2012;7:149.

3. Zhou F, Shang W, Yu X, Tian J. Glypican-3: A promising biomarker for hepatocellular carcinoma diagnosis and treatment. Med Res Rev. 2018;38(2):741-67.
4. El-Saadany S, El-Demerdash T, Helmy A, Mayah WW, El-Sayed Hussein B, Hassanien M, et al. Diagnostic value of glypican-3 for hepatocellular carcinomas. Asian Pac J Cancer Prev. 2018;19(3):811-7.
5. Johnson PJ. Role of alpha-fetoprotein in the diagnosis and management of hepatocellular carcinoma. J Gastroenterol Hepatol. 1999;14(Suppl): S32-6.
6. Choi WT, Kakar S. Immunohistochemistry in the diagnosis of hepatocellular carcinoma. Gastroenterol Clin North Am. 2017;46(2):311-25.
7. Ibrahim TR, Abdel-Raouf SM. Immunohistochemical study of glypican-3 and HepPar-1 in differentiating hepatocellular carcinoma from metastatic carcinomas in FNA of the liver. Pathol Oncol Res. 2015;21(2):379-87.
8. Hooks KB, Audoux J, Fazli H, Lesjean S, Ernault T, Dugot-Senant N, et al. New insights into diagnosis and therapeutic options for proliferative hepatoblastoma. Hepatology. 2018;68(1):89-102.
9. Lendvai G, Szekerczés T, Illyés I, Dóra R, Kontsek E, Gógl A, et al. Cholangiocarcinoma: Classification, histopathology and molecular carcinogenesis. Pathol Oncol Res. 2020;26(1):3-15.
10. Olnes MJ, Erlich R. A review and update on cholangiocarcinoma. Oncology. 2004;66(3):167-79.

CHAPTER 15

Immunocytochemistry of Pancreatic Tumors

INTRODUCTION

The tumors of the pancreatic ductal epithelial and acinar markers include:

- CAM5.2
- CK8
- CK7
- Carcinoembryonic antigen (CEA)
- PDX-1

The following markers are used for pancreatic endocrine tumors:

- CK
- Chromogranin
- Synaptophysin
- Other endocrine enzyme markers: Glucagon, insulin, gastrin, vasoactive intestinal polypeptides
- Ki67

PANCREATIC AND DUODENAL HOMEOBOX-1

Pancreatic and duodenal homeobox-1 (PDX-1), a transcription factor, helps in the development of the pancreas. It is positive in the endocrine cells of the pancreas[1] (**Table 1**).

S100P

S100P is a member of the S100 protein group. It is expressed in normal placenta and also in the epithelial cells of the gastrointestinal tract (GIT). S100P is positive in pancreatic ductal adenocarcinoma[2] and breast carcinoma (**Table 2**).

TABLE 1: Pancreatic and duodenal homeobox-1 (PDX-1) expression.

Marker	Positivity in the pancreas	Other tumors that show positivity	Localization
PDX-1	• Pancreatobiliary adenocarcinoma • Neuroendocrine tumor of the duodenum	• Pancreatic acinar cell carcinoma • Adenocarcinoma of stomach • Colonic and rectal adenocarcinoma • Adenocarcinoma of prostate	Nuclear

PANCREATIC DUCTAL ADENOCARCINOMA

The immunocytochemistry (ICC) pattern of the pancreatic ductal adenocarcinoma is:

- *Positive*
  - CK7
  - CK19
  - S100P
  - pVHL
  - Maspin
  - IMP3
  - MUC1
  - MUC3
  - MUC5/6
  - CEA
  - Epithelial membrane antigen (EMA)
- *Negative*
  - CK20
  - MUC2

The best panel of pancreatic ductal adenocarcinoma consists of pVHL, Maspin, IMP3, and S100P.

TABLE 2: S100P expression.

Marker	Positivity in the pancreas	Other tumors that show positivity	Localization
S100P	Pancreatic adenocarcinoma	• Ovarian adenocarcinoma • Lung (non-small) • Breast carcinoma	Nuclear and cytoplasmic

ACINAR CELL CARCINOMA

The ICC pattern of pancreatic acinar cell carcinoma[3] is:

- *Positive*
  - Trypsin
  - Bcl-10
  - CK18
  - Chymotrypsin
- *Negative*
  - CK20
  - S100P
  - MUC1 and MUC2

SOLID AND CYSTIC PAPILLARY NEOPLASM OF THE PANCREAS

Solid and cystic papillary neoplasm (SCPN) of the pancreas is also called "solid pseudopapillary neoplasm."

The ICC pattern of SCPN of pancreas[4,5] is (**Box 1**):

- *Positive*
  - Beta-catenin
  - CD10
  - Neuron-specific enolase (NSE)
  - CD56
  - Vimentin
  - Alpha-1 antitrypsin
- *Negative*
  - E-cadherin
  - Chymotrypsin
  - Chromogranin

The most useful panel of SCPN consists of beta-catenin, E-cadherin, and CD10 and chromogranin.

PANCREATOBLASTOMA

The ICC pattern of pancreatoblastoma[6] is:

- Acinar cells
  - *Positive*:
    - CK7
    - CK19
    - EMA
    - Trypsin
    - CAM5.2
- Ductal cells
  - *Positive*:
    - CK7
    - CK19
    - CAM5.2
    - CEA

BOX 1 Solid and cystic papillary neoplasm.

- *Positive*: Beta-catenin, alpha-1 antitrypsin, vimentin, and CD10
- *Negative*: Chromogranin, acinar, and ductal markers

NEUROENDOCRINE TUMORS OF THE PANCREAS

The general markers of all types of neuroendocrine tumors of the pancreas are[7,8] (**Box 2; Figs. 1** to **5**):

- *Positive*
  - Synaptophysin
  - Chromogranin

BOX 2 Neuroendocrine tumors of the pancreas.

- *Positive*: Synaptophysin, chromogranin, neuron-specific enolase (NSE), and CD56
- *Negative*: CK20, CK7, and vimentin
- The tumor cell may be positive for one or more pancreatic hormone (insulin, gastrin, somatostatin, etc.)
- The tumor is graded according to the Ki67 index

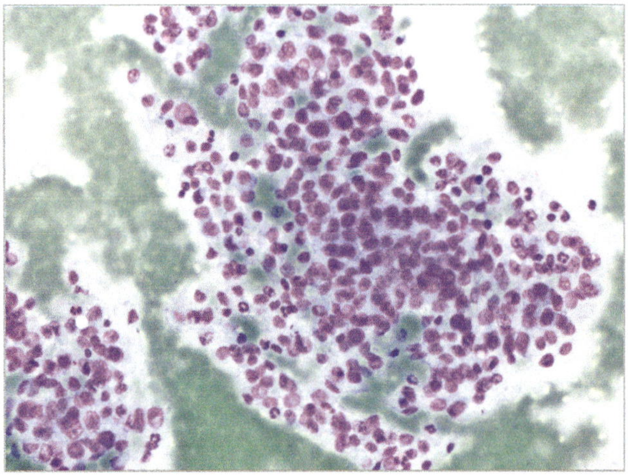

FIG. 1: Neuroendocrine tumor: Fine needle aspiration cytology smear showing round monomorphic cells.

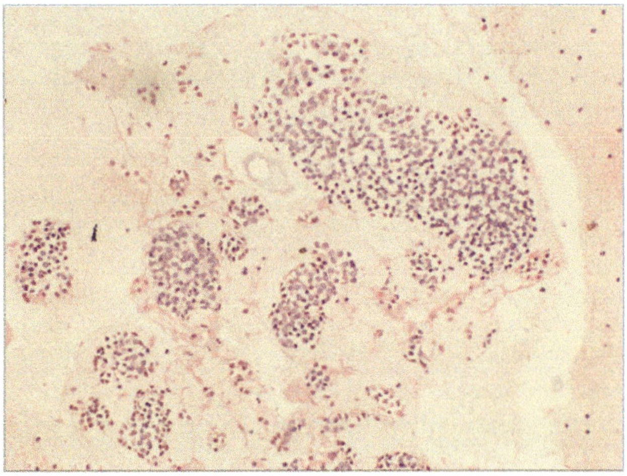

FIG. 2: Neuroendocrine tumor: Cell block section showing islands of tumor cells.

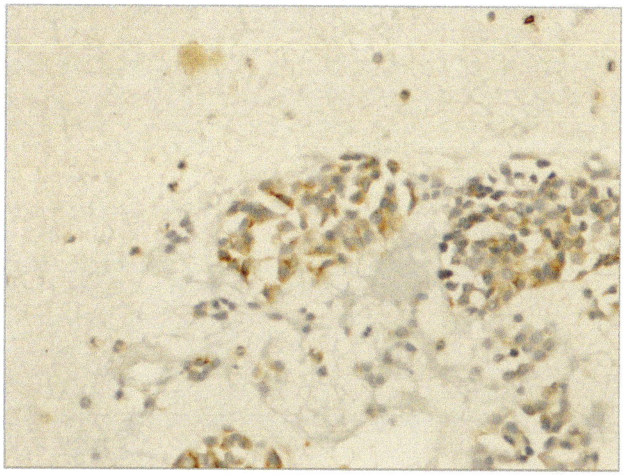

FIG. 3: Neuroendocrine tumor: Tumor cells showing chromogranin positivity.

  - NSE
  - CD56
- *Negative*
  - CK20
  - Vimentin
  - CK7

The grade of the neuroendocrine tumor is based on Ki67 indexing[9] (**Fig. 6**).

- Ki67 index and grading of neuroendocrine tumor
  - Grade I = Less than 2%
  - Grade II = 3–20%
  - Grade III = More than 20%

Further special stains are needed for the exact categorization of the neuroendocrine cells (**Figs. 7** and **8**).

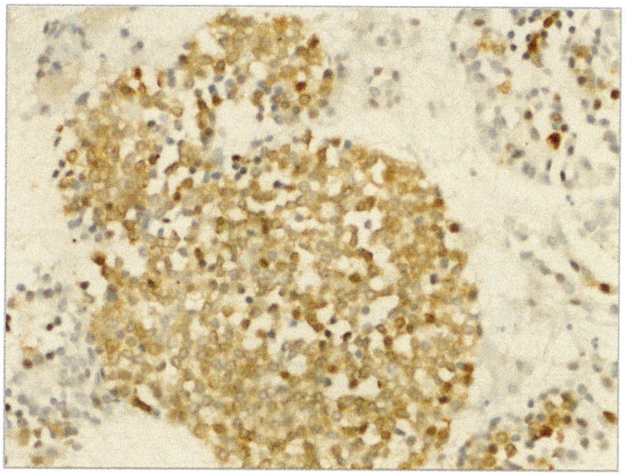

FIG. 4: Neuroendocrine tumor: Tumor cells are positive for neuron-specific enolase (NSE).

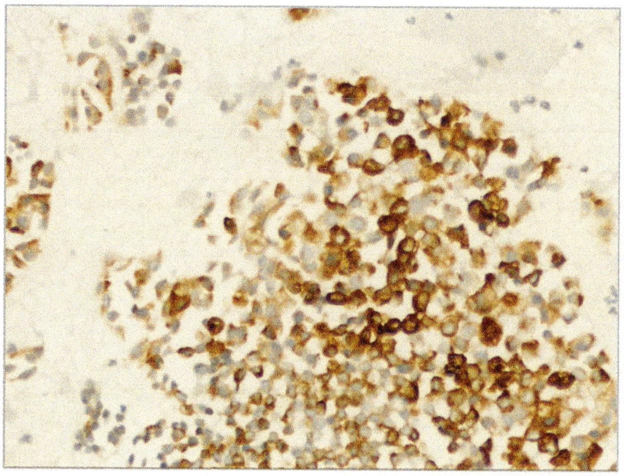

FIG. 5: Neuroendocrine tumor: Membranous positivity of CK.

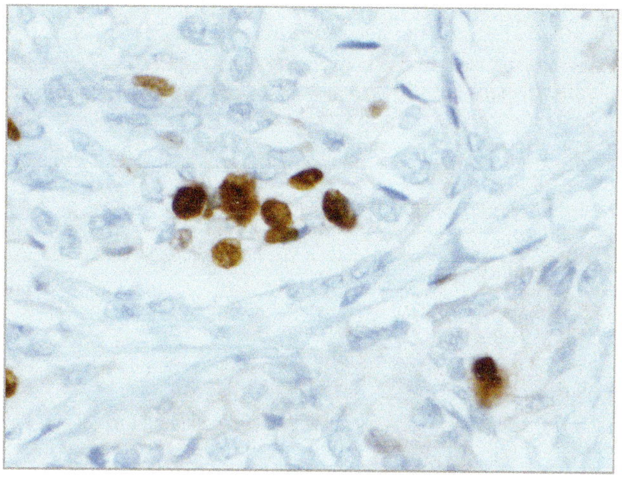

FIG. 6: Neuroendocrine tumor: The tumor cells showing 5% Ki67 indicating grade II tumor.

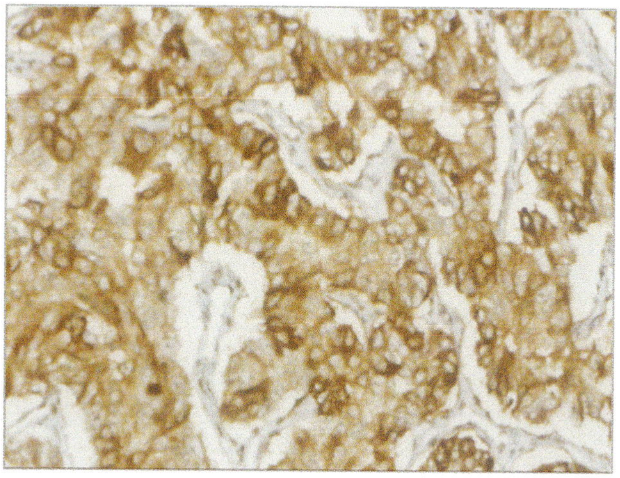

FIG. 7: Neuroendocrine tumor: Chromogranin positive cells in the resected specimen of the pancreas.

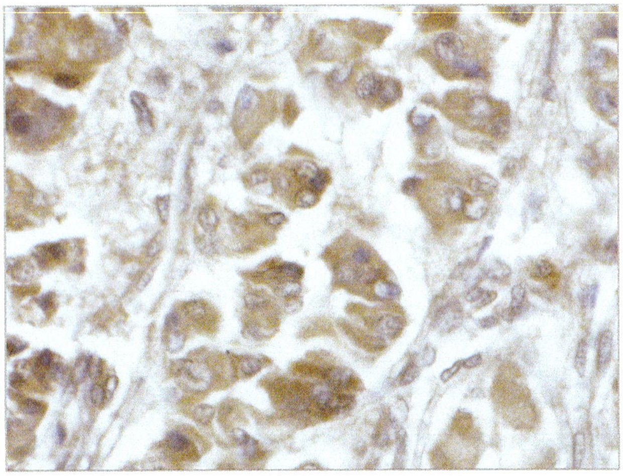

FIG. 8: Neuroendocrine tumor: The tumor cells are positive for gastrin indicating the tumor is a gastrin-producing tumor.

REFERENCES

1. Chan ES, Alexander J, Swanson PE, Jain D, Yeh MM. PDX-1, CDX-2, TTF-1, and CK7: a reliable immunohistochemical panel for pancreatic neuroendocrine neoplasms. Am J Surg Pathol. 2012;36(5):737-43.
2. Aksoy-Altinboga A, Baglan T, Umudum H, Ceyhan K. Diagnostic Value of S100p, IMP3, Maspin, and pVHL in the differential diagnosis of pancreatic ductal adenocarcinoma and normal/chronic pancreatitis in fine needle aspiration biopsy. J Cytol. 2018;35(4):247-51.
3. Ordóñez NG. Pancreatic acinar cell carcinoma. Adv Anat Pathol. 2001;8(3):144-59.
4. Laje P, Bhatti TR, Adzick NS. Solid pseudopapillary neoplasm of the pancreas in children: a 15-year experience and the identification of a unique immunohistochemical marker. J Pediatr Surg. 2013;48(10):2054-60.
5. Ozguven BY, Tuncel D, Polat N, Sakiz D, Kabukcuoglu F, Koksal H, et al. Solid-pseudopapillary neoplasm of the pancreas: Clinicopathologic and immunohistochemical analysis of nine cases. Indian J Pathol Microbiol. 2015;58(3):292-5.
6. al-Kaisi N, Weaver MG, Abdul-Karim FW, Siegler E. Fine needle aspiration cytology of neuroendocrine tumors of the pancreas. A cytologic, immunocytochemical and electron microscopic study. Acta Cytol. 1992;36(5):655-60.
7. Uccella S, Sessa F, La Rosa S. Diagnostic approach to neuroendocrine neoplasms of the gastrointestinal tract and pancreas. Turk Patoloji Derg. 2015;31(Suppl 1):113-27.
8. Guilmette JM, Nosé V. Neoplasms of the neuroendocrine pancreas: An update in the classification, definition, and molecular genetic advances. Adv Anat Pathol. 2019;26(1):13-30.
9. Kim JY, Hong SM, Ro JY. Recent updates on grading and classification of neuroendocrine tumors. Ann Diagn Pathol. 2017;29:11-6.

CHAPTER 16

Immunocytochemistry of Kidney and Urinary Tract Tumors

INTRODUCTION

The panel of diagnostic antibodies in kidney tumors include renal cell carcinoma (RCC) antibody, CD10, paired box gene 8 (PAX8), paired box gene 2 (PAX2), carbonic anhydrase, GATA binding protein 3 (GATA3), human kidney injury molecule-1 (KIM-1), cytokeratin, and DOG1.

Renal Cell Carcinoma Marker

Renal cell carcinoma marker, the glycoprotein, is expressed in the cells of proximal renal tubules and RCC. It is a useful marker for the primary RCC than for the metastatic RCC[1] (**Table 1**).

PAIRED BOX GENE 8

Paired box gene 8 is a transcription factor. It is a member of the paired box gene group. *PAX8* gene is needed for the development and differentiation of the Müllerian system, renal and central nervous systems, and thyroid gland. PAX8 is expressed in follicular and papillary carcinoma of the thyroid, RCC, and ovarian carcinoma.[2] It is widely expressed in the various tumors, and therefore PAX8 should be used along with the other markers (**Table 2**).

PAIRED BOX GENE 2

PAX2 is also a member of the paired box gene family. This is a transcription factor and is associated with the Wolffian and Müllerian duct systems. PAX2 is used as a marker of RCC and tumors developed from the Müllerian duct origin[3,4] (**Table 3**).

TABLE 1: Expression of renal cell carcinoma (RCC) marker.

Markers	Kidney lesion	Other tumors	Localization
RCC	• Clear cell type RCC, • Papillary RCC • Chromophobe adenoma	• Breast carcinoma • Parathyroid adenoma	Membranous and cytoplasmic

CD10

CD10 is also called common acute lymphoblastic leukemia antigen (CALLA). It is membrane metallopeptidase and is expressed in the tubular epithelial cells of the kidney[1] and even various types of non-Hodgkin lymphoma (NHL) (**Table 4**).

TABLE 2: Expression of paired box gene 8 (PAX8).

Markers	Kidney lesion	Other tumors	Localization
PAX8	Renal cell carcinoma (RCC): • Clear cell type • Papillary • Chromophobe • Collecting duct carcinoma • Nephroblastoma	• The tumors developed from the Müllerian system: Serous ovarian carcinoma, endometrial adenocarcinoma • Thyroid carcinomas: Follicular carcinoma, papillary carcinoma • RCC • Thymoma • Thymic carcinoma • Seminoma • Merkel cell carcinoma	Nuclear

TABLE 3: Paired box gene 2 (PAX2) expression.

Markers	Kidney lesion	Other tumors	Localization
PAX2	Renal cell carcinoma except chromophobe type	• The tumors developed from the Müllerian system: Ovarian serous carcinoma, endometrial adenocarcinoma • Lobular carcinoma of breast • Merkel cell carcinoma • Hepatocellular carcinoma	Nuclear

CARBONIC ANHYDRASE IX

Carbonic anhydrase IX (CA IX) is a zinc metalloenzyme associated with hydration of CO_2 and is involved in cellular proliferation and tumor development. It is positive in RCC and helps to differentiate benign renal cells and RCC (**Table 5**).

HUMAN KIDNEY INJURY MOLECULE-1

Human kidney injury molecule-1 is a transmembrane glycoprotein and is expressed in the proximal renal tubular cells after injury. KIM1 is positive in the clear cell type of RCC (**Table 6**).

Renal Oncocytoma

- *Positive*
  - AE1/AE3
  - CD117
  - CD10
  - E-cadherin
- *Negative*
  - Vimentin
  - CA IX
  - RCC

Renal Cell Carcinoma

The immunocytochemistry (ICC) pattern of clear cell carcinoma of the kidney is (**Box 1; Figs. 1** to **6**):

- *Positive*
  - PAX2
  - PAX8

BOX 1 Clear cell renal cell carcinoma (RCC).

- Renal cell carcinoma is positive for: PAX2, PAX8, KIM1, EMA, vimentin, CA IX, and RCC
- Renal cell carcinoma is negative for CK7, CK20, and CD117
- Negative staining of GATA3, CK5/6, and p63 helps to differentiate clear cell RCC from the urothelial cell carcinoma

TABLE 4: CD10 expression.

Markers	Kidney lesion	Other tumors	Localization
CD10	• Renal cell carcinoma • Collecting duct carcinoma	• Follicular lymphoma • Diffuse large B-cell lymphoma • Endometrial stromal sarcoma • Acute lymphoblastic leukemia	Membranous

TABLE 5: Carbonic anhydrase IX (CA IX) expression.

Markers	Kidney lesion	Other tumors	Localization
Carbonic anhydrase IX	Renal cell carcinoma	• Breast carcinoma • Urothelial cell carcinoma • Cervical carcinoma • Endometrial carcinoma • Oral squamous cell carcinoma • Alveolar soft part sarcoma	Membranous and cytoplasmic

TABLE 6: Human kidney injury molecule-1 (KIM-1) expression.

Markers	Kidney lesion	Other tumors	Localization
Human kidney injury molecule-1	Renal cell carcinoma (clear cell type)	• Hepatocellular carcinoma • Clear cell carcinoma of the uterus • Ovarian carcinoma	Membranous and cytoplasmic

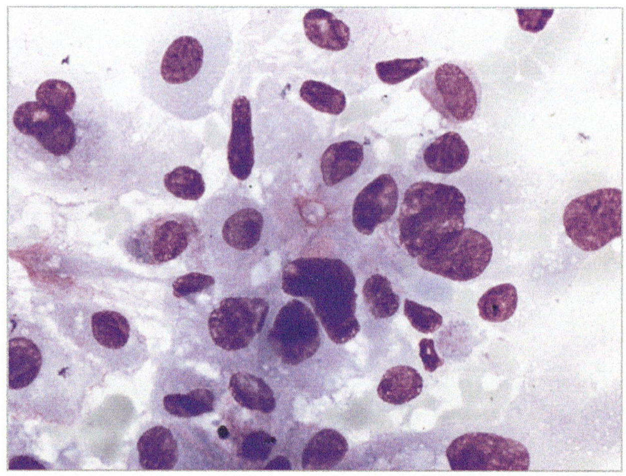

FIG. 1: Renal cell carcinoma (RCC) of clear cell type: Discrete tumor cells with abundant cytoplasm and enlarged centrally placed nuclei.

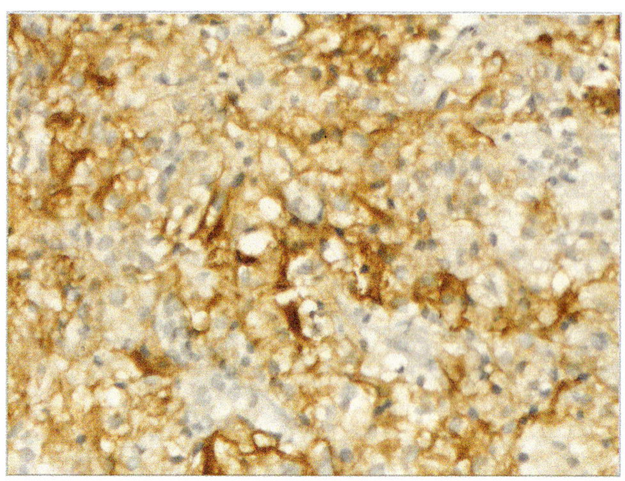

FIG. 2: Renal cell carcinoma (RCC) of clear cell type: CD10 positive cells.

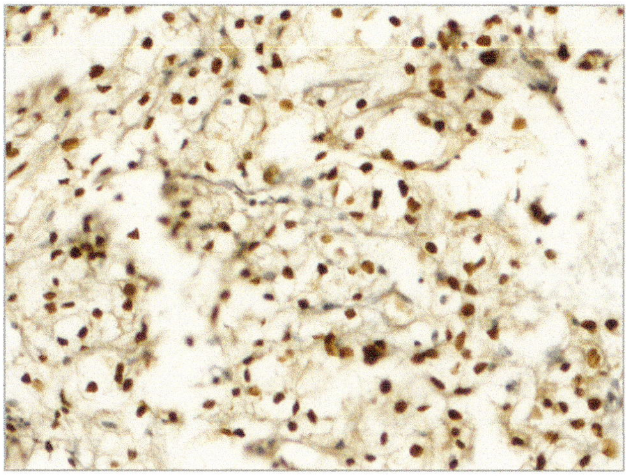

FIG. 3: Renal cell carcinoma (RCC) of clear cell type: Nuclear positive paired box gene 8 (PAX8).

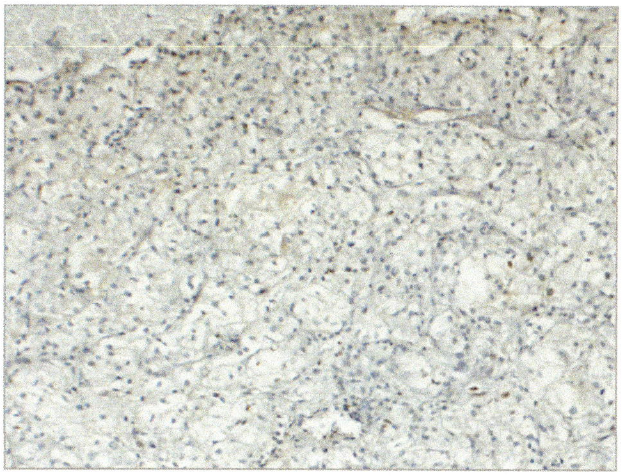

FIG. 6: Renal cell carcinoma (RCC) of clear cell type: Tumor cells are negative for CK20.

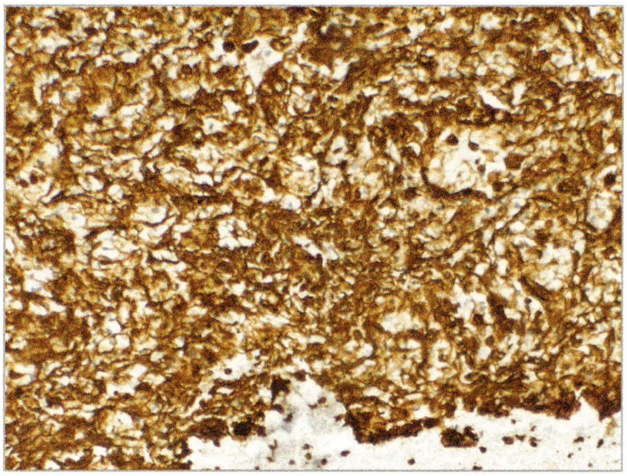

FIG. 4: Renal cell carcinoma (RCC) of clear cell type: Tumor cells are positive for vimentin.

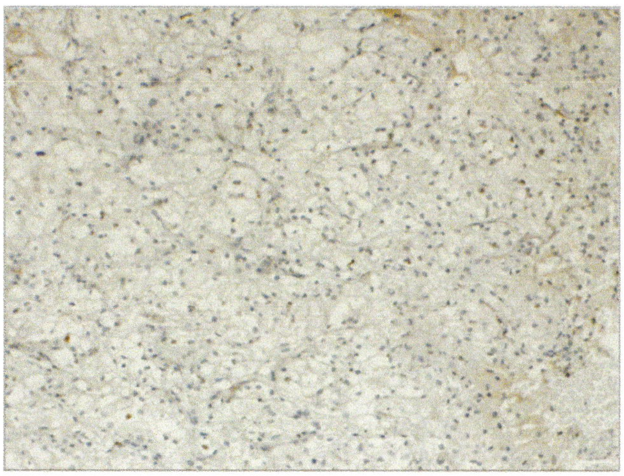

FIG. 5: Renal cell carcinoma (RCC) of clear cell type: Tumor cells are negative for CK7.

  - KIM1
  - EMA
  - Vimentin
  - CA IX
  - RCC
- *Negative*
  - CK7
  - CK20
  - CD117
  - GATA3
  - CK5/6
  - P63

Immunocytochemistry may be helpful to differentiate between different types of clear cell malignancies (**Table 7**).

Papillary Renal Cell Carcinoma

Clear Cell Type

- *Positive*
  - CK7
  - CD10
  - PAX8
  - PAX2
  - CA IX

Chromophobe Type

- *Positive*
  - CK7
  - CK8
  - CD117
  - EMA

Collecting Duct Carcinoma

- *Positive*
  - CK8
  - CK18

TABLE 7: Differentiating between different clear cell malignancies.

Markers	RCC	Clear cell carcinoma of the ovary	Clear cell sarcoma	HCC	Adrenocortical carcinoma
PAX8	Positive	Positive	Negative	Negative	Negative
CD10	Positive	Negative	Negative	Positive	Negative
Arginase	Negative	Negative	Negative	Positive	Negative
HMB45	Negative	Negative	Positive	Negative	Negative
Inhibin	Negative	Negative	Negative	Negative	Positive
Calcitonin	Negative	Negative	Negative	Negative	Positive
RCC	Positive	Negative	Negative	Negative	Negative

(HCC: hepatocellular carcinoma; RCC: renal cell carcinoma)

TABLE 8: Immunocytochemistry (ICC) to differentiate between urothelial cell carcinoma (UCC) of renal pelvis and renal cell carcinoma (RCC).

Markers	Urothelial cell carcinoma	Renal cell carcinoma
GATA3	Positive	Negative
S100P	Positive	Negative
PAX8	Negative	Positive
PAX2	Negative	Positive
p63	Positive	Negative
RCC	Negative	Positive
CK7	Positive	Negative
CK20	Positive	Negative

- CK19
- Lectin
- PAX8
- E-cadherin

- *Negative*
    - CD10
    - KIM1
    - RCC

Urothelial Cell Carcinoma of Renal Pelvis

- *Positive*
    - CK7
    - CK18
    - CK19
    - GATA3
    - Thrombomodulin (TM)
    - S100P

The differential diagnosis of urothelial cell carcinoma (UCC) of renal pelvis and RCC is highlighted in **Table 8**.

Angiomyolipoma of Kidney

The ICC pattern of angiomyolipoma of the kidney is described below (**Box 2; Figs. 7** to **12**).

- *Positive*
    - HMB45

BOX 2 Angiomyolipoma of the kidney.

- Angiomyolipoma of the kidney is a member of the PEComa group of tumors
- The tumor consists of adipose tissue, smooth muscle cells, and vascular tissue
- *Positive:* HMB45, Melan-A, Desmin, and Calponin
- *Negative:* CD10, PAX8, EMA, renal cell carcinoma (RCC), and S100
- Differential diagnosis: RCC is positive for PAX8, CD10, RCC, and EMA

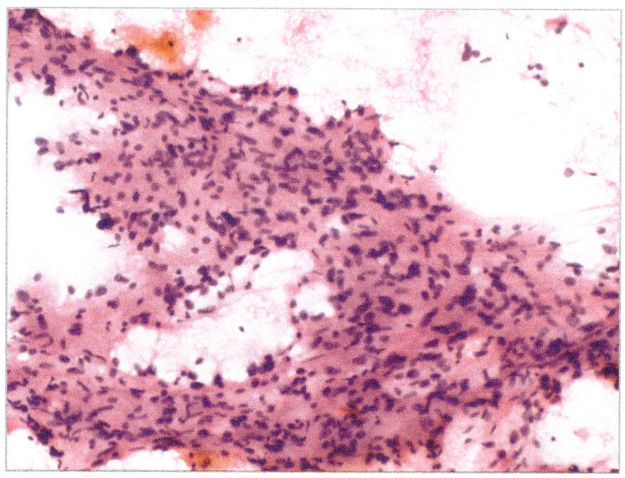

FIG. 7: Angiomyolipoma: Fine needle aspiration cytology smear showing clusters of oval to spindle cells (hematoxylin and eosin stain).

    - Melan-A
    - Calponin
    - Desmin
    - SMA
- *Negative*
    - CD10
    - PAX8 and PAX2
    - EMA
    - RCC
    - S100

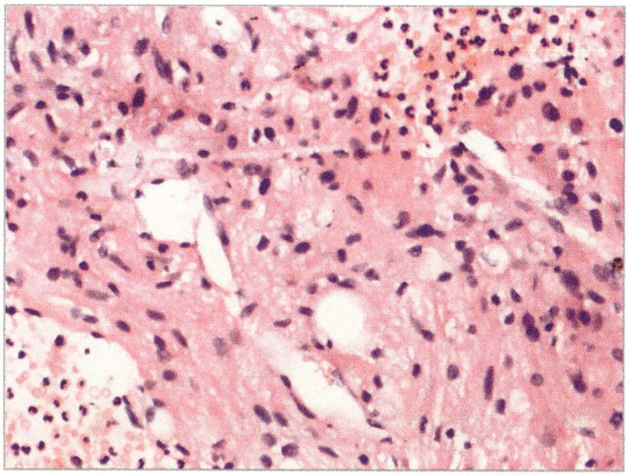

FIG. 8: Angiomyolipoma: Cell block section of the same showing vascular channels and spindle cells.

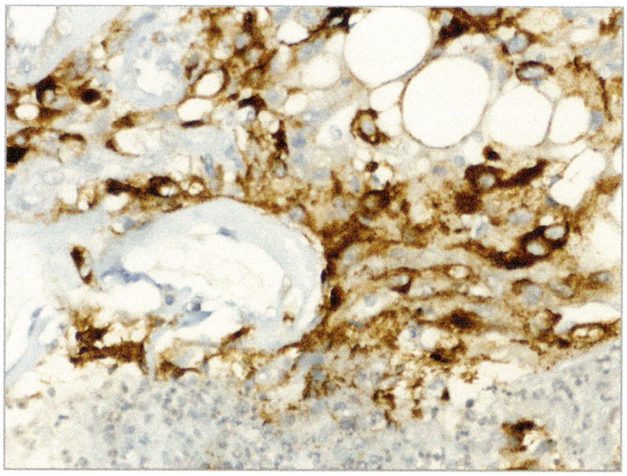

FIG. 9: Angiomyolipoma: Tumor cells are positive for HMB45.

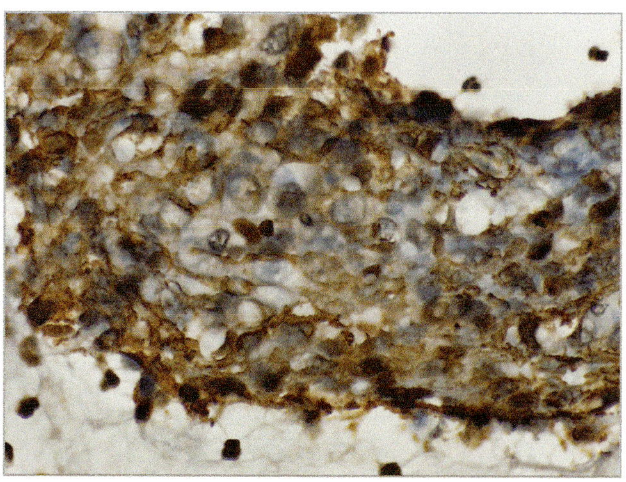

FIG. 10: Angiomyolipoma: Tumor cells are positive for smooth muscle actin (SMA).

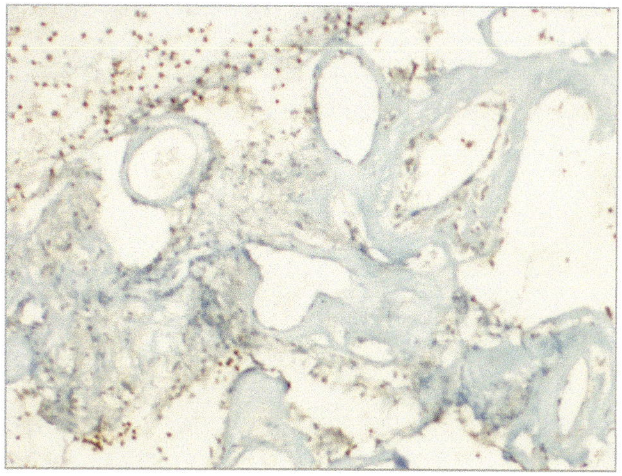

FIG. 11: Angiomyolipoma: Tumor cells are negative for CD10.

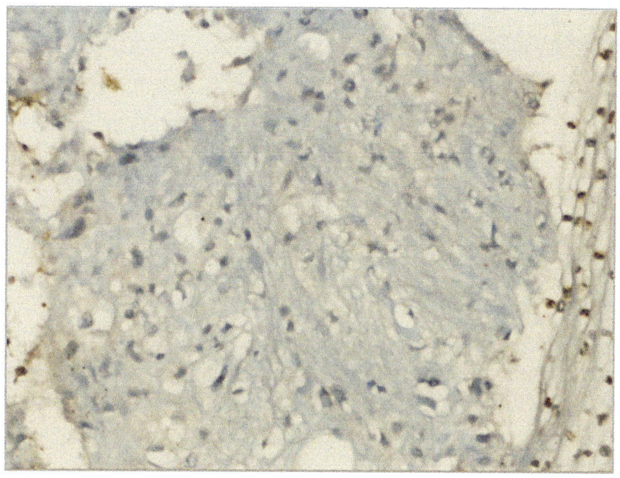

FIG. 12: Angiomyolipoma: Tumor cells are negative for paired box gene 8 (PAX8).

BOX 3 Nephroblastoma.

- The most common renal tumor of the children
- The blastemal cells are positive for: WT1, CD56, PAX2/PAX8, focally positive for NSE, and occasionally CK markers
- Blastemal cells are consistently negative for CD99

Nephroblastoma (Wilms' Tumor)

The ICC pattern of nephroblastoma is described below (**Box 3**):

- *Positive*
  - WT1
  - CD56
  - PAX2/PAX8
  - CK8
  - CK18
- *Negative*
  - CD99

- Myogenin
- MyoD1

Clear Cell Sarcoma of the Kidney

- *Positive*
  - Vimentin
  - Bcl2
  - Cyclin D1

Rhabdoid Tumor of the Kidney

- Vimentin
- Pan cytokeratin
- EMA

TUMORS OF THE URINARY TRACT

The ICC markers of urinary tract tumors include GATA3, uroplakin, TM, CK, p63, and S100P.

Uroplakin

Uroplakin III is a transmembrane protein which is present in the apical part of the urothelial cells. It is a specific marker of urothelial carcinomas with relatively low sensitivity. Out of the four subtypes of uroplakins (Ia, Ib, II, and III), the specific markers are uroplakin Ia and II (**Table 9**).

TABLE 9: Uroplakin expression.

Markers	Kidney lesion	Other tumors	Localization
Uroplakin	Urothelial cell carcinoma	• Uroplakin III is occasionally positive in malignant Brenner's tumor	Membranous

Thrombomodulin

Thrombomodulin is a glycoprotein that acts as an anticoagulant. It is present in the endothelial cells of the blood vessels, mesothelial cells, and urothelial cells. Other than UCC, TM is also expressed in mesothelioma, vascular neoplasm, and squamous cell carcinoma (**Table 10**). One of the important uses of TM is to differentiate UCC from the prostatic adenocarcinoma.

TABLE 10: Thrombomodulin (TM) expression.

Markers	Kidney lesion	Other tumors	Localization
TM	Urothelial cell carcinoma	• Mesothelioma • Vascular neoplasm • Squamous cell carcinoma	Membranous and cytoplasmic

GATA Binding Protein 3

GATA binding protein 3 is a zinc finger binding transcription protein which plays an essential role in cell proliferation and differentiation of the breast, urothelium, and trophoblasts. It is considered as one of the useful markers of UCC. However, as other tumors are also positive for GATA3, a panel of the marker is needed for the definite interpretation of GATA3 (**Table 11**).

Urothelial Cell Carcinoma

The ICC pattern of UCC is (**Box 4; Figs. 13** to **17**):

- *Positive*
  - CK5
  - CK7

TABLE 11: GATA binding protein 3 (GATA3) expression.

Markers	Kidney lesion	Other tumors	Localization
GATA3	• Urothelial cell carcinoma	• Breast carcinoma: Ductal and lobular • Paraganglioma • Choriocarcinoma • Yolk sac tumor • Pheochromocytoma • Salivary duct carcinoma • Mammary analogue secretory carcinoma	Nuclear

BOX 4 Urothelial cell carcinoma (UCC).

- Urothelial cell carcinomas are positive for: CK7, CK20 occasionally, p63, CK19, GATA3, uroplakin III, and thrombomodulin
- Immunocytochemistry is particularly helpful in:
  - Identifying secondary involvement by local tumors such as rectal or prostate
  - Metastatic UCC
  - Identifying rare variety of UCC
- The combined panel of markers consisting of CK7, CK20, GATA3, and thrombomodulin is helpful in identifying metastatic UCC

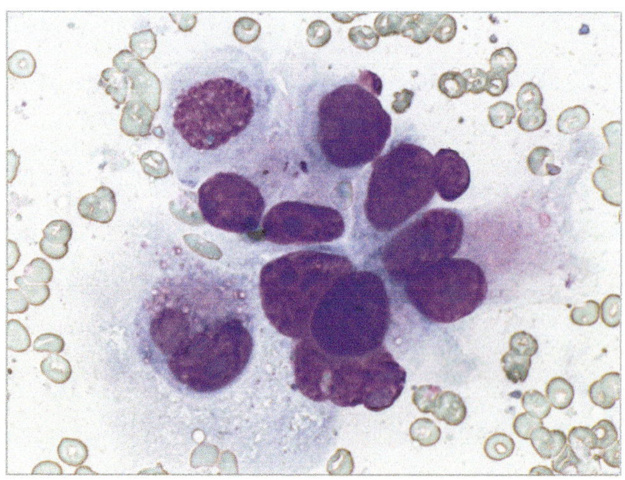

FIG. 13: Urothelial cell carcinoma (UCC): Fine needle aspiration cytology smear showing clusters of tumor cells in a pelvic mass (May–Grunwald–Giemsa stain).

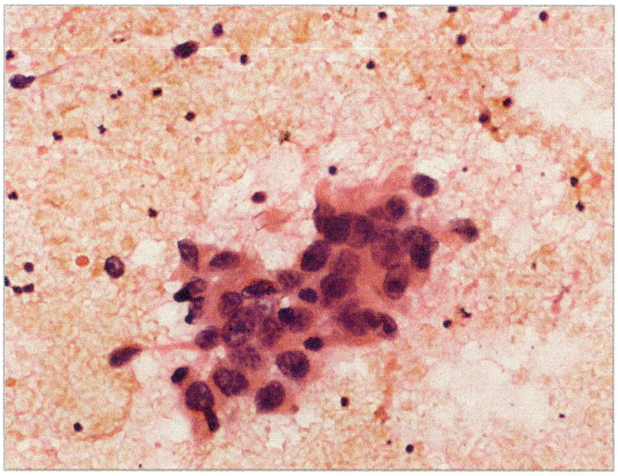

FIG. 14: Urothelial cell carcinoma (UCC): Fine needle aspiration cytology smear showing oval to elongated hematoxylin and eosin stain.

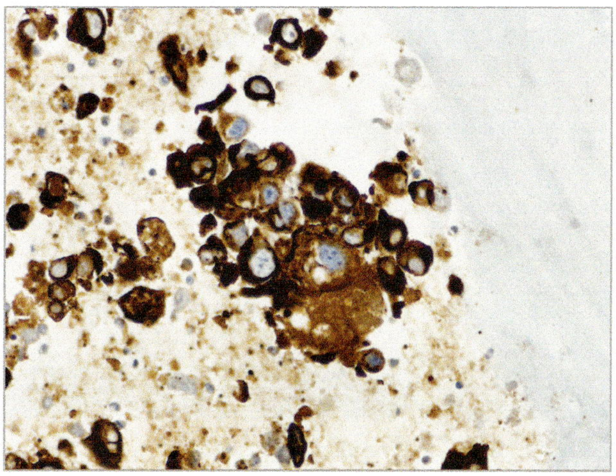

FIG. 15: Urothelial cell carcinoma (UCC): Tumor cells are positive for CK7.

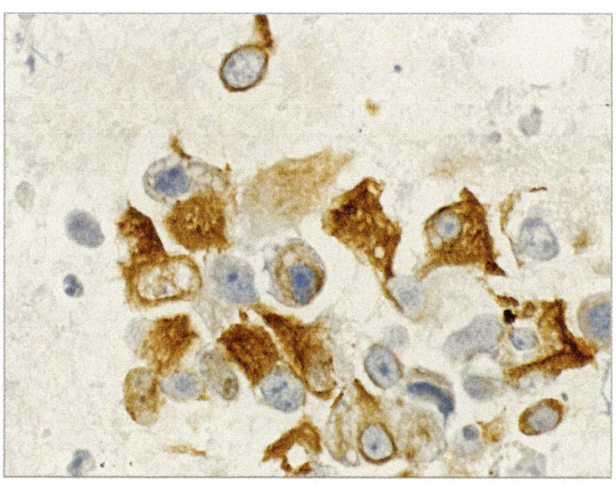

FIG. 16: Urothelial cell carcinoma (UCC): Tumor cells are positive for CK20.

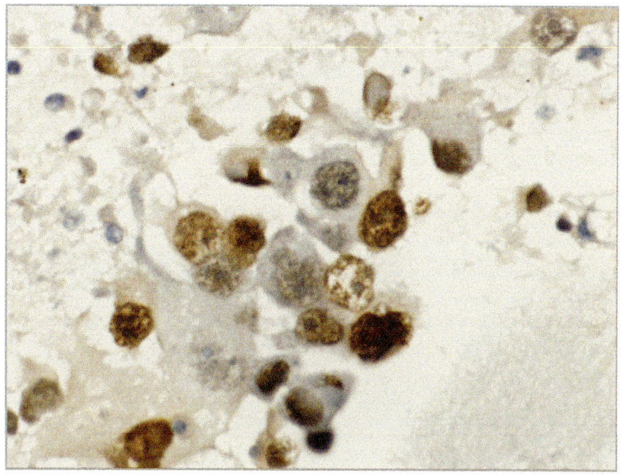

FIG. 17: Urothelial cell carcinoma (UCC): Tumor cells are positive for GATA binding protein 3 (GATA3).

TABLE 12: Immunocytochemistry (ICC) to distinguish urothelial cell carcinoma (UCC) from prostatic carcinoma.

Markers	Urothelial cell carcinoma	Prostatic carcinoma
GATA3	Positive	Negative
p63	Positive	Negative
Thrombomodulin	Positive	Negative
PSA	Negative	Positive

- CK20 (occasionally)
- CK19
- GATA3
- TM
- P63

The ICC markers to differentiate UCC and prostatic adenocarcinoma are highlighted in **Table 12**.

REFERENCES

1. Avery AK, Beckstead J, Renshaw AA, Corless CL. Use of antibodies to RCC and CD10 in the differential diagnosis of renal neoplasms. Am J Surg Pathol. 2000;24(2):203-10.
2. Cimino-Mathews A, Sharma R, Netto GJ. Diagnostic use of PAX8, CAIX, TTF-1, and TGB in metastatic renal cell carcinoma of the thyroid. Am J Surg Pathol. 2011;35(5):757-61.
3. Mazal PR, Stichenwirth M, Koller A, Blach S, Haitel A, Susani M. Expression of aquaporins and PAX-2 compared to CD10 and cytokeratin 7 in renal neoplasms: a tissue microarray study. Mod Pathol. 2005;18(4):535-40.
4. Daniel L, Lechevallier E, Giorgi R, Sichez H, Zattara-Cannoni H, Figarella-Branger D, et al. Pax-2 expression in adult renal tumors. Hum Pathol. 2001;32(3):282-7.

CHAPTER 17

Immunocytochemistry of Ovarian Tumors

INTRODUCTION

The panel of markers for surface epithelial ovarian tumors includes cytokeratin (CK), PAX8, Wilms tumor 1 (WT1), p53, carcinoembryonic antigen (CEA), CA125, p16, and GATA3.

- *Cytokeratin:* The details of CK have been described in Chapter 8.
- *PAX8:* The details of PAX8 have been described in Chapter 16.
- *GATA3:* It has been described in Chapter 16.
- *WT1:* It has been discussed in Chapter 7.
- *CA125:* It has been discussed in Chapter 6 as MUC-16.

DIAGNOSTIC PANEL OF OVARIAN GERM CELL TUMORS

The diagnostic panel of ovarian germ cell tumors includes Octamer-binding transcription factor 4 (OCT4), Sal-like protein 4 (SALL4), placental-like alkaline phosphatase (PLAP), SOX2, and CD117.

Octamer-binding Transcription Factor 4

Octamer-binding transcription factor 4, a transcription factor, is needed for the self-renewal of the pluripotent germ cells and primordial germ cells. OCT4 is a reliable marker of germ cell with high sensitivity and specificity. It is positive in the seminoma/dysgerminoma and embryonal carcinoma (**Table 1**).[1]

Sal-like Protein 4

Sal-like protein 4, a zinc-finger transcriptional factor, is required for the initiation and maintenance of pluripotent germ cells. It is a reliable sensitive and specific marker of germ cell tumors. SALL4 is positive in a wide range of germ cell tumors.[2,3] However, various other tumors also show SALL4 positivity (**Table 2**).

TABLE 1: Octamer-binding transcription factor 4 (OCT4) expression.

Marker	Positivity in ovarian tumors	Other tumors showing positivity	Localization
OCT4	• Dysgerminoma • Embryonal carcinoma • Gonadoblastoma	• Diffuse large B-cell lymphoma • Occasionally in breast carcinoma	Nuclear

Placental-like Alkaline Phosphatase

Placental-like alkaline phosphatase is a metalloenzyme and is expressed in the syncytiotrophoblasts of the placenta. It a highly sensitive marker (100%) for germ cell tumors (**Table 3**).[4]

TABLE 2: Sal-like protein 4 (SALL4) expression.

Marker	Positivity in ovarian tumors	Other tumors showing positivity	Localization
SALL4	• Seminoma/ dysgerminoma • Embryonal carcinoma • Choriocarcinoma • Yolk sac tumor	• Adenocarcinoma of colon • Adenocarcinoma of esophagus • Adenocarcinoma of endocervix • Urothelial carcinoma • Rhabdoid tumors • Nephroblastoma • Acute myeloid leukemia • Serous adenocarcinoma of ovary	Nuclear

TABLE 3: Placental-like alkaline phosphatase (PLAP) expression.

Marker	Positivity in ovarian tumors	Other tumors showing positivity	Localization
PLAP	• Seminoma • Embryonal carcinoma • Focally positive: Choriocarcinoma, yolk sac tumor	• Renal cell carcinoma • Breast carcinoma • Lung carcinoma	Cytoplasmic and membranous

CD117

It is a tyrosine kinase receptor and is expressed in the epithelial cells of breast, germ cells, immature cells of myeloid series, and melanocytes. It is mainly expressed in dysgerminoma (**Table 4**).[5,6]

SOX2

It is a transcription factor that belongs to the SOX family. It is related to the self-renewal of the stem cells.[7] It is a useful marker of embryonal carcinoma (**Table 5**).

DIAGNOSTIC PANEL OF OVARIAN SEX CORD TUMORS

The diagnostic panel of ovarian sex cord tumors includes inhibin, forkhead box transcription factor L2 (FOXL2), Melan A, and CD56.[8]

Inhibin

Inhibin, a growth and differentiation factor, is expressed in ovarian theca cells and adrenal gland. It is a reliable marker of ovarian sex cord tumor (**Table 6**).

Forkhead Box Transcription Factor L 2

This is a transcription factor and is related to the development of the ovary. It is a sensitive and specific marker of sex cord-stromal tumor of the ovary (**Table 7**).[9,10]

TABLE 4: CD117 expression.

Marker	Positivity in ovarian tumors	Other tumors showing positivity	Localization
CD117	Dysgerminoma	• Gastrointestinal stromal tumor • Small cell carcinoma of lung • Ewing's sarcoma • Thyroid follicular and papillary carcinoma • Hepatocellular carcinoma • Neuroblastoma • Merkel cell carcinoma	Cytoplasmic and membranous

TABLE 5: SOX2 expression.

Marker	Positivity in ovarian tumors	Other tumors showing positivity	Localization
SOX2	Embryonal carcinoma	• Carcinoma of prostate • Squamous cell carcinoma	Nuclear

Melan A (MART1)

It is a transmembrane protein that is associated with the maturation of the melanosome and regulation of pigment formation. It is a marker of the melanocytic tumor. However, Melan A is expressed in steroid cell tumor (**Table 8**).[10]

IMMUNOCYTOCHEMISTRY PATTERN OF OVARIAN TUMORS

The immunocytochemistry (ICC) patterns of the different surface epithelial tumors of the ovary are described here.[11,12]

Serous Epithelial Tumors of the Ovary

The ICC patterns of serous epithelial tumors of the ovary are (**Box 1; Figs. 1** to **6**):

- *Positive*
  - AE1/AE3
  - CK7

TABLE 6: Inhibin expression.

Marker	Positivity in ovarian tumors	Other tumors showing positivity	Localization
Inhibin	Sex cord-stromal tumor	• Adrenocortical tumor • Adenocarcinoma of prostate • Choriocarcinoma • Mesothelioma	Cytoplasmic

TABLE 7: Forkhead box transcription factor L2 (FOXL2) expression.

Marker	Positivity in ovarian tumors	Other tumors showing positivity	Localization
FOXL2	• Sex cord-stromal tumor	• Carcinoma of breast • Pituitary adenoma	Nuclear

TABLE 8: Melan A expression.

Marker	Positivity in ovarian tumors	Other tumors showing positivity	Localization
Melan A	Steroid cell tumor	• Melanoma • Adrenocortical tumor • Angiomyolipoma • Osteosarcoma	Cytoplasmic

BOX 1 Serous epithelial tumors of ovary.

- All the serous epithelial tumors are positive for CK7
- The tumors are positive for: PAX8, AE1/AE3, EMA, BerEP4, CA125, and WT1
- Negative for: CDX2 and CK20
- PAX8 is not entirely specific for serous epithelial ovarian carcinoma

(CK: cytokeratin; EMA: epithelial membrane antigen; WT1: Wilms tumor 1)

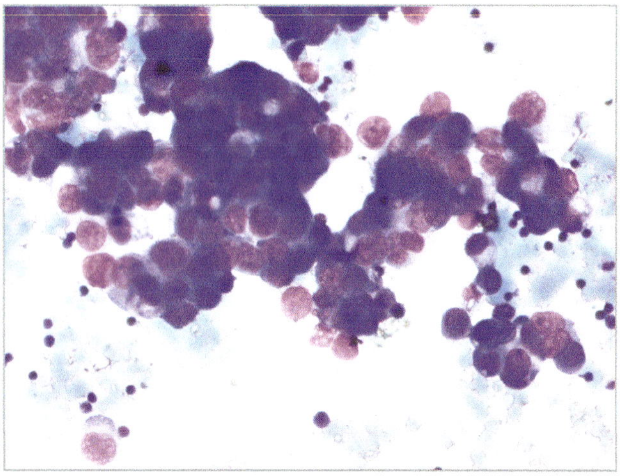

FIG. 1: Serous adenocarcinoma of ovary: Fine needle aspiration cytology smear showing multiple clusters of malignant cells.

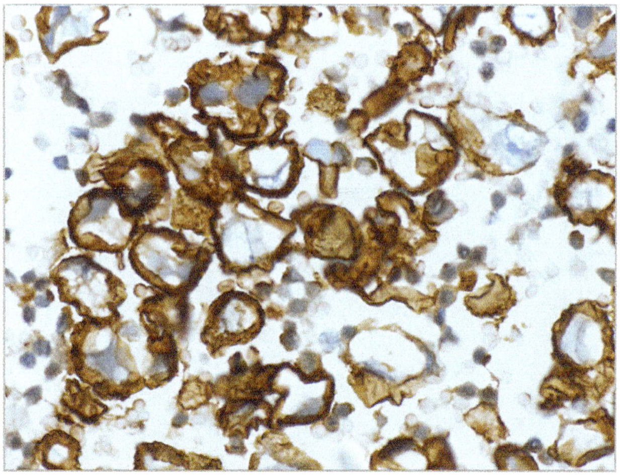

FIG. 2: Serous adenocarcinoma of ovary: Membranous positive CA125.

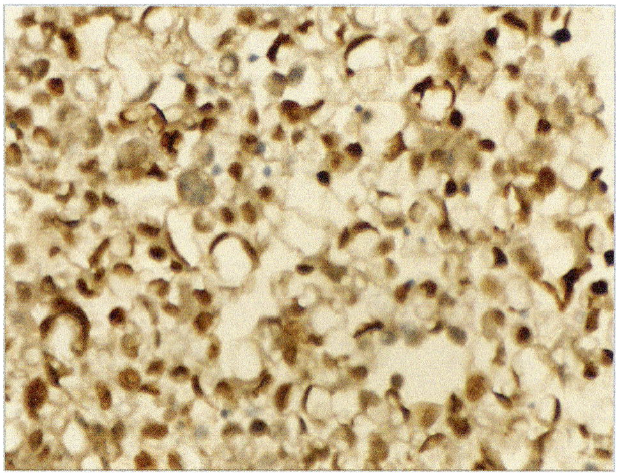

FIG. 3: Serous adenocarcinoma of ovary: Nuclear positive PAX8.

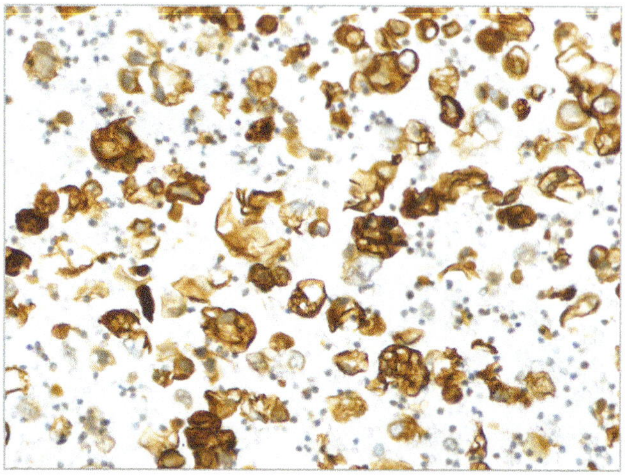

FIG. 4: Serous adenocarcinoma of ovary: Strong membranous positive cytokeratin 7 (CK7).

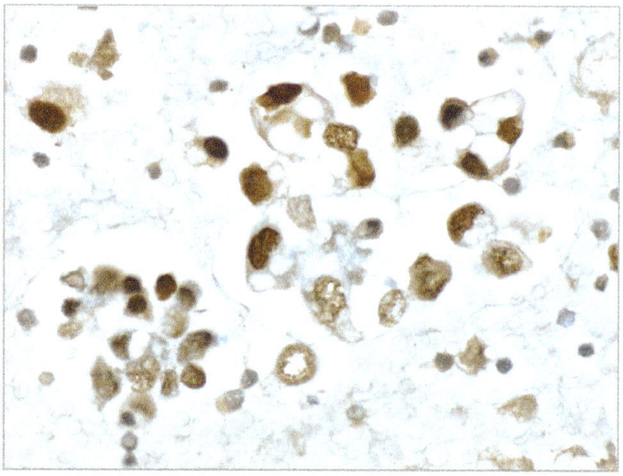

FIG. 5: Serous adenocarcinoma of ovary: Tumor cells are positive for p53.

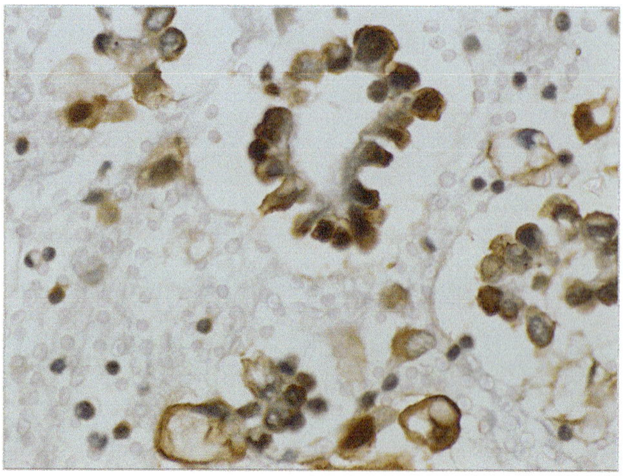

FIG. 6: Serous adenocarcinoma of ovary: Tumor cells are positive for Wilms tumor 1 (WT1).

- Epithelial membrane antigen (EMA)
- BerEP4
- CA125
- WT1
- PAX8

- *Negative*
  - CK20
  - CDX2
  - D2-40
  - OCT4
  - p16

The most reliable markers of the epithelial tumors are CK7, PAX8, MOC31, and BerEP4.

Mucinous Tumors of the Ovary[12,13]

The ICC pattern of mucinous neoplasms of the ovary is described below (**Box 2**; **Figs. 7** to **11**):

- *Positive*
  - AE1/AE3
  - CK7
  - EMA
  - BerEP4
  - MOC31
  - MUC1
  - MUC5AC

BOX 2 Mucinous tumors of ovary.

- Mucinous tumors of ovary are positive for CK7 and may also show CK20 positivity
- The intestinal type of primary mucinous tumor of ovary may often show CDX2 positivity
- Strongly positive CK7 along with negative CDX2 and CK20 indicates a primary mucinous tumor of ovary

(CK: cytokeratin)

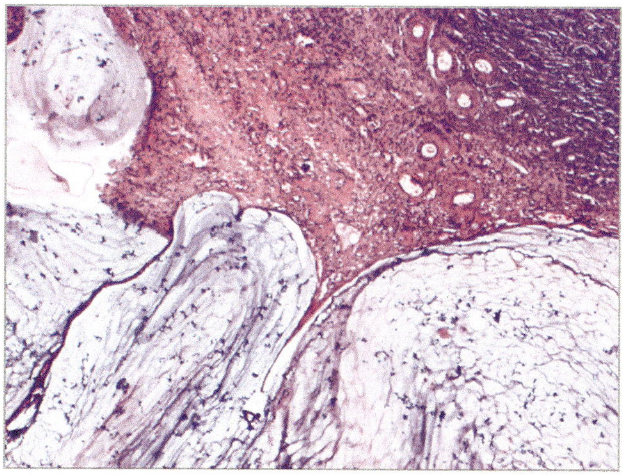

FIG. 7: Metastatic mucinous tumor in the ovary from colon: Multiple cystic mucin-containing spaces focally lined by mucinous epithelium.

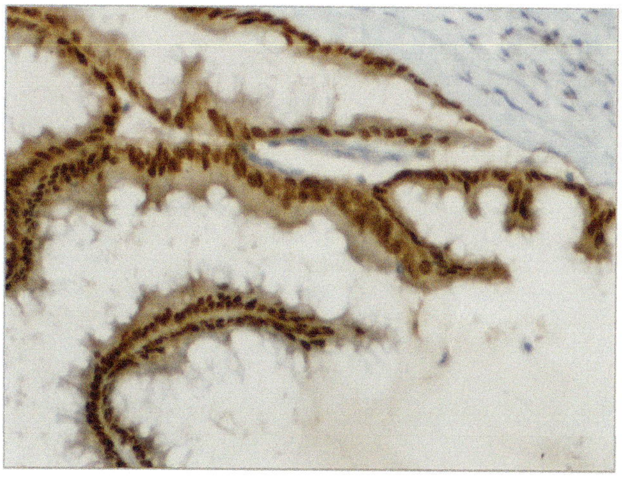

FIG. 8: Metastatic mucinous tumor in the ovary from colon: Positive for CDX2.

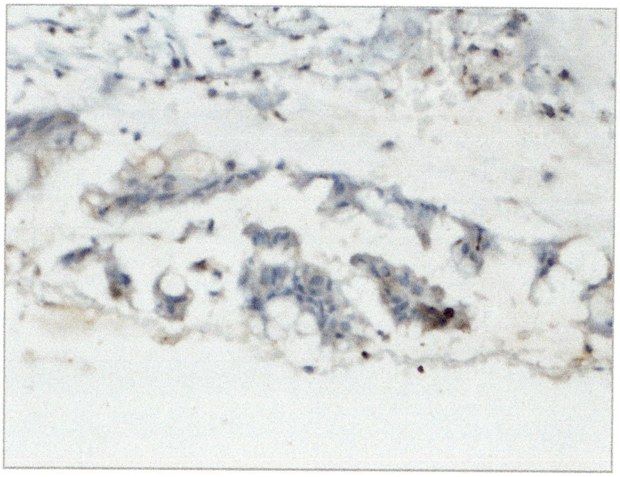

FIG. 9: Metastatic mucinous tumor in the ovary from colon: Negative for PAX8.

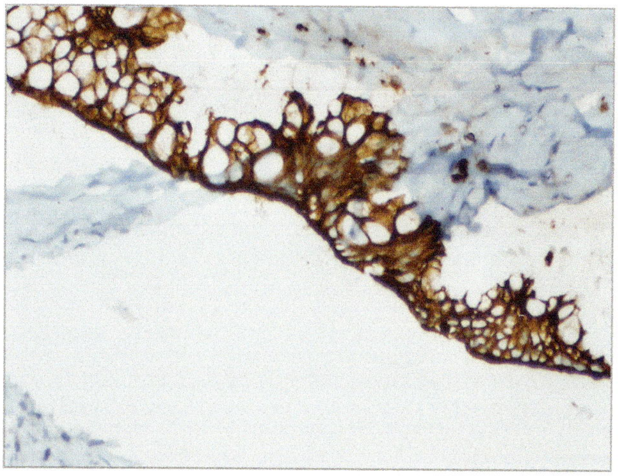

FIG. 10: Metastatic mucinous tumor in the ovary from colon: Positive for cytokeratin 20 (CK20).

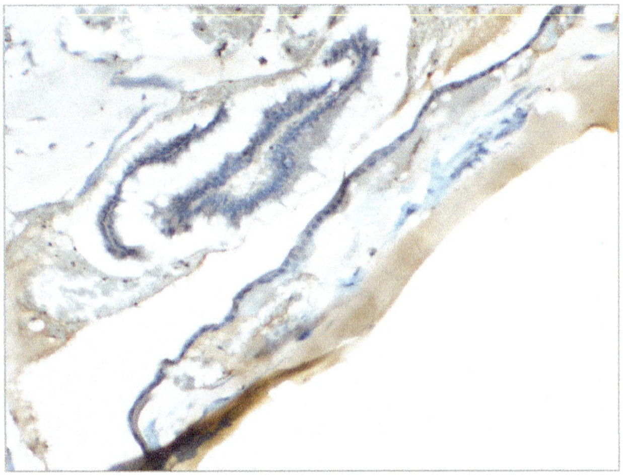

FIG. 11: Metastatic mucinous tumor in the ovary from colon: Negative for cytokeratin 7 (CK7).

- *Negative*
    - CK20
    - PAX8 (variably expressed)
    - OCT4

The mucinous tumor of the ovary may not express PAX8. It expresses AE1/AE3, CK7, MOC31, and BerEP4. Occasionally, the mucinous tumors of the ovary may express patchy positivity of CK20. However, the strong and diffuse positivity of CK20 which is seen in the colonic adenocarcinoma is usually absent in mucinous tumor of ovary.

Endometrioid Tumors of the Ovary[11]

The ICC pattern of endometrioid tumors of the ovary is described below (**Box 3**):

- *Positive*
    - AE1/AE3
    - CK7
    - CAM5.2
    - EMA
    - CA125
    - PAX8
    - ER
    - PR
    - Patchy positive p16
- *Negative*
    - WT1
    - P53

The most important differentiating marker of serous versus endometrioid carcinoma is WT1 as serous carcinoma is positive for WT1, whereas endometrioid carcinoma is negative for WT1.

Clear Cell Carcinoma of the Ovary[11]

The ICC pattern of clear cell carcinoma of the ovary is described below (**Box 4**):

- *Positive*
    - CK7
    - PAX8
    - HNF1β (hepatocyte nuclear factor 1β)
    - Napsin A
    - EMA
- *Negative*
    - P53
    - CA125
    - ER

The following ICC markers are helpful to differentiate serous adenocarcinoma from clear cell carcinoma of the ovary (**Table 9**).

BOX 3 Endometrioid tumors of ovary.

- The tumors are often positive for ER, PR, and PAX8
- Like other surface epithelial tumors, endometrioid tumors are also positive for CK7
- The tumors also show beta catenin positivity (38%)

(ER: estrogen receptor; PR: progesterone receptor; CK: cytokeratin)

BOX 4 Clear cell carcinoma of ovary.

- The tumors are positive for PAX8, CK7, and EMA
- These tumors are characteristically positive for HNF1β (hepatocyte nuclear factor 1β) and Napsin A
- Napsin A is a superior marker than HNF1β

(EMA: epithelial membrane antigen; CK: cytokeratin)

TABLE 9: Differentiating serous carcinoma from clear cell carcinoma.

Markers	Serous carcinoma	Clear cell carcinoma
HNF1β	Negative	Positive
WT1	Positive	Negative
KIM1	Negative	Positive
pVHL	Negative	Positive

(HNF1β: hepatocyte nuclear factor 1β; KIM1: kidney injury molecule-1; pVHL: von Hippel–Lindau gene product; WT1: Wilms tumor 1)

Brenner Tumor of the Ovary

- *Positive*
    - CK7
    - EMA
    - P63
    - Uroplakin III

Sex Cord-stromal Tumor of the Ovary[14-16]

The diagnostic ICC pattern of sex cord-stromal tumor of the ovary is described in **Box 5**.

Granulosa Cell Tumor

- *Positive*
    - Alpha-inhibin
    - Calretinin

BOX 5 Sex cord-stromal tumor of ovary.

- Sex cord-stromal tumors of ovary are always negative for EMA
- These tumors are characteristically positive for calretinin, inhibin, SF1, and forkhead box transcription factor L2 (FOXL2)
- Inhibin and FOXL2 are the specific markers
- Calretinin and SF1 are the relatively sensitive markers with less specificity

(EMA: epithelial membrane antigen)

  - FOXL2
  - Vimentin
  - CD99
  - SF1
- *Negative*
  - CK7
  - EMA

Thecoma/Fibroma

- *Positive*
  - Alpha-inhibin
  - Calretinin
  - FOXL2
  - Vimentin
  - SF1
- *Negative*
  - AE1/AE3
  - EMA

Sclerosing Stromal Tumor

- *Positive*
  - Alpha-inhibin
  - Calretinin
  - FOXL2
  - Vimentin
  - CD10
- *Negative*
  - AE1/AE3
  - EMA

Sertoli–Leydig Cell Tumor

- *Positive*
  - Alpha-inhibin
  - Calretinin
  - FOXL2 (occasionally negative)
  - WT1
  - Melan A
  - Vimentin
  - SF1
- *Negative*
  - CEA
  - EMA

The differentiating ICC markers of Sertoli–Leydig cell tumor and endometrioid carcinoma of the ovary are highlighted in **Table 10**.

TABLE 10: Differentiating Sertoli–Leydig cell tumor from endometrioid carcinoma of the ovary.

Markers	Sertoli–Leydig cell tumor	Endometrioid carcinoma of the ovary
Alpha-inhibin	Positive	Negative
WT1	Positive	Negative
SF1	Positive	Negative
PR	Negative	Positive
EMA	Negative	Positive

(EMA: epithelial membrane antigen; PR: progesterone receptor; WT1: Wilms tumor 1; SF1: splicing factor 1)

Germ Cell Tumor of the Ovary[1,2,10]

Dysgerminoma

- *Positive*
  - OCT4
  - SALL4
  - PLAP
  - CD117
  - NANOG
- *Negative*
  - AE1/AE3
  - EMA

Embryonal Carcinoma

- *Positive*
  - OCT4
  - SALL4
  - PLAP
  - SOX2
  - NANOG
  - CD30
  - CK7
- *Negative*
  - EMA
  - CD117

Yolk Sac Tumor

- *Positive*
  - AFP
  - SALL4
  - Glypican3
  - AE1/AE3
- *Negative*
  - EMA
  - CK7

Choriocarcinoma

Syncytiotrophoblast

- *Positive*
  - β-HCG
  - Inhibin

- CD10
- AE1/AE3

Cytotrophoblast

- *Positive*
  - Inhibin
  - CD10
  - AE1/AE3

Intermediate Trophoblast

- *Positive*
  - Human placental lactogen (HPL)
  - Inhibin
  - CD10
  - AE1/AE3

Box 6 highlights certain essential points regarding germ cell tumor of the ovary.

The differentiating features between syncytiotrophoblast, cytotrophoblast, and intermediate trophoblast are highlighted in **Table 11**.

Immature Teratoma

- *Positive*
  - GFAP: Neuroepithelial cells
  - S100: Neuroepithelial cells
  - CD56
  - OCT4: Immature cells
  - SALL4: Immature cells

BOX 6 Germ cell tumor of ovary.

- All germ cell tumors are negative for epithelial membrane antigen expression
- Majority of the germ cell tumors are positive for placental-like alkaline phosphatase
- Dysgerminoma: Octamer-binding transcription factor 4 (OCT4) and CD117 positive
- Embryonal carcinoma: CD30 and OCT4
- Yolk sac tumor: AFP, SALL4, glypican 3
- Choriocarcinoma: Beta-HCG

(SALL4: sal-like protein 4; HCG: human chorionic gonadotropin)

TABLE 11: Immunocytochemistry (ICC) pattern of trophoblasts.

Markers	Syncytiotrophoblast	Intermediate trophoblast	Cytotrophoblast
HCG	Positive	Negative	Negative
HPL	Negative	Positive	Negative
Inhibin	Positive	Positive	Positive
CD10	Positive	Positive	Positive

(HCG: human chorionic gonadotropin; HPL: human placental lactogen)

OTHER TUMORS

Carcinoid of Ovary[17]

- *Positive*
  - Chromogranin
  - Synaptophysin
  - CD56
  - Neuron-specific enolase (NSE)

Small Cell Carcinoma Pulmonary Type (Figs. 12 to 20)

- *Positive*
  - CD56
  - NSE

Small Cell Carcinoma: Hypercalcemic Type

- *Positive*
  - WT1
  - EMA

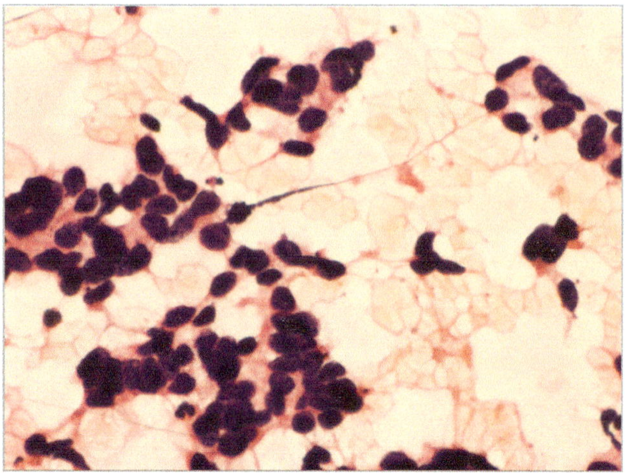

FIG. 12: Small cell carcinoma of the ovary: Fine needle aspiration cytology smear showing small clusters of monomorphic tumor cells.

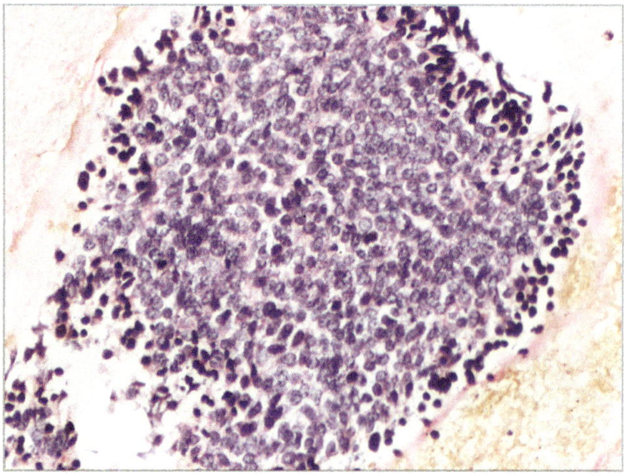

FIG. 13: Small cell carcinoma of the ovary: Cell block section showing small monomorphic tumor cells.

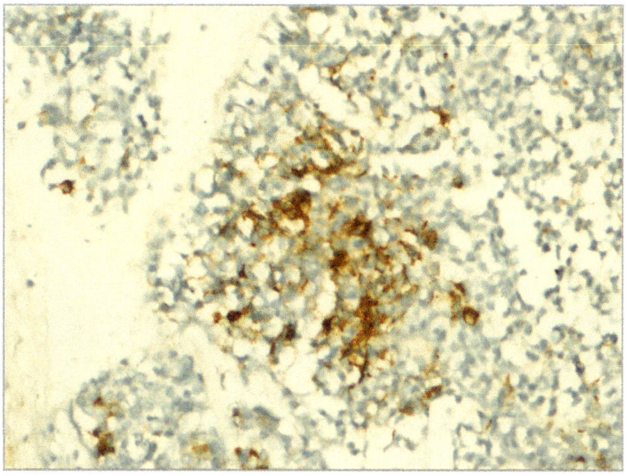

FIG. 14: Small cell carcinoma of the ovary: The cells are positive for chromogranin.

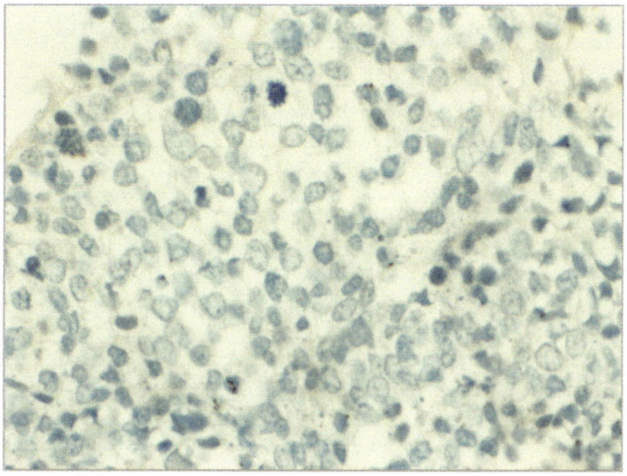

FIG. 15: Small cell carcinoma of the ovary: The tumor cells are negative for thyroid transcription factor 1 (TTF1).

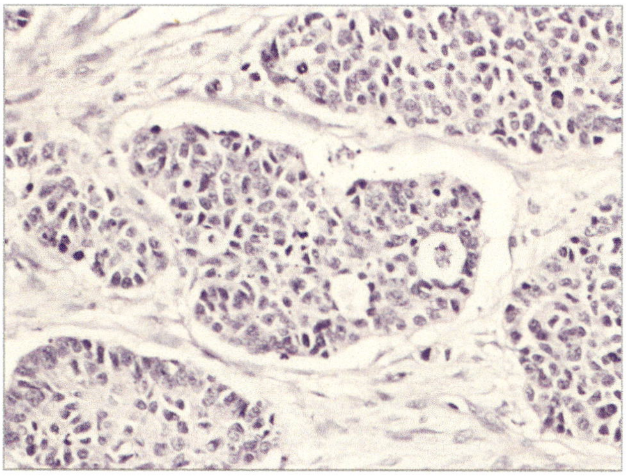

FIG. 16: Small cell carcinoma of the ovary: Multiple islands of tumor cells in the resected specimen of the ovary.

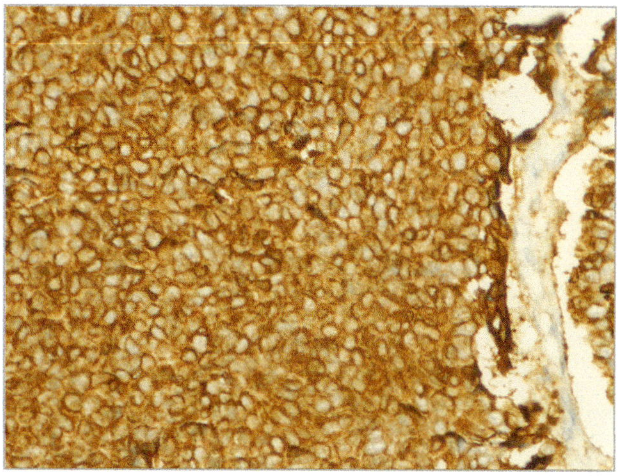

FIG. 17: Small cell carcinoma of the ovary: The tumor cells are positive for chromogranin in the resected specimen of the ovary.

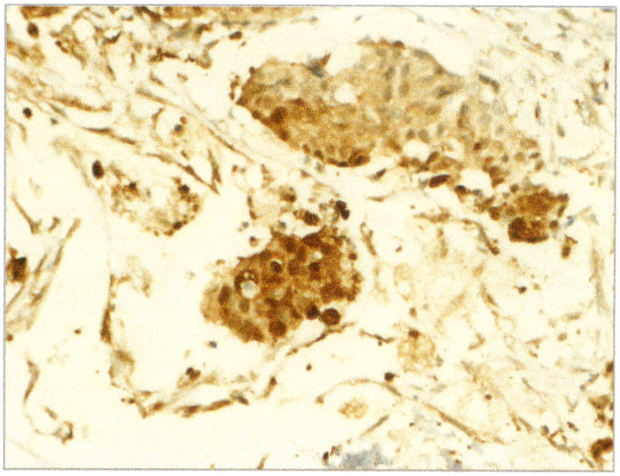

FIG. 18: Small cell carcinoma of the ovary: The tumor cells are positive for neuron-specific enolase (NSE).

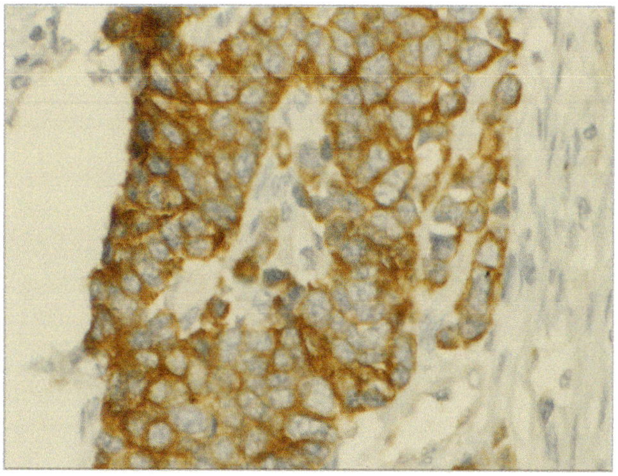

FIG. 19: Small cell carcinoma of the ovary: The tumor cells are positive for pan cytokeratin (CK).

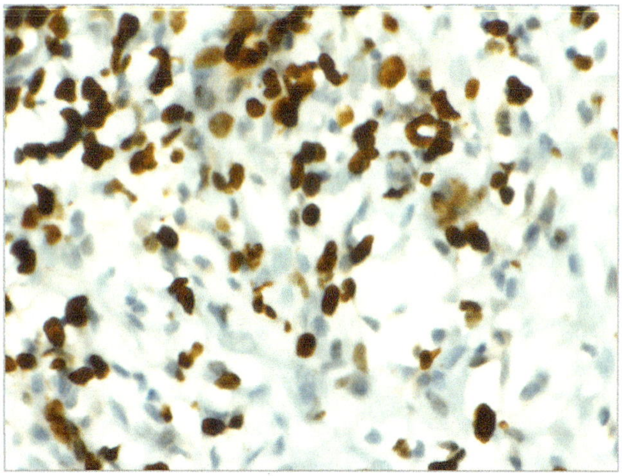

FIG. 20: Small cell carcinoma of the ovary: High Ki67 positive tumor cells (65% index).

REFERENCES

1. Cheng L, Sung MT, Cossu-Rocca P, Jones TD, MacLennan GT, De Jong J, et al. OCT4: biological functions and clinical applications as a marker of germ cell neoplasia. J Pathol. 2007;211(1):1-9.
2. Cao D, Guo S, Allan RW, Molberg KH, Peng Y. SALL4 is a novel sensitive and specific marker of ovarian primitive germ cell tumors and is particularly useful in distinguishing yolk sac tumor from clear cell carcinoma. Am J Surg Pathol. 2009;33(6):894-904.
3. Liu A, Cheng L, Du J, Peng Y, Allan RW, Wei L, et al. Diagnostic utility of novel stem cell markers SALL4, OCT4, NANOG, SOX2, UTF1, and TCL1 in primary mediastinal germ cell tumors. Am J Surg Pathol. 2010;34(5):697-706.
4. Hustin J, Gillerot Y, Collette J, Franchimont P. Placental alkaline phosphatase in developing normal and abnormal gonads and in germ-cell tumours. Virchows Arch A Pathol Anat Histopathol. 1990;417(1):67-72.
5. Gibson PC, Cooper K. CD117 (KIT): a diverse protein with selective applications in surgical pathology. Adv Anat Pathol. 2002;9(1):65-9.
6. Miettinen M, Sobin LH, Sarlomo-Rikala M. Immunohistochemical spectrum of GISTs at different sites and their differential diagnosis with a reference to CD117 (KIT). Mod Pathol. 2000;13(10):1134-42.
7. Novak D, Hüser L, Elton JJ, Umansky V, Altevogt P, Utikal J. SOX2 in development and cancer biology. Semin Cancer Biol. 2019;S1044-579X(18)30185-8.
8. Hanley KZ, Mosunjac MB. Practical review of ovarian sex cord-stromal tumors. Surg Pathol Clin. 2019;12(2):587-620.
9. Fuller PJ, Leung D, Chu S. Genetics and genomics of ovarian sex cord-stromal tumors. Clin Genet. 2017;91(2):285-291.
10. Rabban JT, Zaloudek CJ. A practical approach to immunohistochemical diagnosis of ovarian germ cell tumours and sex cord-stromal tumours. Histopathology. 2013;62(1):71-88
11. Ramalingam P. Morphologic, immunophenotypic, and molecular features of epithelial ovarian cancer. Oncology (Williston Park). 2016;30(2):166-76.
12. McCluggage WG, Young RH. Immunohistochemistry as a diagnostic aid in the evaluation of ovarian tumors. Semin Diagn Pathol. 2005;22(1):3-32.
13. Baker PM, Oliva E. Immunohistochemistry as a tool in the differential diagnosis of ovarian tumors: an update. Int J Gynecol Pathol. 2005;24(1):39-55.
14. Young RH. Sex cord-stromal tumors of the ovary and testis: their similarities and differences with consideration of selected problems. Mod Pathol. 2005;18:S81-98.
15. Young RH. Ovarian sex cord-stromal tumours and their mimics. Pathology. 2018;50(1):5-15.
16. Deavers MT, Malpica A, Liu J, Broaddus R, Silva EG. Ovarian sex cord-stromal tumors: an immunohistochemical study including a comparison of calretinin and inhibin. Mod Pathol. 2003;16(6):584-90.
17. Howitt BE, Kelly P, McCluggage WG. Pathology of neuroendocrine tumours of the female genital tract. Curr Oncol Rep. 2017;19(9):59.

CHAPTER 18

Immunocytochemistry of Soft-tissue Tumors

INTRODUCTION

The commonly used markers in soft-tissue tumors include cytokeratin, CD34, calponin, CD99, CD117, desmin, h-caldesmon, HMB45, MyoD1, myogenin, NB84, osteocalcin, SOX-9, vimentin, and vWF.

The common marker of all soft-tissue tumors is vimentin.

VIMENTIN

Vimentin is expressed in all soft-tissue tumors. Initially, vimentin is positive in the primitive cells, and slowly the vimentin is replaced by other types of intermediate filaments in the process of embryogenesis.[1,2] Other than sarcoma, vimentin is also expressed by carcinomas (**Table 1**).

MARKERS OF SKELETAL MUSCLE

The markers of skeletal muscle are desmin, myoglobin, and MyoD1.

Desmin

Desmin is positive in the skeletal, cardiac, and smooth muscle. In skeletal muscle, it is located in the Z disc and in smooth muscle it is seen as cytoplasmic dense bodies. Desmin is a relatively specific marker of all sorts of myogenic tumor (**Table 2**).[3,4]

Myoglobin

Myoglobin is a polypeptide that is related to the transport of oxygen. It is expressed only in the skeletal muscle (**Table 3**).[5] This antigen is not detectable in the embryonic tissue.

MyoD1 and Myogenin

Both MyoD1 and myogenin are transcription factors and the members of the MyoD family. They are related to the activation and proliferation of the stem cells of skeletal muscle. Both these markers are specific for rhabdomyosarcoma (**Table 4**).[6]

TABLE 1: Vimentin expression.

Marker	Expression in soft tissue	Other tumors expressing this antigen	Localization	Comments
Vimentin	All soft-tissue tumors	• Melanoma • Lymphoma • Renal cell carcinoma • Thyroid follicular carcinoma • Endometrial adenocarcinoma • Salivary gland carcinoma • Metaplastic carcinoma • Hepatocellular carcinoma	Cytoplasmic	• Wide variety of tumors express vimentin • Vimentin can only be used along with other markers

TABLE 2: Desmin expression.

Marker	Expression in myogenic tumor	Other tumors expressing this antigen	Localization	Comments
Desmin	• Rhabdomyosarcoma • Rhabdomyoma • Leiomyoma • Leiomyosarcoma	• Alveolar soft-part sarcoma • Dedifferentiated liposarcoma • Desmoplastic round cell tumor • Inflammatory myofibroblastic tumor	Cytoplasmic	• Wide variety of tumors express desmin • At least one other myogenic marker should be used along with desmin

TABLE 3: Myoglobin expression.

Marker	Expression in myogenic tumor	Other tumors expressing this antigen	Localization	Comments
Myoglobin	• Rhabdomyosarcoma • Rhabdomyoma	Breast, colorectal and prostatic carcinomas	Cytoplasmic	Wide variety of tumors express myoglobin

TABLE 4: Myogenic determination factor 1 (MyoD1) and myogenin expression.

Marker	Expression in myogenic tumor	Other tumors expressing this antigen	Localization	Comments
MyoD1 and myogenin	Rhabdomyosarcoma	Nephroblastoma	Nuclear	• The only nuclear stain should be considered as positive • Regenerative muscle cells also express them

IMMUNOCYTOCHEMISTRY PROFILE OF SKELETAL MUSCLE TUMORS

Rhabdomyosarcoma[6-8]

The immunocytochemistry (ICC) pattern of embryonal rhabdomyosarcoma is (**Box 1; Figs. 1** to **8**):

- *Positive*
  - Desmin
  - Myogenin
  - MyoD1

The specific markers of rhabdomyosarcoma are MyoD1 and myogenin.

BOX 1 Rhabdomyosarcoma.

- Rhabdomyosarcoma is positive for desmin, myogenin, and myogenic determination factor 1 (MyoD1)
- MyoD1 is the most specific marker of this tumor and is consistently present in rhabdomyosarcoma
- Occasionally, the undifferentiated cells of rhabdomyosarcoma may be negative for desmin

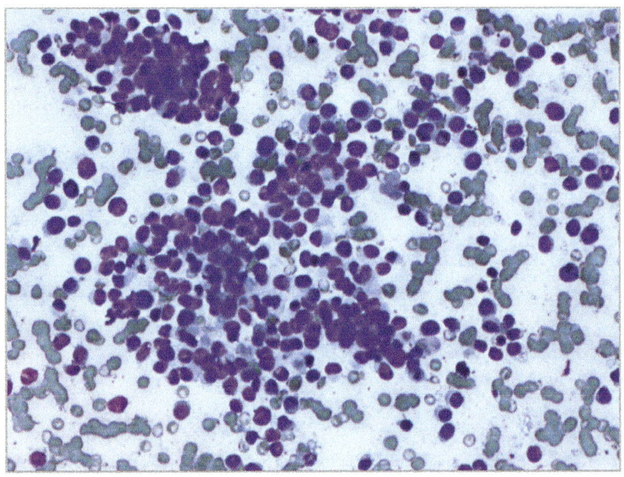

FIG. 1: Rhabdomyosarcoma: Fine needle aspiration cytology smear of a paratesticular swelling showing loose clusters of tumor cells having round nuclei and scanty cytoplasm.

Rhabdomyoma

- *Positive*
  - Desmin
  - Myoglobin

MARKERS OF SMOOTH MUSCLE

The markers of smooth muscle include desmin, smooth muscle actin, h-caldesmon, calponin, and steroid hormone receptors.

Smooth Muscle Actin

Actin is an intermediate filaments with six major isotypes: alpha, beta and gamma subtypes. Alpha actin has three isotypes: alpha actin 1, 2 and 3. Alpha actin 2 is commonly known as smooth muscle actin (SMA). SMA is expressed in smooth muscle tumors and tumors of myofibroblasts (**Table 5**).

h-Caldesmon

h-Caldesmon, a calcium-binding cytoplasmic protein, has two different isoforms: Low-molecular-weight isoform and high-molecular-weight isoform. The low-molecular-weight isoform of caldesmon is associated with cell proliferation and cytoskeleton modification. The high-molecular-weight isoform (h-caldesmon) is attached with actin, tropomyosin, and various phospholipids. h-Caldesmon is positive in the smooth muscle cells. The myofibroblasts do not express h-caldesmon (**Table 6**).[9]

Calponin

Calponin is a protein that binds with calcium and controls the contraction of the smooth muscle. It is expressed in both smooth muscle cells and myoepithelial cells (**Table 7**). Gastrointestinal stromal tumor (GIST) does not show calponin positivity.

Transgelin

Transgelin is an actin-binding protein. It is expressed in smooth muscle tumors (**Table 8**).[10,11]

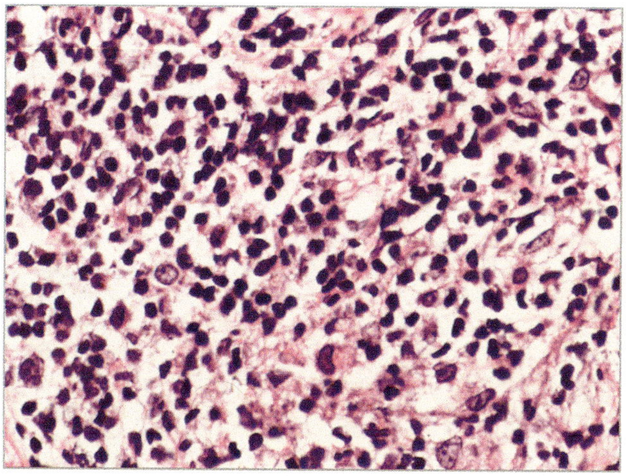

FIG. 2: Rhabdomyosarcoma: Cell block section showing diffuse sheets of cells.

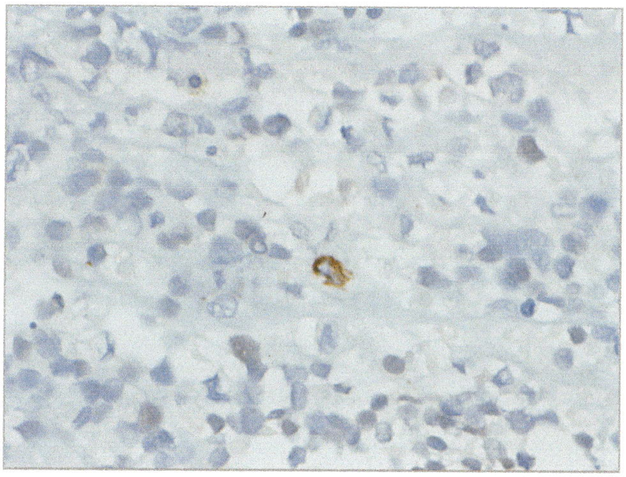

FIG. 3: Rhabdomyosarcoma: The tumor cells are negative for leukocyte common antigen (LCA).

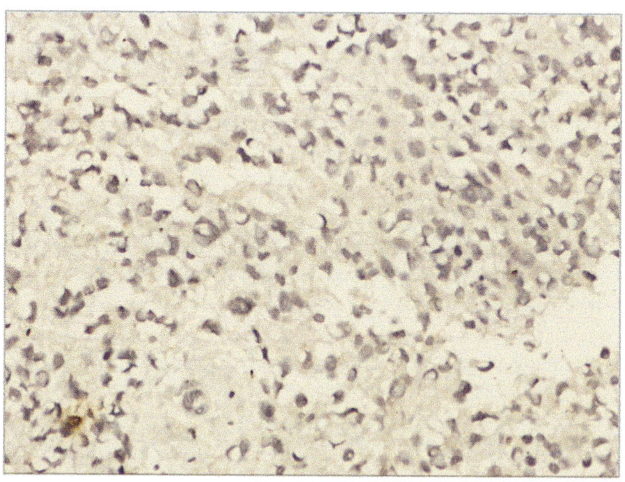

FIG. 4: Rhabdomyosarcoma: The tumor cells are negative for pan cytokeratin.

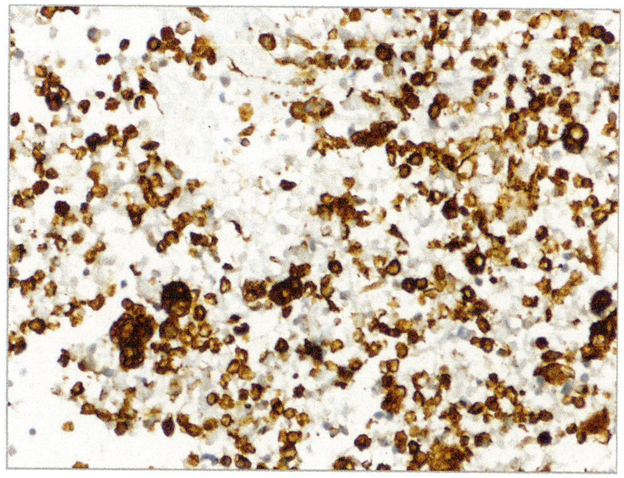

FIG. 5: Rhabdomyosarcoma: The tumor cells are positive for vimentin.

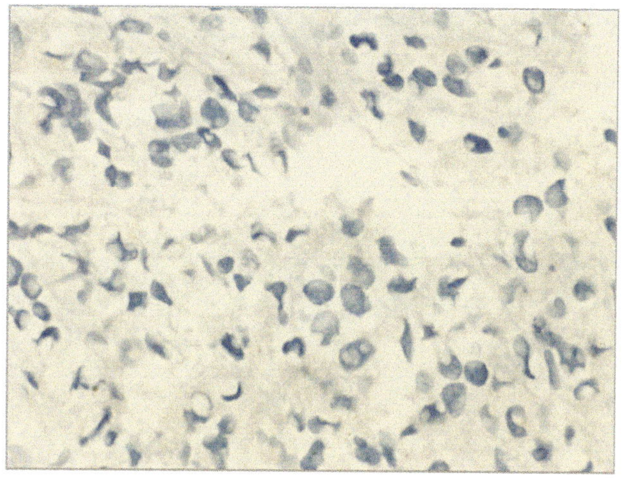

FIG. 6: Rhabdomyosarcoma: The tumor cells are negative for OCT4.

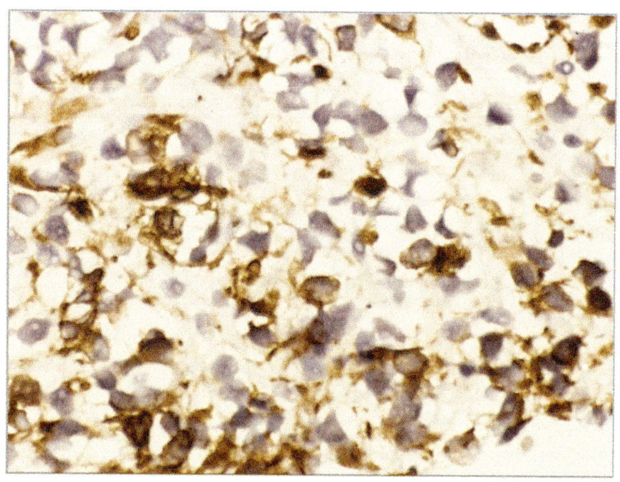

FIG. 7: Rhabdomyosarcoma: The tumor cells are positive for desmin.

Leiomyoma and Leiomyosarcoma

The ICC profile of smooth muscle tumors—leiomyoma and leiomyosarcoma—is described below (**Box 2; Figs. 9** and **10**):[9-12]

- *Positive*
  - h-Caldesmon
  - SMA
  - Calponin
  - Desmin
  - Transgelin
  - Estrogen receptor (ER)
  - Progesterone receptor (PR)

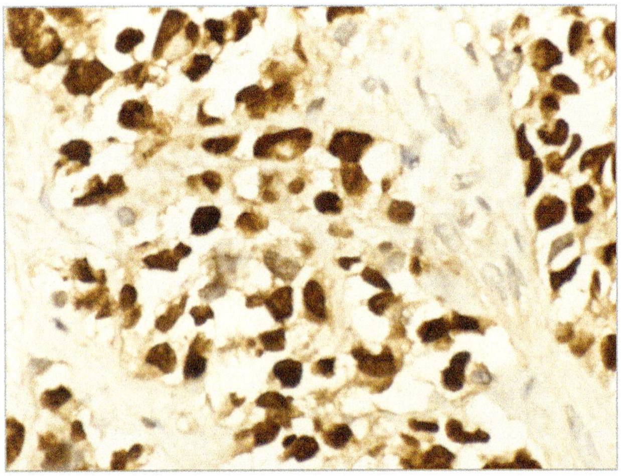

FIG. 8: Rhabdomyosarcoma: The tumor cells are positive for myogenin.

MARKERS OF PERIPHERAL NERVE SHEATH TUMORS

The markers of peripheral nerve sheath tumors are S100, myelin basic protein, glial fibrillary acidic protein, and neurofilaments.

Neurofilament Proteins[13]

Neurofilament proteins (NP) are the members of intermediate filaments, and they form the skeleton of axon and dendrites of the nervous system. NPs are the markers of tumors derived from the neurons and ganglionic cells (**Table 9**).

S100 Protein

S100 protein is one of the members of low molecular weight proteins. The name S100 protein is so because it was for the first time isolated by 100% saturated ammonium sulfate. S100 protein is positive in a large range of tissues such as neurons, glial cells, chondrocytes, melanocytes, Langerhans

TABLE 5: Smooth muscle actin (SMA) expression.

Marker	Expression in myogenic tumor	Other tumors expressing this antigen	Localization	Comments
SMA	Leiomyoma and leiomyosarcoma	• Myofibroblastic tumor • Gastrointestinal stromal tumor • Endometrial stromal sarcoma	Cytoplasmic	As myofibroblastic and myoepithelial tumors also show SMA positivity so it should be carefully interpreted

TABLE 6: h-Caldesmon expression.

Marker	Expression in myogenic tumor	Other tumors expressing this antigen	Localization	Comments
h-Caldesmon	Leiomyoma and leiomyosarcoma	Epithelioid mesothelioma • Gastrointestinal stromal tumor (GIST) • Glomus tumor	Cytoplasmic	h-Caldesmon is also positive in other nonsmooth muscle tumors and therefore should be interpreted carefully

TABLE 7: Calponin expression.

Marker	Expression in myogenic tumor	Other tumors expressing this antigen	Localization	Comments
Calponin	Leiomyoma and leiomyosarcoma	• Myoepithelial cells • Myofibroblastic tumor	Cytoplasmic	Calponin is not expressed by gastrointestinal stromal tumor (GIST)

TABLE 8: Transgelin expression.

Marker	Expression in myogenic tumor	Other tumors expressing this antigen	Localization	Comments
Transgelin	Visceral and vascular smooth muscle tumors	• Triple-negative breast carcinoma • Malignant nerve sheath tumor (a subset)	Cytoplasmic	Transgelin is expressed in the myofibroblast and fibroblast and also some epithelial neoplasm

BOX 2 Leiomyoma and leiomyosarcoma.

- The markers of smooth muscle tumor are h-caldesmon, smooth muscle actin, and desmin
- Morphology of the tumor along with the above-mentioned markers support the diagnosis
- Desmin is not a specific marker of this tumor, and various other tumors may also show desmin expression

cells, and epithelial cells of breast, pancreas, and sweat gland (**Table 10**).[14]

SOX10

SOX10, a transcription factor, helps in the differentiation and development of Schwann cells and melanocytes. It is positive in nerve sheath tumors and melanoma (**Table 11**).[14]

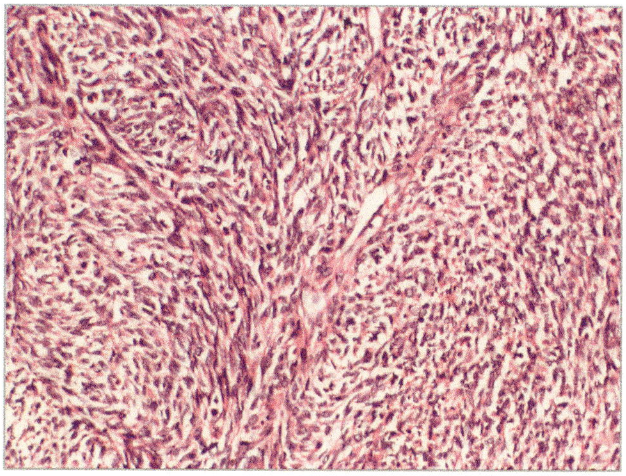

FIG. 9: Leiomyosarcoma of uterus: Histopathology section showing fascicles of spindle cells.

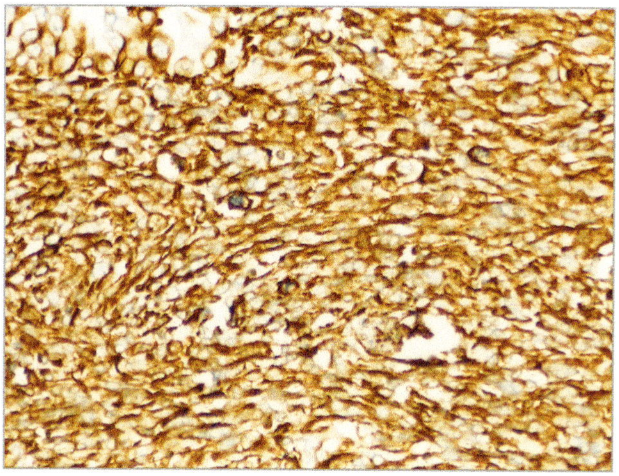

FIG. 10: Leiomyosarcoma of uterus: Desmin positive cells.

TABLE 9: Neurofilament proteins expression.

Marker	Expression in nerve sheath tumors	Other tumors expressing this antigen	Localization	Comments
Neurofilament proteins	• Neurofibroma • Paraganglioma • Neuroblastoma • Retinoblastoma	• Pancreatic endocrine tumors • Merkel cell carcinoma • Parathyroid tumors • Pheochromocytoma	Cytoplasmic	Other than nerve sheath tumor, many tumors are positive for neurofilament proteins

TABLE 10: S100 proteins expression.

Marker	Expression in nerve sheath tumors	Other tumors expressing this antigen	Localization	Comments
S100 proteins	• Neurofibroma • Malignant peripheral nerve sheath tumors	• Melanoma • Chondrocytic tumors • Lipomatous tumors • Myoepithelial tumors • Langerhans cell histiocytosis	Cytoplasmic and nuclear	A wide variety of tumors express S100 protein

TABLE 11: SOX10 expression.

Marker	Expression in nerve sheath tumors	Other tumors expressing this antigen	Localization	Comments
SOX10	• Neurofibroma • Malignant peripheral nerve sheath tumors	• Melanoma • Granular cell tumor • Myoepithelial tumors • Clear cell sarcoma	Nuclear	A wide variety of tumors express S100 protein

IMMUNOCYTOCHEMISTRY PATTERN OF NERVE SHEATH TUMORS[13,14]

The key diagnostic markers of nerve sheath tumors are described in **Box 3**.

Neurofibroma

- *Positive*
  - SOX10
  - CD34
  - S100
  - Epithelial membrane antigen (EMA)

Schwannoma

- *Positive*
  - SOX10
  - S100
  - D2-40
  - Calretinin

Paraganglioma

- *Positive*
  - Neuron-specific enolase (NSE)
  - CD56
  - Synaptophysin
  - S100

Melanotic Neuroectodermal Tumor of Infancy

- *Positive*
  - HMB45
  - Pan cytokeratin

Malignant Peripheral Nerve Sheath Tumor

- *Positive*
  - Myelin basic protein
  - S100 (occasionally negative)
  - SOX10 (occasionally negative)

BOX 3 Nerve sheath tumors.

- Neurofibroma, schwannoma, paraganglioma, and malignant peripheral nerve sheath tumor (MPNST) are all positive for S100 protein
- There are no definite diagnostic markers of MPNST. However, these tumors are often positive for S100, glial fibrillary acidic protein (GFAP), SOX10, and CD56

Granular Cell Tumor

- *Positive*
  - SOX10
  - S100
  - NSE
  - CD56

MARKERS OF VASCULAR TUMORS

The markers of vascular tumors include CD31, CD34, Factor VIII, ERG, D2-40, claudin 5, and thrombomodulin.

CD31

CD31 is also called "platelet-endothelial cell adhesion molecule" (PECAM-1). This transmembrane glycoprotein is positive in platelets, megakaryocytes, endothelial cells, monocytes, and granulocytes. It is a relatively specific marker of the vascular tumor (**Table 12**).

CD34

CD34 is a glycoprotein related to cell surface adhesion and is expressed in the hematopoietic progenitor cells of the myeloid and lymphoid system and also endothelial cells. CD34 is mainly used as a marker of the vascular tumor and endothelial cells of the blood vessels. It is also positive in other nonvascular tumors such as GIST and dermatofibrosarcoma protuberans (**Table 13**).

Factor VIII Related Antigen (von Willebrand Factor)

Factor VIII related antigen is also known as "von Willebrand factor" (vWF). It is synthesized by endothelial cells and megakaryocytes and stored in the Weibel–Palade bodies of the endothelial cells. vWF is a large polymeric protein with three subunits. After proteolysis, vWF breaks down into multiple fragments and are present in the circulating plasma. The function of vWF is mainly to help in the intrinsic pathway of coagulation and platelet aggregation. vWF is a useful marker of the vascular tumor (**Table 14**). As it is stored in the endothelial cells, vWF is also a marker of the endothelial cells.

V-ETS Avian Erythroblastosis Virus Oncogene Homolog (ERG)

ERG, a transcription factor, controls the differentiation of the endothelial cells and angiogenesis. ERG is the marker of the

TABLE 12: CD31 expression.

Marker	Expression in vascular tumors	Other tumors expressing this antigen	Localization	Comments
CD31	Vascular tumor	• Langerhans cell histiocytosis • Plasmacytoma • Granulocytic sarcoma • Ewing's sarcoma	Membranous and cytoplasmic	Many nonvascular tumors also express this marker

TABLE 13: CD34 expression.

Marker	Expression in vascular tumors	Other tumors expressing this antigen	Localization	Comments
CD34	Vascular tumor	• Dermatofibrosarcoma protuberans • Spindle cell lipoma • Gastrointestinal stromal tumor (GIST) • Solitary fibrous tumor • Granulocytic sarcoma • Liposarcoma • Peripheral nerve sheath tumor	Membranous and cytoplasmic	Sensitive but less specific than CD31 as a marker of vascular tumor

TABLE 14: von Willebrand factor.

Marker	Expression in vascular tumors	Other tumors expressing this antigen	Localization	Comments
von Willebrand factor	Vascular tumor	No	Cytoplasmic	Highly specific but difficult to interpret the result so it is not used presently

TABLE 15: ERG factor expression.

Marker	Expression in vascular tumors	Other tumors expressing this antigen	Localization	Comments
ERG	Vascular tumor	• Prostatic adenocarcinoma • Ewing's sarcoma • Epithelioid sarcoma • Meningioma	Nuclear	It is considered as a more sensitive and specific marker than CD31 and CD34

endothelial cells with nearly 100% specificity. However, other tumors such as prostatic carcinoma and Ewing's sarcoma (ES) may also show ERG expression (**Table 15**).[16,17]

Podoplanin (D2-40)

Podoplanin (D2-40) has been discussed in Chapter 7.

Claudin 5

Claudin 5 has been discussed in Chapter 6.

Thrombomodulin

Thrombomodulin has been described in Chapter 16.

IMMUNOCYTOCHEMISTRY PATTERN OF VASCULAR TUMORS[15-17]

The key diagnostic features of vascular soft-tissue tumors have been mentioned in **Box 4**.

BOX 4 Vascular soft-tissue tumors.

- The vascular soft-tissue tumors are consistently positive for ERG, CD31, and podoplanin
- ERG is the most specific for vascular soft-tissue tumors
- Angiosarcomas are always positive for ERG and Fli-1

Hemangioma

- *Positive*
  - CD31
  - CD34
  - Claudin 5
  - ERG
  - vWF

Retiform Hemangioendothelioma

- *Positive*
  - CD31
  - CD34
  - vWF
  - ERG
  - Claudin 5
  - D2-40

Epithelioid Hemangioendothelioma

- *Positive*
  - CD31
  - CD34
  - vWF
  - Claudin 5

Lymphangioma

- *Positive*
  - Podoplanin (D2-40)
  - CD31
  - CD34
  - vWF

Kaposi's Sarcoma

- *Positive*
  - Podoplanin (D2-40)
  - HHV-8
  - CD31
  - CD34
  - Claudin 5
  - ERG

Angiosarcoma

- *Positive*
  - CD31
  - CD34
  - ERG
  - Claudin 5
  - Fli-1

MARKERS OF HISTIOCYTIC AND DENDRITIC GROUPS OF TUMORS

The various markers of the fibrohistiocytic group of tumors include CD1a, CD68, CD207, and podoplanin.

CD1a

CD1a is a glycoprotein and is expressed in the Langerhans cells, dendritic cells, immature T cells, and cortical thymocytes. It is mainly used as a marker of Langerhans cell histiocytosis (**Table 16**).

CD68

CD68 is a glycoprotein which is present in the lysosomes. It is a good marker of histiocytes. Melanocytic lesions also express this marker (**Table 17**).

CD207 (Langerin)

CD207 (Langerin) is a calcium-dependent glycoprotein. It is responsible for the generation of Birbeck granules within the Langerhans cells. CD207 is considered as a reliable marker of Langerhans cell histiocytosis (LCH) (**Table 18**).[18]

IMMUNOCYTOCHEMISTRY PATTERN OF HISTIOCYTIC AND DENDRITIC NEOPLASMS

Follicular Dendritic Cell Sarcoma

- *Positive*
  - CD21
  - CD23
  - CD35
  - Podoplanin
  - Fascin
  - D2-40

TABLE 16: CD1a expression.

Marker	Main use	Other tumors expressing this antigen	Localization	Comments
CD1a	Langerhans cell histiocytosis (LCH)	• T-cell lymphoma • Myeloid leukemia • Mycosis fungoides	Membranous	It is a useful marker of LCH

TABLE 17: CD68 expression.

Marker	Main use in soft-tissue tumor	Other tumors expressing this antigen	Localization	Comments
CD68	• Benign and malignant fibrous histiocytoma • Nodular fasciitis • Granular cell tumor • Giant cell tumors	• Dendritic cell tumor • Mast cell tumor • Acute myeloid leukemia (AML-M4)	Cytoplasmic and membranous	This marker is widely expressed in many tumors and so it should be carefully interpreted

TABLE 18: CD207 (Langerin) expression.

Marker	Main use	Other tumors expressing this antigen	Localization	Comments
CD207 (Langerin)	Langerhans cell histiocytosis (LCH)	No other tumor	Membranous	This is a specific marker of LCH. It should be used along with CD1a

    - EMA
    - S100
    - CD68
- *Negative*
    - CD1a
    - CD3
    - CD34

Histiocytic Sarcoma

- *Positive*
    - CD68
    - HLA-DR

Langerhans Cell Histiocytosis[19] (Box 5; Figs. 11 to 14)

- *Positive*
    - CD1a
    - CD207
    - S100

IMMUNOCYTOCHEMISTRY PATTERN OF FIBROBLASTIC TUMORS

Nodular Fasciitis

- *Positive*
    - Vimentin
    - SMA
    - Calponin
    - CD68

BOX 5 Langerhans cell histiocytosis.

- Positive: CD1a, S100, and CD207 (langerin)
- Langerin is a sensitive marker of Langerhans cell histiocytosis
- Electron microscopy: Birbeck granule

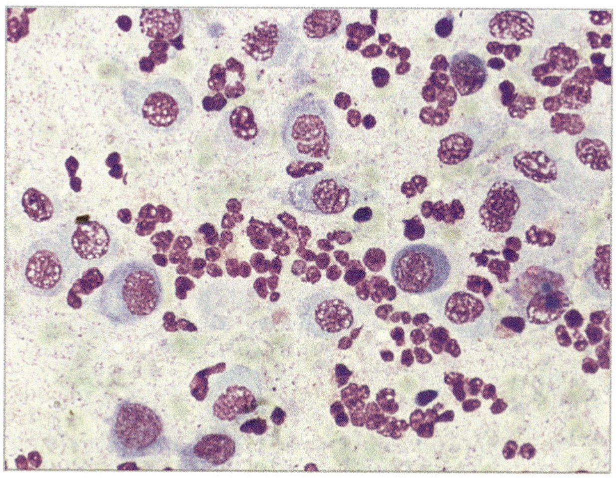

FIG. 11: Langerhans cell histiocytosis: Fine needle aspiration cytology smear showing many mononuclear cells and eosinophils. The cells show nuclear grooves.

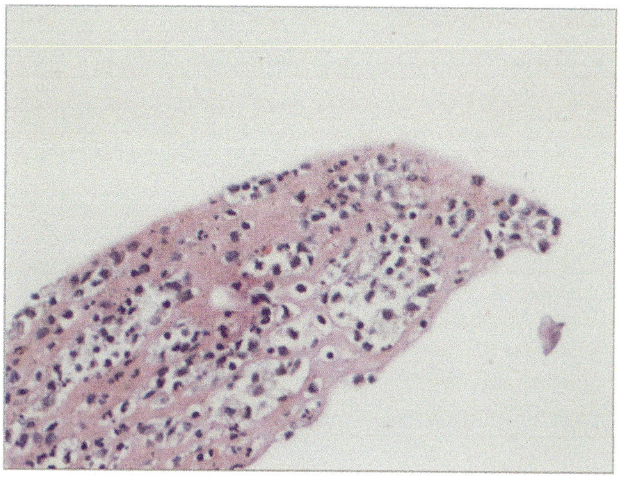

FIG. 12: Langerhans cell histiocytosis: Cell block section showing histiocytes and eosinophils.

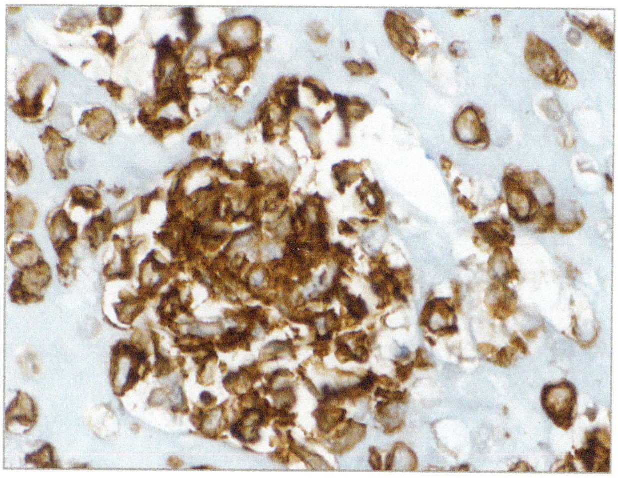

FIG. 13: Langerhans cell histiocytosis: The tumor cells are positive for CD1a.

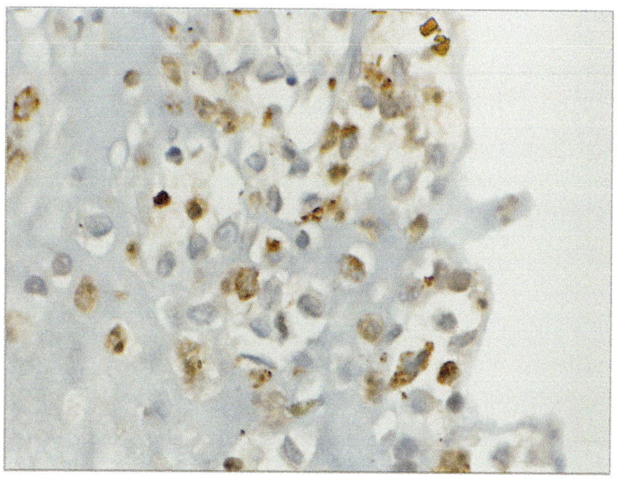

FIG. 14: Langerhans cell histiocytosis: The tumor cells are negative for CD68.

Fibrous Hamartoma of Infancy

- *Positive*
  - SMA
  - S100
  - CD34
  - Bcl2

Myofibroblastoma

- *Positive*
  - Vimentin
  - SMA
  - Calponin
  - Desmin

Desmoid Tumor

- *Positive*
  - Beta catenin
  - Vimentin
  - Desmin

Fibrosarcoma

- *Positive*
  - Vimentin

IMMUNOCYTOCHEMISTRY PATTERN OF FIBROHISTIOCYTIC TUMORS

Fibrous Histiocytoma

- *Positive*
  - Vimentin
  - Alpha 1-antitrypsin

Dermatofibrosarcoma Protuberans

- *Positive*
  - Vimentin
  - CD34
  - PDGF

Giant Cell Tumor of the Soft Parts

- *Positive*
  - Vimentin
  - CD68

MARKERS OF LIPOMATOUS TUMORS

The markers of lipomatous tumors include CD34, MDM2, S100, and p16.

CD34

CD34 has been discussed under 'Vascular Tumor'.

S100

S100 has been discussed under 'Nerve Sheath Tumors'.

Murine Double Minute 2

Murine double minute 2 (MDM2) is an enzyme that inhibits the function of p53 and thereby prevents apoptosis and regulates the cell cycle. The MDM2 protein is overexpressed in liposarcoma.[20] Many other tumors also express MDM2 such as myxofibrosarcoma, malignant peripheral nerve sheath tumor (MPNST) and embryonal rhabdomyosarcoma (**Table 19**).[21]

p16

p16 is a protein that plays a vital role in tumor suppression. It is mainly related to human papillomavirus (HPV) infection and is used as a surrogate marker of HPV infection (**Table 20**). In soft-tissue tumor, p16 is expressed in well-differentiated liposarcoma.[22]

IMMUNOCYTOCHEMISTRY PATTERN OF LIPOMATOUS TUMORS[20-22]

Lipoma

- *Positive*
  - Vimentin
  - CD34

TABLE 19: Murine double minute 2 (MDM2) expression.

Marker	Expression in lipomatous tumors	Other tumors expressing this antigen	Localization	Comments
MDM2	Vascular tumor	• Myxofibrosarcoma • Malignant peripheral nerve sheath tumor (MPNST) • Embryonal rhabdomyosarcoma • Desmoplastic small round cell tumor • Osteosarcoma	Nuclear	MDM2 may be cross-reactive with cytokeratin (CK)

TABLE 20: p16 expression.

Marker	Expression in lipomatous tumors	Other tumors expressing this antigen	Localization	Comments
p16	Well-differentiated liposarcoma	• Cervical dysplasia • Squamous cell carcinoma of the cervix	Nuclear	It is not a specific marker of liposarcoma

Well-differentiated Liposarcoma

- *Positive*
  - Vimentin
  - CD34
  - MDM2
  - S100
  - p16

Myxoid Liposarcoma

- *Positive*
  - Vimentin
  - TLS
  - MDM2 (occasionally positive)
  - S100
  - p16

Pleomorphic Liposarcoma

- *Positive*
  - Vimentin
  - S100

The ICC pattern of various liposarcomas is highlighted in **Box 6**.

MARKERS OF MELANOCYTIC TUMORS

The markers of melanocytic tumors include HMB45, MART1, SOX10, and tyrosinase.

Human Melanoma Black 45

Human Melanoma Black 45 is a glycoprotein that is present in the immature melanosomes. It is associated with the maturation of melanosomes. HMB45 is expressed in the melanoma and also in lesions with melanocytic differentiation (**Table 21**).

Melan A (MART1)

Melan A is also called MART1. It is associated with the development of melanosomes and controls the generation of pigment. Melan A is expressed in the melanocytes and adrenal cortical cells.[23,24] It is a relatively less specific marker than HMB45 to identify the melanocytes (**Table 22**).

Tyrosinase

Tyrosinase is a copper-containing catalytic enzyme which is involved in the production of melanin pigment from tyrosine. It is a specific marker of melanoma (**Table 23**).

BOX 6 Liposarcoma.

- Liposarcomas are overall positive for MDM2 and S100
- MDM2 is usually negative in myxoid liposarcoma
- Lipoblasts: Positive for S100

(MDM2: murine double minute 2)

TABLE 21: HMB45 expression.

Marker	Expression in melanocytic tumors	Other tumors expressing this antigen	Localization	Comments
HMB45	• Malignant melanoma • Clear cell sarcoma of soft parts	• Cellular blue nevi • Spitz nevi • PEComa • Angiomyolipoma	Cytoplasmic	• One tenth of malignant melanoma may not express HMB45 • A cocktail of different melanoma markers may be helpful to detect more such cases

(HMB45: Human Melanoma Black 45)

TABLE 22: Melan A expression.

Marker	Expression in melanocytic tumors	Other tumors expressing this antigen	Localization	Comments
Melan A	• Malignant melanoma • Clear cell sarcoma of soft parts	• Angiomyolipoma • Adrenal cortical tumors • Sex cord stromal tumor	Cytoplasmic	Less specific marker for melanoma

TABLE 23: Tyrosinase expression.

Marker	Expression in melanocytic tumors	Other tumors expressing this antigen	Localization	Comments
Tyrosinase	• Malignant melanoma • Clear cell sarcoma of soft parts	Angiomyolipoma	Cytoplasmic	Highly specific marker for melanoma

IMMUNOCYTOCHEMISTRY PATTERN OF MELANOCYTIC TUMORS

Melanoma

- *Positive*
    - HMB45
    - Melan A
    - Sox10
    - Tyrosinase

MARKERS OF PERIPHERAL NEUROECTODERMAL TUMORS

The markers of peripheral neuroectodermal tumors include CD99, Fli-1, CD56, and chromogranin.

CD99

CD99 is also called Mic2. It is a glycoprotein that is expressed in cortical thymocytes and also certain T and B lymphocytes. It is involved in the differentiation of neuroectodermal cells and apoptosis of T lymphocytes. CD99 is a well-recognized marker of ES/peripheral neuroectodermal tumors (PNET).[25] It is widely expressed in various other tumors, and therefore its specificity is low in the diagnosis of ES (**Table 24**).

Friend Leukemia Virus Integration Site 1

Friend leukemia virus integration site 1 (Fli-1) is a member of the ETS transcription factor family. It is expressed during the embryogenic development of the blood vessels. Fli-1 is related to cell proliferation and tumorigenesis. Fli-1 is a specific marker for ES/PNET (**Table 25**).[26,27]

CD56

CD56 is responsible for the growth of neurites, cell-to-cell adhesion, migration, proliferation, and cell death by apoptosis. CD56 is expressed in NK cells, neuroendocrine cells, subsets of T cells, glial cells, and skeletal muscle. CD56 is present in various soft-tissue tumor as mentioned in **Table 26**.

TABLE 24: CD99 expression.

Marker	Expression in Ewing's tumors	Other tumors expressing this antigen	Localization	Comments
CD99	Ewing's sarcoma/peripheral neuroectodermal tumors	• T-cell lymphoma • Large cell lymphoma • Gastrointestinal stromal tumor (GIST) • Hepatocellular carcinoma • Breast carcinoma • Prostatic carcinoma • Sex-cord stromal tumor of the ovary • Merkel cell carcinoma • Wilms' tumor	Membranous	• Less specific marker for Ewing's sarcoma • Neuroblastoma is negative for CD99, and so it may be helpful to distinguish neuroblastoma from Ewing's sarcoma/ peripheral neuroectodermal tumors

TABLE 25: Fli-1 expression.

Marker	Expression in Ewing's tumors	Other tumors expressing this antigen	Localization	Comments
Fli-1	• Ewing's sarcoma/peripheral neuroectodermal tumors	• Lymphoblastic lymphoma • Merkel cell carcinoma • Melanoma • Desmoplastic small round cell tumor	Nuclear	• Relatively good marker for Ewing's sarcoma • Acute lymphoblastic lymphoma also expresses Fli-1

(Fli-1: Friend leukemia virus integration site 1)

TABLE 26: CD56 expression.

Marker	Expression in soft-tissue tumors	Other tumors expressing this antigen	Localization	Comments
CD56	• Ewing's sarcoma/peripheral neuroectodermal tumors • Angiosarcoma, synovial sarcoma, alveolar rhabdomyosarcoma, schwannoma	• Neuroendocrine tumors • Angiosarcoma • Synovial sarcoma • Alveolar rhabdomyosarcoma • Schwannoma • Melanoma	Membranous and cytoplasmic	A wide variety of tumor express CD56, so this marker should be used along with other markers to have a specific diagnosis

Ewing's Sarcoma/Peripheral Neuroectodermal Tumors[25-27]

The ICC pattern of ES/PNET is (**Box 7; Figs. 15** to **19**):

- *Positive*
  - CD99
  - Fli-1
  - Vimentin
- *Negative*
  - Desmin
  - Chromogranin

BOX 7	Ewing's sarcoma (ES)/peripheral neuroectodermal tumors (PNETs).

- ES/PNET tumors are consistently positive for CD99
- These tumors are also positive for Fl-i1 and neuron-specific enolase
- CD99 and Fli-1 are not specific markers of ES/PNET
- These tumors are negative for desmin, CD45, and chromogranin

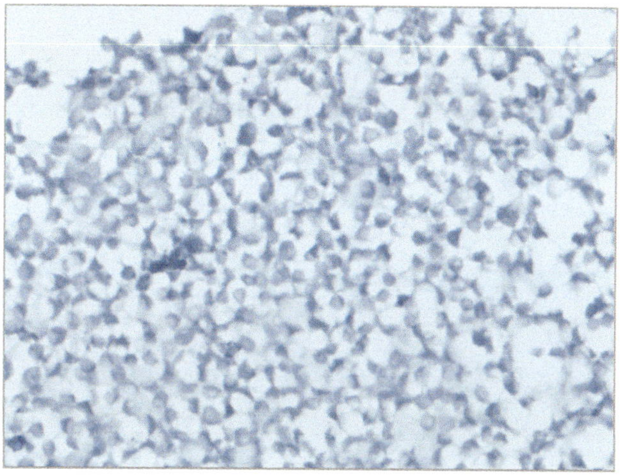

FIG. 17: Ewing's sarcoma: The tumor cells are negative for CD45.

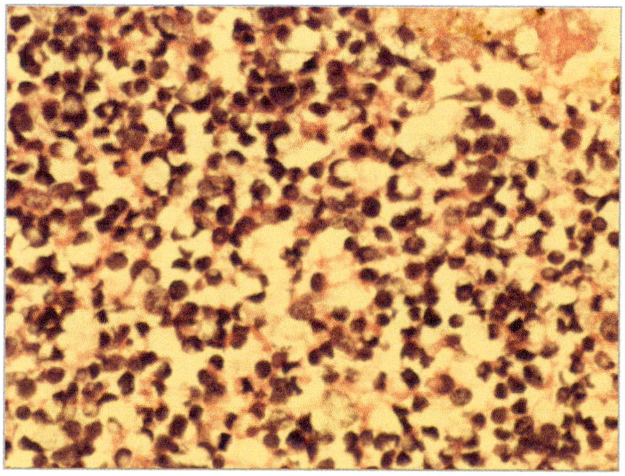

FIG. 15: Ewing's sarcoma: Cell block section showing malignant round cells.

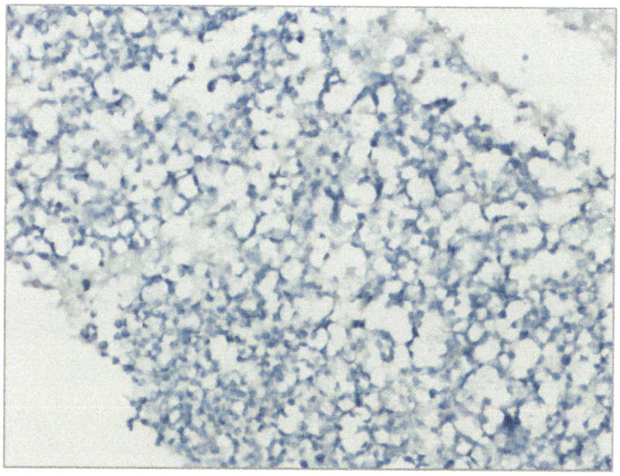

FIG. 18: Ewing's sarcoma: The tumor cells are negative for desmin.

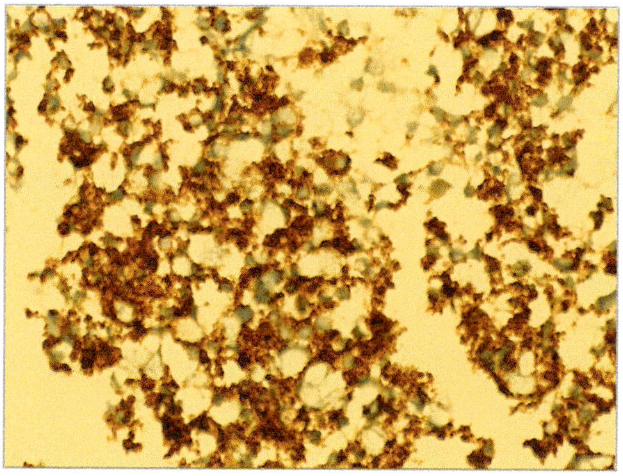

FIG. 16: Ewing's sarcoma: The tumor cells are positive for Mic2.

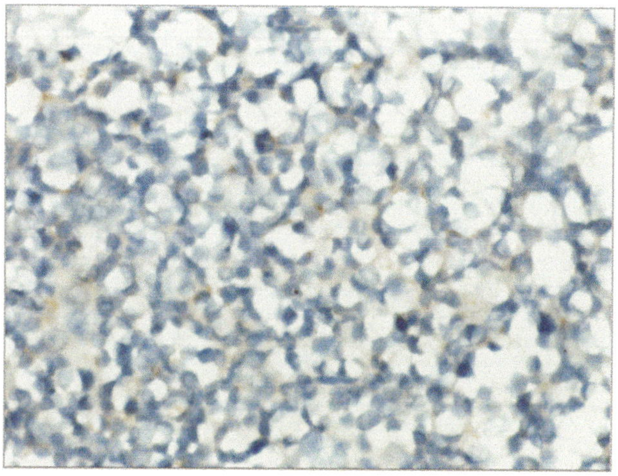

FIG. 19: Ewing's sarcoma: The tumor cells are negative for CD56.

Neuroblastoma

- *Positive*
  - CD56
  - NSE
  - NB84
  - Synaptophysin
  - Neurofilament
  - GATA-3
  - S100

Merkel Cell Carcinoma [28,29] (Figs. 20 to 22)

- *Positive*
  - CK20
  - EMA
  - NSE
  - Chromogranin
  - Synaptophysin

MARKERS OF MISCELLANEOUS SOFT-TISSUE TUMORS

Transducer-like Enhancer of Split 1

Transducer-like enhancer of split 1 (TLE-1) is a transcriptional repressor protein, and it represses cellular differentiation. TLE-1 expression is upregulated in synovial sarcoma. However, various other tumors may also overexpress TLE-1 (**Table 27**).[30,31]

Immunocytochemistry Pattern of Synovial Sarcoma[30,31] (Box 8; Figs. 23 to 25)

Epithelial Cell Component

- *Positive*
  - TLE-1
  - Pan cytokeratin
  - EMA
  - CK7
  - CK19

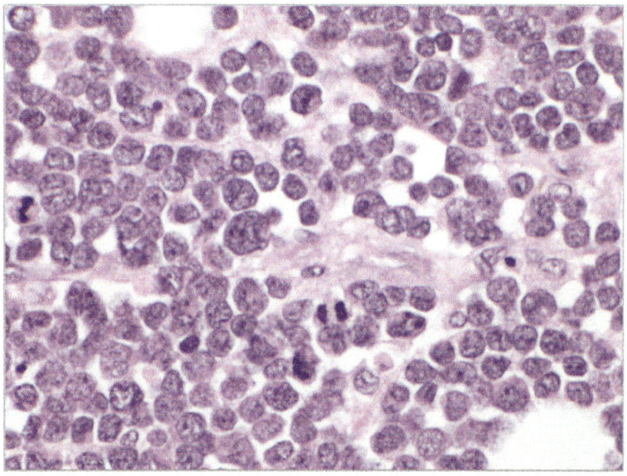

FIG. 21: Merkel cell carcinoma: Cell block section of the same showing sheets of malignant round cells.

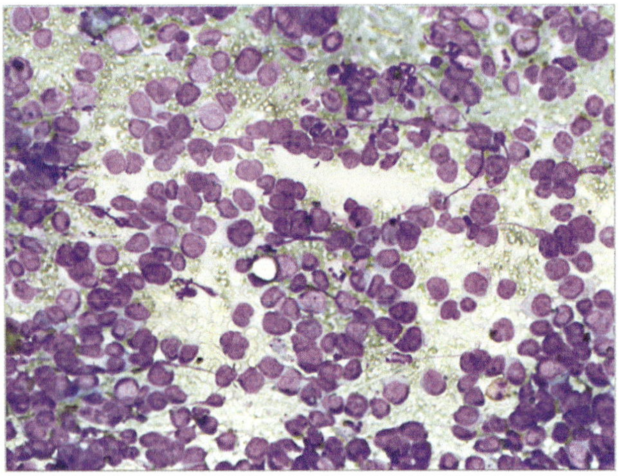

FIG. 20: Merkel cell carcinoma: Fine needle aspiration cytology of a swelling over the forehead. The smear shows round monomorphic cells.

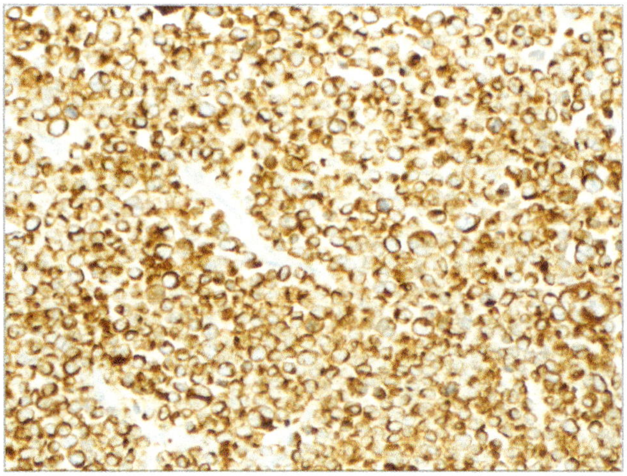

FIG. 22: Merkel cell carcinoma: The tumor cells are strongly positive for CK20.

TABLE 27: TLE-1 expression.

Marker	Mainly used	Other tumors expressing this antigen	Localization	Comments
TLE-1	Synovial sarcoma	• Endometrial stromal sarcoma • Solitary fibrous tumor • Epithelioid sarcoma • Mesothelioma	Nuclear	TLE-1 is widely expressed in many tumors

(TLE-1: transducer-like enhancer of split 1)

Spindle Cell Component

- *Positive*
  - TLE-1
  - Vimentin
  - Calponin
  - CD99

BOX 8 Synovial sarcoma.

- Positive: Both pan cytokeratin and vimentin
- EMA is positive in both monophasic and biphasic—synovial sarcoma
- The tumor is also positive for CD99, Bcl2, and TLE-1
- TLE-1 is positive in 90% cases of synovial sarcoma; however, it is not a specific marker

(EMA: epithelial membrane antigen; TLE-1: transducer-like enhancer of split 1)

Clear Cell Sarcoma (Figs. 26 to 29)

- *Positive*
  - HMB45
  - S100
  - Tyrosinase
  - MITF
  - Melan A
  - Vimentin

Epithelioid Sarcoma[32]

- *Positive*
  - AE1/AE3
  - EMA
  - D2-40
  - Fli-1
  - Vimentin

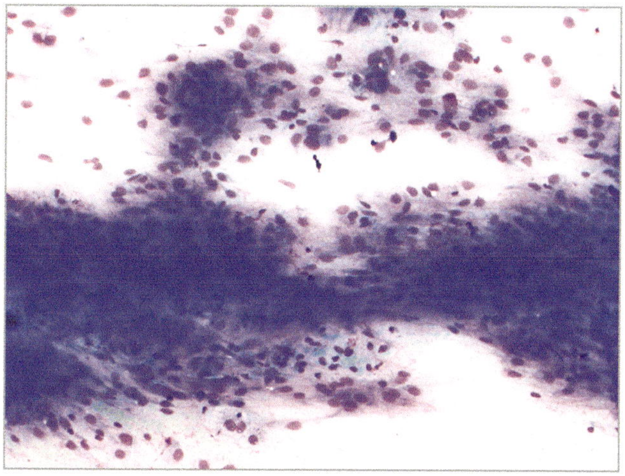

FIG. 23: Synovial sarcoma: Fine needle aspiration cytology of swelling over the chest wall showing loose clusters and fascicles of spindle cells.

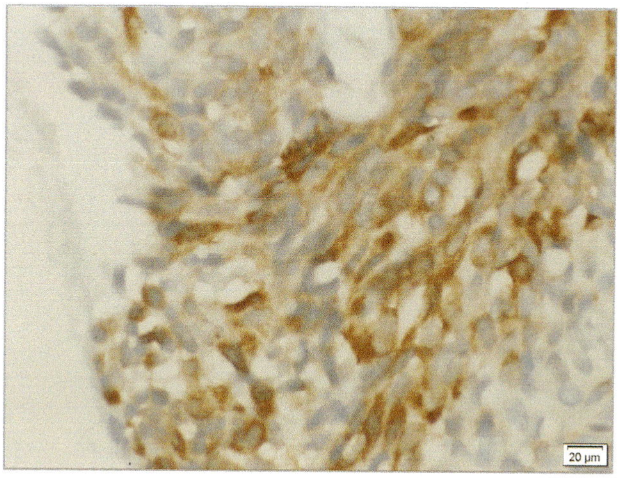

FIG. 25: Synovial sarcoma: The tumor cells are also positive for Bcl2.

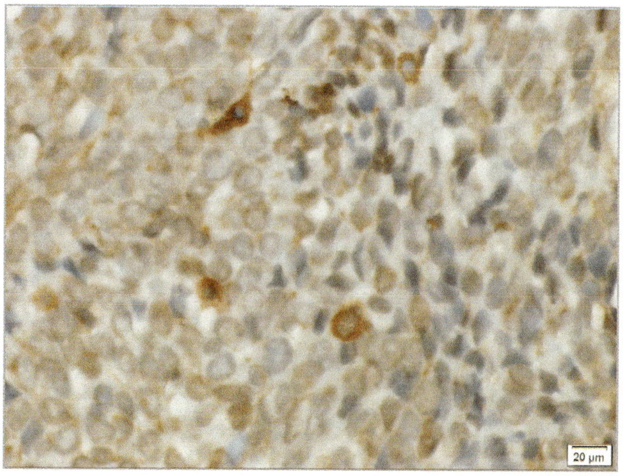

FIG. 24: Synovial sarcoma: The tumor cells showing TLE-1 positive cells.

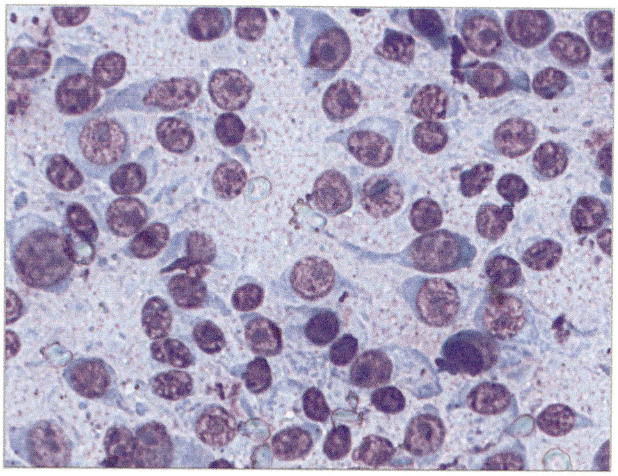

FIG. 26: Clear cell sarcoma: Fine needle aspiration cytology of swelling over the leg showing discrete cells with a moderate amount of clear cytoplasm.

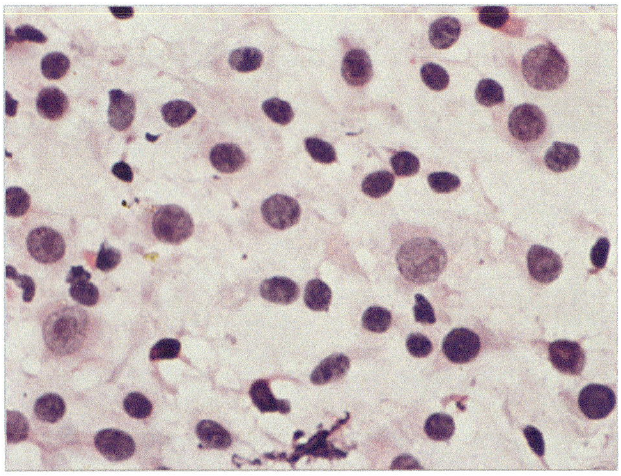

FIG. 27: Clear cell sarcoma: Fine needle aspiration cytology of swelling showing cells with round nuclei having large prominent nucleoli.

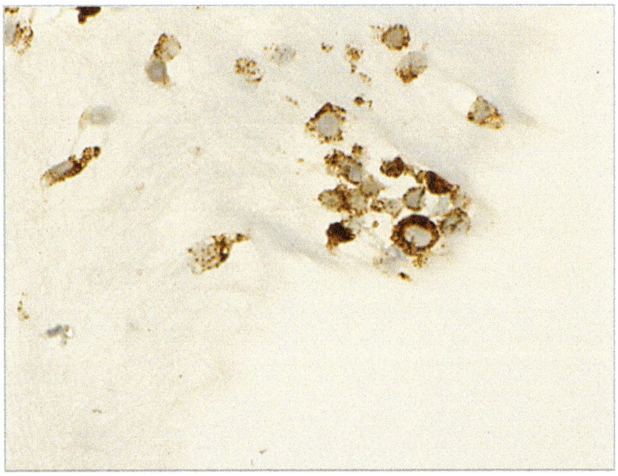

FIG. 28: Clear cell sarcoma: The tumor cells are positive for Human Melanoma Black 45 (HMB45).

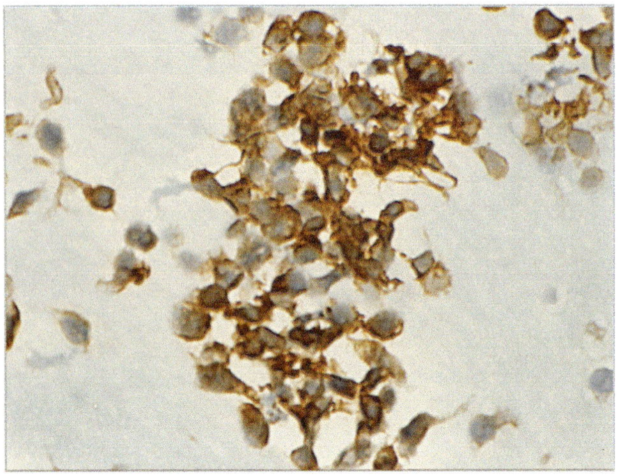

FIG. 29: Clear cell sarcoma: The tumor cells are positive for vimentin.

Desmoplastic Small Round Cell Tumor[33]

- *Positive*
  - AE1/AE3
  - EMA
  - MOC31
  - BER-EP4
  - WT1
  - Vimentin
  - Desmin

Rhabdoid Tumor

- *Positive*
  - AE1/AE3
  - Vimentin
  - CD99

Alveolar Soft-part Sarcoma

- *Positive*
  - TFE3
  - Vimentin

Chordoma[34]

- *Positive*
  - Brachyury
  - NSE
  - Pan CK
  - EMA
  - CEA
  - S100
  - Vimentin

REFERENCES

1. Sharma P, Alsharif S, Fallatah A, Chung BM. intermediate filaments as effectors of cancer development and metastasis: A focus on keratins, vimentin, and nestin. Cells. 2019;8(5).pii: E497.
2. Dey P, Togra J, Mitra S. Intermediate filament: structure, function, and applications in cytology. Diagn Cytopathol. 2014;42(7):628-35.
3. Truong LD, Rangdaeng S, Cagle P, Ro JY, Hawkins H, Font RL. The diagnostic utility of desmin. A study of 584 cases and review of the literature. Am J Clin Pathol. 1990;93(3):305-14.
4. Ordóñez NG. Antidesmin antibodies. Their use in diagnostic pathology. Am J Clin Pathol. 1990;93(3):430-1.
5. Hosoi H, Sugimoto T, Hayashi Y, Inaba T, Horii Y, Morioka H, et al. Differential expression of myogenic regulatory genes, MyoD1 and myogenin, in human rhabdomyosarcoma sublines. Int J Cancer. 1992;50(6):977-83.
6. Wang NP, Marx J, McNutt MA, Rutledge JC, Gown AM. Expression of myogenic regulatory proteins (myogenin and MyoD1) in small blue round cell tumors of childhood. Am J Pathol. 1995;147(6):1799-810.
7. Furlong MA, Mentzel T, Fanburg-Smith JC. Pleomorphic rhabdomyosarcoma in adults: a clinicopathologic study of 38 cases with emphasis on morphologic variants and recent skeletal muscle-specific markers. Mod Pathol. 2001;14(6):595-603.
8. Parham DM, Webber B, Holt H, Williams WK, Maurer H. Immunohistochemical study of childhood rhabdomyosarcomas and related neoplasms. Results of an intergroup rhabdomyosarcoma study project. Cancer. 1991;67(12):3072-80.

9. Watanabe K, Kusakabe T, Hoshi N, Saito A, Suzuki T. h-Caldesmon in leiomyosarcoma and tumors with smooth muscle cell-like differentiation: its specific expression in the smooth muscle cell tumor. Hum Pathol. 1999;30(4):392-6.
10. Tawfik O, Rao D, Nothnick WB, Graham A, Mau B, Fan F. Transgelin, a novel marker of smooth muscle differentiation, effectively distinguishes endometrial stromal tumors from uterine smooth muscle tumors. Int J Gynecol Obstet Reprod Med Res. 2014;1(1):26-31.
11. Robin YM, Penel N, Pérot G, Neuville A, Vélasco V, Ranchère-Vince D, et al. Transgelin is a novel marker of smooth muscle differentiation that improves diagnostic accuracy of leiomyosarcomas: a comparative immunohistochemical reappraisal of myogenic markers in 900 soft tissue tumors. Mod Pathol. 2013;26(4):502-10.
12. Lee CH, Turbin DA, Sung YC, Espinosa I, Montgomery K, van de Rijn M, et al. A panel of antibodies to determine site of origin and malignancy in smooth muscle tumors. Mod Pathol. 2009;22(12):1519-3.13. Yuan A, Rao MV, Veeranna, Nixon RA. Neurofilaments and neurofilament proteins in health and disease. Cold Spring Harb Perspect Biol. 2017;9(4):a018309
14. Karamchandani JR, Nielsen TO, van de Rijn M, West RB. Sox10 and S100 in the diagnosis of soft-tissue neoplasms. Appl Immunohistochem Mol Morphol. 2012;20(5):445-50.
15. Yaskiv O, Rubin BP, He H, Falzarano S, Magi-Galluzzi C, Zhou M. ERG protein expression in human tumors detected with a rabbit monoclonal antibody. Am J Clin Pathol. 2012;138(6):803-10.
16. Miettinen M, Wang ZF, Paetau A, Tan SH, Dobi A, Srivastava S, et al ERG transcription factor as an immunohistochemical marker for vascular endothelial tumors and prostatic carcinoma. Am J Surg Pathol. 2011;35(3): 432-41.
17. Naeem N, Mushtaq S, Akhter N, Hussain M, Hassan U. Effectiveness of vascular markers (immunohistochemical stains) in soft tissue sarcomas. J Coll Physicians Surg Pak. 2018;28(5):352-6.
18. Chikwava K, Jaffe R. Langerin (CD207) staining in normal pediatric tissues, reactive lymph nodes, and childhood histiocytic disorders. Pediatr Dev Pathol. 2004;7(6):607-14.
19. El Demellawy D, Young JL, de Nanassy J, Chernetsova E, Nasr A. Langerhans cell histiocytosis: a comprehensive review. Pathology. 2015;47(4):294-301.
20. Binh MB, Sastre-Garau X, Guillou L, de Pinieux G, Terrier P, Lagacé R, et al. MDM2 and CDK4 immunostainings are useful adjuncts in diagnosing well-differentiated and dedifferentiated liposarcoma subtypes: a comparative analysis of 559 soft tissue neoplasms with genetic data. Am J Surg Pathol. 2005;29(10):1340-7.
21. Wallander ML, Tripp S, Layfield LJ. MDM2 amplification in malignant peripheral nerve sheath tumors correlates with p53 protein expression. Arch Pathol Lab Med. 2012;136(1):95-9.
22. Kammerer-Jacquet SF, Thierry S, Cabillic F, Lannes M, Burtin F, Henno S, et al. Differential diagnosis of atypical lipomatous tumor/well-differentiated liposarcoma and dedifferentiated liposarcoma: utility of p16 in combination with MDM2 and CDK4 immunohistochemistry. Hum Pathol. 2017;59:34-40.
23. Shidham VB, Qi DY, Acker S, Kampalath B, Chang CC, George V, et al. Evaluation of micrometastases in sentinel lymph nodes of cutaneous melanoma: higher diagnostic accuracy with Melan-A and MART-1 compared with S-100 protein and HMB-45. Am J Surg Pathol. 2001;25(8): 1039-46.
24. Zubovits J, Buzney E, Yu L, Duncan LM. HMB-45, S-100, NK1/C3, and MART-1 in metastatic melanoma. Hum Pathol. 2004;35(2):217-23.
25. Folpe AL, Goldblum JR, Rubin BP, Shehata BM, Liu W, Dei Tos AP, et al. Morphologic and immunophenotypic diversity in Ewing family tumors: a study of 66 genetically confirmed cases. Am J Surg Pathol. 2005;29(8):1025-33.
26. Lee AF, Hayes MM, Lebrun D, Espinosa I, Nielsen GP, Rosenberg AE, et al. FLI-1 distinguishes Ewing sarcoma from small cell osteosarcoma and mesenchymal chondrosarcoma. Appl Immunohistochem Mol Morphol. 2011;19(3):233-8.
27. Folpe AL, Hill CE, Parham DM, O'Shea PA, Weiss SW. Immunohistochemical detection of FLI-1 protein expression: a study of 132 round cell tumors with emphasis on CD99-positive mimics of Ewing's sarcoma/primitive neuroectodermal tumor. Am J Surg Pathol. 2000;24(12):1657-62.
28. Harms PW. Update on Merkel cell carcinoma. Clin Lab Med. 2017;37(3): 485-501.
29. Pasternak S, Carter MD, Ly TY, Doucette S, Walsh NM. Immunohistochemical profiles of different subsets of Merkel cell carcinoma. Hum Pathol. 2018;82:232-8.
30. Lin G, Doyle LA. An update on the application of newly described immunohistochemical markers in soft tissue pathology. Arch Pathol Lab Med. 2015;139(1):106-21.
31. Zaccarini DJ, Deng X, Tull J, Maciak C, Valente AL, Zhang S. Expression of TLE-1 and CD99 in Carcinoma: Pitfalls in diagnosis of synovial sarcoma. Appl Immunohistochem Mol Morphol. 2018;26(6):368-73.
32. Armah HB, Parwani AV. Epithelioid sarcoma. Arch Pathol Lab Med. 2009;133(5):814-9.
33. Ordóñez NG. Desmoplastic small round cell tumor: II: an ultrastructural and immunohistochemical study with emphasis on new immunohistochemical markers. Am J Surg Pathol. 1998;22(11):1314-27.
34. Sun HI, Guduk M, Gucyetmez B, Yapicier O, Pamir MN. Chordoma: Immunohistochemical analysis of Brachury. Turk Neurosurg. 2018;28(2):174-8.

INDEX

Page numbers followed by *b* refer to box, *f* refer to figure, and *t* refer to table.

L

M

N

O

P

R

S

T

U

V

W

Y